# Advanced Control Engineering

In fond memory of
my mother

# Advanced Control Engineering

**Roland S. Burns**

*Professor of Control Engineering*
*Department of Mechanical and Marine Engineering*
*University of Plymouth, UK*

OXFORD   AUCKLAND   BOSTON   JOHANNESBURG   MELBOURNE   NEW DELHI

Butterworth-Heinemann
Linacre House, Jordan Hill, Oxford OX2 8DP
225 Wildwood Avenue, Woburn, MA 01801-2041
A division of Reed Educational and Professional Publishing Ltd

A member of the Reed Elsevier plc group

First published 2001
Transferred to digital printing 2004
© Roland S. Burns 2001

**British Library Cataloguing in Publication Data**
A catalogue record for this book is available from the British Library
**Library of Congress Cataloguing in Publication Data**
A catalogue record for this book is available from the Library of Congress

ISBN 0 7506 5100 8

Typeset in India by Integra Software Services Pvt. Ltd.,
Pondicherry, India 605 005, www.integra-india.com

FOR EVERY TITLE THAT WE PUBLISH, BUTTERWORTH-HEINEMANN
WILL PAY FOR BTCV TO PLANT AND CARE FOR A TREE.

# Contents

# List of Tables

# Preface and acknowledgements

The material presented in this book is as a result of four decades of experience in the field of control engineering. During the 1960s, following an engineering apprenticeship in the aircraft industry, I worked as a development engineer on flight control systems for high-speed military aircraft. It was during this period that I first observed an unstable control system, was shown how to frequency-response test a system and its elements, and how to plot a Bode and Nyquist diagram. All calculations were undertaken on a slide-rule, which I still have. Also during this period I worked in the process industry where I soon discovered that the incorrect tuning for a PID controller on a 100 m long drying oven could cause catastrophic results.

On the 1st September 1970 I entered academia as a lecturer (Grade II) and in that first year, as I prepared my lecture notes, I realized just how little I knew about control engineering. My professional life from that moment on has been one of discovery (currently termed 'life-long learning'). During the 1970s I registered for an M.Phil. which resulted in writing a FORTRAN program to solve the matrix Riccati equations and to implement the resulting control algorithm in assembler on a minicomputer.

In the early 1980s I completed a Ph.D. research investigation into linear quadratic Gaussian control of large ships in confined waters. For the past 17 years I have supervised a large number of research and consultancy projects in such areas as modelling the dynamic behaviour of moving bodies (including ships, aircraft missiles and weapons release systems) and extracting information using state estimation techniques from systems with noisy or incomplete data. More recently, research projects have focused on the application of artificial intelligence techniques to control engineering projects. One of the main reasons for writing this book has been to try and capture four decades of experience into one text, in the hope that engineers of the future benefit from control system design methods developed by engineers of my generation.

The text of the book is intended to be a comprehensive treatment of control engineering for any undergraduate course where this appears as a topic. The book is also intended to be a reference source for practising engineers, students undertaking Masters degrees, and an introductory text for Ph.D. research students.

One of the fundamental aims in preparing the text has been to work from basic principles and to present control theory in a way that is easily understood and applied. For most examples in the book, all that is required to obtain a solution is a calculator. However, it is recognized that powerful software packages exist to aid control system design. At the time of writing, MATLAB, its Toolboxes and SIMULINK have emerged as becoming the industry standard control system design package. As a result, Appendix 1 provides script file source code for most examples presented in the main text of the book. It is suggested however, that these script files be used to check hand calculation when used in a tutorial environment.

Depending upon the structure of the undergraduate programme, it is suggested that content of Chapters 1, 2 and 3 be delivered in Semester 3 (first Semester, year two), where, at the same time, Laplace Transforms and complex variables are being studied under a Mathematics module. Chapters 4, 5 and 6 could then be studied in Semester 4 (second Semester, year two). In year 3, Chapters 7 and 8 could be studied in Semester 5 (first Semester) and Chapters 9 and 10 in Semester 6 (second Semester). However, some of the advanced material in Chapters 9 and 10 could be held back and delivered as part of a Masters programme.

When compiling the material for the book, decisions had to be made as to what should be included, and what should not. It was decided to place the emphasis on the control of continuous and discrete-time linear systems. Treatment of nonlinear systems (other than linearization) has therefore not been included and it is suggested that other works (such as Feedback Control Systems, Phillips and Harbor (2000)) be consulted as necessary.

I would wish to acknowledge the many colleagues, undergraduate and postgraduate students at the University of Plymouth (UoP), University College London (UCL) and the Open University (OU) who have contributed to the development of this book. I am especially indebted to the late Professor Tom Lambert (UCL), the late Professor David Broome (UCL), ex-research students Dr Martyn Polkinghorne, Dr Paul Craven and Dr Ralph Richter. I would like to thank also my colleague Dr Bob Sutton, Reader in Control Systems Engineering, in stimulating my interest in the application of artificial intelligence to control systems design. Thanks also go to OU students Barry Drew and David Barrett for allowing me to use their T401 project material in this book. Finally, I would like to express my gratitude to my family. In particular, I would like to thank Andrew, my son, and Janet my wife, for not only typing the text of the book and producing the drawings, but also for their complete support, without which the undertaking would not have been possible.

Roland S. Burns

# Introduction to control engineering

## 1.1 Historical review

Throughout history mankind has tried to control the world in which he lives. From the earliest days he realized that his puny strength was no match for the creatures around him. He could only survive by using his wits and cunning. His major asset over all other life forms on earth was his superior intelligence. Stone Age man devised tools and weapons from flint, stone and bone and discovered that it was possible to train other animals to do his bidding – and so the earliest form of control system was conceived. Before long the horse and ox were deployed to undertake a variety of tasks, including transport. It took a long time before man learned to replace animals with machines.

Fundamental to any control system is the ability to measure the output of the system, and to take corrective action if its value deviates from some desired value. This in turn necessitates a sensing device. Man has a number of 'in-built' senses which from the beginning of time he has used to control his own actions, the actions of others, and more recently, the actions of machines. In driving a vehicle for example, the most important sense is sight, but hearing and smell can also contribute to the driver's actions.

The first major step in machine design, which in turn heralded the industrial revolution, was the development of the steam engine. A problem that faced engineers at the time was how to control the speed of rotation of the engine without human intervention. Of the various methods attempted, the most successful was the use of a conical pendulum, whose angle of inclination was a function (but not a linear function) of the angular velocity of the shaft. This principle was employed by James Watt in 1769 in his design of a flyball, or centrifugal speed governor. Thus possibly the first system for the automatic control of a machine was born.

The principle of operation of the Watt governor is shown in Figure 1.1, where change in shaft speed will result in a different conical angle of the flyballs. This in turn results in linear motion of the sleeve which adjusts the steam mass flow-rate to the engine by means of a valve.

Watt was a practical engineer and did not have much time for theoretical analysis. He did, however, observe that under certain conditions the engine appeared to hunt,

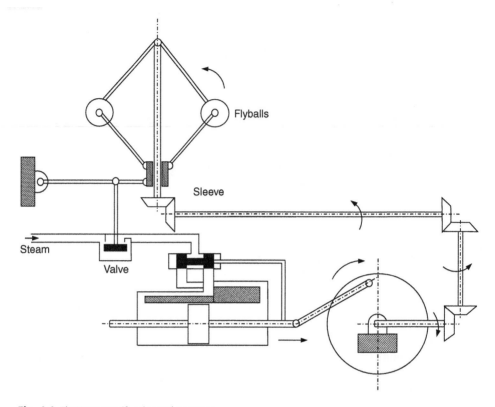

**Fig. 1.1** The Watt centrifugal speed governor.

where the speed output oscillated about its desired value. The elimination of hunting, or as it is more commonly known, instability, is an important feature in the design of all control systems.

In his paper 'On Governors', Maxwell (1868) developed the differential equations for a governor, linearized about an equilibrium point, and demonstrated that stability of the system depended upon the roots of a characteristic equation having negative real parts. The problem of identifying stability criteria for linear systems was studied by Hurwitz (1875) and Routh (1905). This was extended to consider the stability of nonlinear systems by a Russian mathematician Lyapunov (1893). The essential mathematical framework for theoretical analysis was developed by Laplace (1749–1827) and Fourier (1758–1830).

Work on feedback amplifier design at Bell Telephone Laboratories in the 1930s was based on the concept of frequency response and backed by the mathematics of complex variables. This was discussed by Nyquist (1932) in his paper 'Regeneration Theory', which described how to determine system stability using frequency domain methods. This was extended by Bode (1945) and Nichols during the next 15 years to give birth to what is still one of the most commonly used control system design methodologies.

Another important approach to control system design was developed by Evans (1948). Based on the work of Maxwell and Routh, Evans, in his Root Locus method, designed rules and techniques that allowed the roots of the characteristic equation to be displayed in a graphical manner.

The advent of digital computers in the 1950s gave rise to the state-space formulation of differential equations, which, using vector matrix notation, lends itself readily to machine computation. The idea of optimum design was first mooted by Wiener (1949). The method of dynamic programming was developed by Bellman (1957), at about the same time as the maximum principle was discussed by Pontryagin (1962). At the first conference of the International Federation of Automatic Control (IFAC), Kalman (1960) introduced the dual concept of controllability and observability. At the same time Kalman demonstrated that when the system dynamic equations are linear and the performance criterion is quadratic (LQ control), then the mathematical problem has an explicit solution which provides an optimal control law. Also Kalman and Bucy (1961) developed the idea of an optimal filter (Kalman filter) which, when combined with an optimal controller, produced linear-quadratic-Gaussian (LQG) control.

The 1980s saw great advances in control theory for the robust design of systems with uncertainties in their dynamic characteristics. The work of Athans (1971), Safanov (1980), Chiang (1988), Grimble (1988) and others demonstrated how uncertainty can be modelled and the concept of the H∞ norm and $\mu$-synthesis theory.

The 1990s has introduced to the control community the concept of intelligent control systems. An intelligent machine according to Rzevski (1995) is one that is able to achieve a goal or sustained behaviour under conditions of uncertainty. Intelligent control theory owes much of its roots to ideas laid down in the field of Artificial Intelligence (AI). Artificial Neural Networks (ANNs) are composed of many simple computing elements operating in parallel in an attempt to emulate their biological counterparts. The theory is based on work undertaken by Hebb (1949), Rosenblatt (1961), Kohonen (1987), Widrow-Hoff (1960) and others. The concept of fuzzy logic was introduced by Zadeh (1965). This new logic was developed to allow computers to model human vagueness. Fuzzy logic controllers, whilst lacking the formal rigorous design methodology of other techniques, offer robust control without the need to model the dynamic behaviour of the system. Workers in the field include Mamdani (1976), Sugeno (1985) Sutton (1991) and Tong (1978).

## 1.2  Control system fundamentals

### 1.2.1  Concept of a system

Before discussing the structure of a control system it is necessary to define what is meant by a system. Systems mean different things to different people and can include purely physical systems such as the machine table of a Computer Numerically Controlled (CNC) machine tool or alternatively the procedures necessary for the purchase of raw materials together with the control of inventory in a Material Requirements Planning (MRP) system.

However, all systems have certain things in common. They all, for example, require inputs and outputs to be specified. In the case of the CNC machine tool machine table, the input might be the power to the drive motor, and the outputs might be the position, velocity and acceleration of the table. For the MRP system inputs would include sales orders and sales forecasts (incorporated in a master

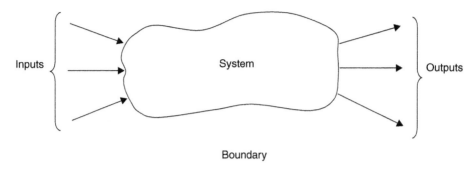

**Fig. 1.2** The concept of a system.

production schedule), a bill of materials for component parts and subassemblies, inventory records and information relating to capacity requirements planning. Material requirements planning systems generate various output reports that are used in planning and managing factory operations. These include order releases, inventory status, overdue orders and inventory forecasts. It is necessary to clearly define the boundary of a system, together with the inputs and outputs that cross that boundary. In general, a system may be defined as a collection of matter, parts, components or procedures which are included within some specified boundary as shown in Figure 1.2. A system may have any number of inputs and outputs.

In control engineering, the way in which the system outputs respond in changes to the system inputs (i.e. the system response) is very important. The control system design engineer will attempt to evaluate the system response by determining a mathematical model for the system. Knowledge of the system inputs, together with the mathematical model, will allow the system outputs to be calculated.

It is conventional to refer to the system being controlled as the plant, and this, as with other elements, is represented by a block diagram. Some inputs, the engineer will have direct control over, and can be used to control the plant outputs. These are known as control inputs. There are other inputs over which the engineer has no control, and these will tend to deflect the plant outputs from their desired values. These are called disturbance inputs.

In the case of the ship shown in Figure 1.3, the rudder and engines are the control inputs, whose values can be adjusted to control certain outputs, for example heading and forward velocity. The wind, waves and current are disturbance inputs and will induce errors in the outputs (called controlled variables) of position, heading and forward velocity. In addition, the disturbances will introduce increased ship motion (roll, pitch and heave) which again is not desirable.

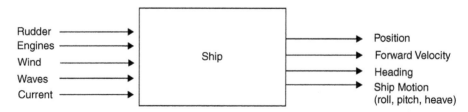

**Fig. 1.3** A ship as a dynamic system.

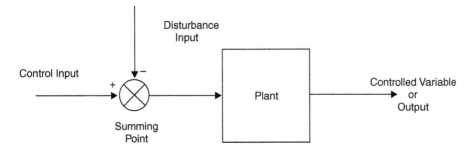

**Fig. 1.4** Plant inputs and outputs.

Generally, the relationship between control input, disturbance input, plant and controlled variable is shown in Figure 1.4.

## 1.2.2 Open-loop systems

Figure 1.4 represents an open-loop control system and is used for very simple applications. The main problem with open-loop control is that the controlled variable is sensitive to changes in disturbance inputs. So, for example, if a gas fire is switched on in a room, and the temperature climbs to 20 °C, it will remain at that value unless there is a disturbance. This could be caused by leaving a door to the room open, for example. Or alternatively by a change in outside temperature. In either case, the internal room temperature will change. For the room temperature to remain constant, a mechanism is required to vary the energy output from the gas fire.

## 1.2.3 Closed-loop systems

For a room temperature control system, the first requirement is to detect or sense changes in room temperature. The second requirement is to control or vary the energy output from the gas fire, if the sensed room temperature is different from the desired room temperature. In general, a system that is designed to control the output of a plant must contain at least one sensor and controller as shown in Figure 1.5.

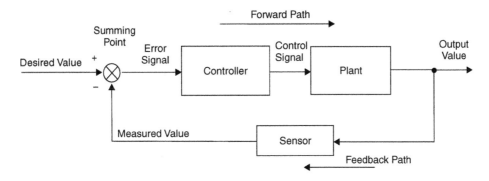

**Fig. 1.5** Closed-loop control system.

Figure 1.5 shows the generalized schematic block-diagram for a closed-loop, or feedback control system. The controller and plant lie along the forward path, and the sensor in the feedback path. The measured value of the plant output is compared at the summing point with the desired value. The difference, or error is fed to the controller which generates a control signal to drive the plant until its output equals the desired value. Such an arrangement is sometimes called an error-actuated system.

## 1.3 Examples of control systems

### 1.3.1 Room temperature control system

The physical realization of a system to control room temperature is shown in Figure 1.6. Here the output signal from a temperature sensing device such as a thermocouple or a resistance thermometer is compared with the desired temperature. Any difference or error causes the controller to send a control signal to the gas solenoid valve which produces a linear movement of the valve stem, thus adjusting the flow of gas to the burner of the gas fire. The desired temperature is usually obtained from manual adjustment of a potentiometer.

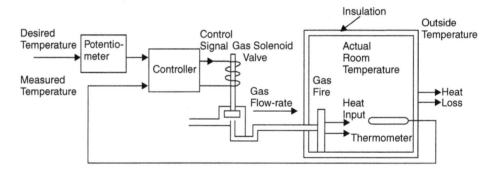

**Fig. 1.6** Room temperature control system.

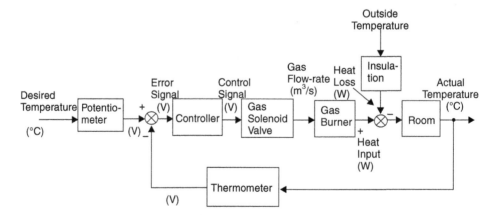

**Fig. 1.7** Block diagram of room temperature control system.

A detailed block diagram is shown in Figure 1.7. The physical values of the signals around the control loop are shown in brackets.

Steady conditions will exist when the actual and desired temperatures are the same, and the heat input exactly balances the heat loss through the walls of the building.

The system can operate in two modes:

(a) *Proportional control*: Here the linear movement of the valve stem is proportional to the error. This provides a continuous modulation of the heat input to the room producing very precise temperature control. This is used for applications where temperature control, of say better than 1 °C, is required (i.e. hospital operating theatres, industrial standards rooms, etc.) where accuracy is more important than cost.

(b) *On–off control*: Also called thermostatic or bang-bang control, the gas valve is either fully open or fully closed, i.e. the heater is either on or off. This form of control produces an oscillation of about 2 or 3 °C of the actual temperature about the desired temperature, but is cheap to implement and is used for low-cost applications (i.e. domestic heating systems).

## 1.3.2 Aircraft elevator control

In the early days of flight, control surfaces of aircraft were operated by cables connected between the control column and the elevators and ailerons. Modern high-speed aircraft require power-assisted devices, or servomechanisms to provide the large forces necessary to operate the control surfaces.

Figure 1.8 shows an elevator control system for a high-speed jet.

Movement of the control column produces a signal from the input angular sensor which is compared with the measured elevator angle by the controller which generates a control signal proportional to the error. This is fed to an electrohydraulic servovalve which generates a spool-valve movement that is proportional to the control signal,

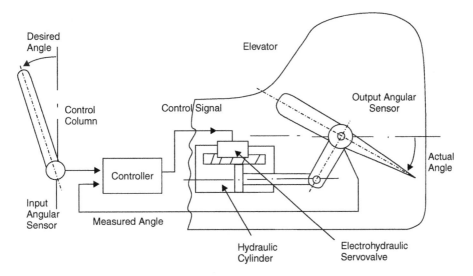

**Fig. 1.8** Elevator control system for a high-speed jet.

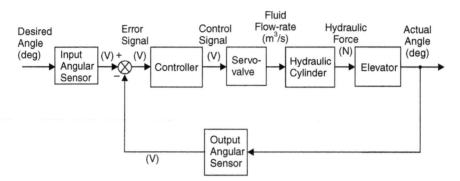

**Fig. 1.9** Block diagram of elevator control system.

thus allowing high-pressure fluid to enter the hydraulic cylinder. The pressure difference across the piston provides the actuating force to operate the elevator.

Hydraulic servomechanisms have a good power/weight ratio, and are ideal for applications that require large forces to be produced by small and light devices.

In practice, a 'feel simulator' is attached to the control column to allow the pilot to sense the magnitude of the aerodynamic forces acting on the control surfaces, thus preventing excess loading of the wings and tail-plane. The block diagram for the elevator control system is shown in Figure 1.9.

### 1.3.3 Computer Numerically Controlled (CNC) machine tool

Many systems operate under computer control, and Figure 1.10 shows an example of a CNC machine tool control system.

Information relating to the shape of the work-piece and hence the motion of the machine table is stored in a computer program. This is relayed in digital format, in a sequential form to the controller and is compared with a digital feedback signal from the shaft encoder to generate a digital error signal. This is converted to an analogue

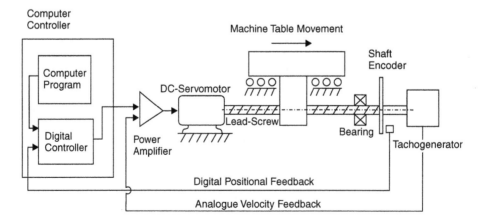

**Fig. 1.10** Computer numerically controlled machine tool.

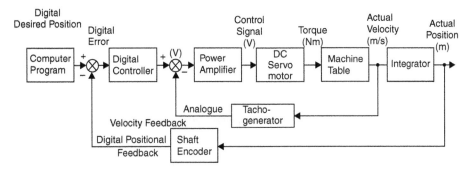

**Fig. 1.11** Block diagram of CNC machine-tool control system.

control signal which, when amplified, drives a d.c. servomotor. Connected to the output shaft of the servomotor (in some cases through a gearbox) is a lead-screw to which is attached the machine table, the shaft encoder and a tachogenerator. The purpose of this latter device, which produces an analogue signal proportional to velocity, is to form an inner, or minor control loop in order to dampen, or stabilize the response of the system.

The block diagram for the CNC machine tool control system is shown in Figure 1.11.

## 1.3.4 Ship autopilot control system

A ship autopilot is designed to maintain a vessel on a set heading while being subjected to a series of disturbances such as wind, waves and current as shown in Figure 1.3. This method of control is referred to as course-keeping. The autopilot can also be used to change course to a new heading, called course-changing. The main elements of the autopilot system are shown in Figure 1.12.

The actual heading is measured by a gyro-compass (or magnetic compass in a smaller vessel), and compared with the desired heading, dialled into the autopilot by the ship's master. The autopilot, or controller, computes the demanded rudder angle and sends a control signal to the steering gear. The actual rudder angle is monitored by a rudder angle sensor and compared with the demanded rudder angle, to form a control loop not dissimilar to the elevator control system shown in Figure 1.8.

The rudder provides a control moment on the hull to drive the actual heading towards the desired heading while the wind, waves and current produce moments that may help or hinder this action. The block diagram of the system is shown in Figure 1.13.

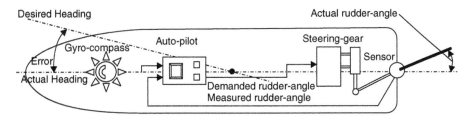

**Fig. 1.12** Ship autopilot control system.

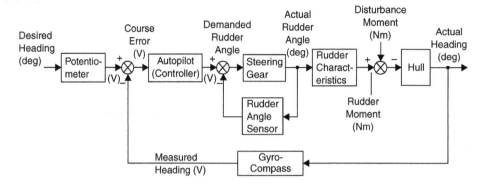

**Fig. 1.13** Block diagram of ship autopilot control system.

## 1.4  Summary

In order to design and implement a control system the following essential generic elements are required:

- *Knowledge of the desired value*: It is necessary to know what it is you are trying to control, to what accuracy, and over what range of values. This must be expressed in the form of a performance specification. In the physical system this information must be converted into a form suitable for the controller to understand (analogue or digital signal).
- *Knowledge of the output or actual value*: This must be measured by a feedback sensor, again in a form suitable for the controller to understand. In addition, the sensor must have the necessary resolution and dynamic response so that the measured value has the accuracy required from the performance specification.
- *Knowledge of the controlling device*: The controller must be able to accept measurements of desired and actual values and compute a control signal in a suitable form to drive an actuating element. Controllers can be a range of devices, including mechanical levers, pneumatic elements, analogue or digital circuits or microcomputers.
- *Knowledge of the actuating device*: This unit amplifies the control signal and provides the 'effort' to move the output of the plant towards its desired value. In the case of the room temperature control system the actuator is the gas solenoid valve and burner, the 'effort' being heat input (W). For the ship autopilot system the actuator is the steering gear and rudder, the 'effort' being turning moment (Nm).
- *Knowledge of the plant*: Most control strategies require some knowledge of the static and dynamic characteristics of the plant. These can be obtained from measurements or from the application of fundamental physical laws, or a combination of both.

### 1.4.1  Control system design

With all of this knowledge and information available to the control system designer, all that is left is to design the system. The first problem to be encountered is that the

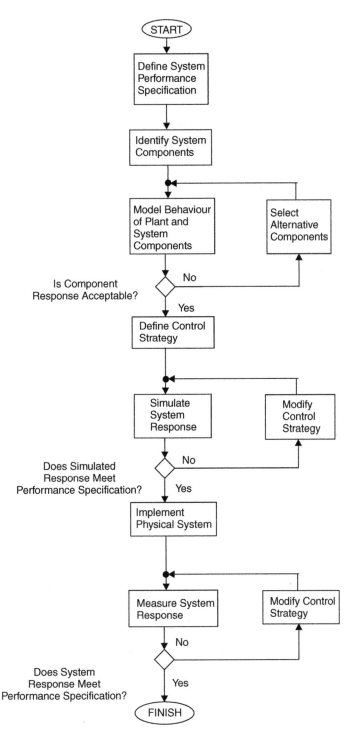

**Fig. 1.14** Steps in the design of a control system.

knowledge of the system will be uncertain and incomplete. In particular, the dynamic characteristics of the system may change with time (time-variant) and so a fixed control strategy will not work. Due to fuel consumption for example, the mass of an airliner can be almost half that of its take-off value at the end of a long haul flight.

Measurements of the controlled variables will be contaminated with electrical noise and disturbance effects. Some sensors will provide accurate and reliable data, others, because of difficulties in measuring the output variable may produce highly random and almost irrelevant information.

However, there is a standard methodology that can be applied to the design of most control systems. The steps in this methodology are shown in Figure 1.14.

The design of a control system is a mixture of technique and experience. This book explains some tried and tested, and some more recent approaches, techniques and methods available to the control system designer. Experience, however, only comes with time.

# 2

# System modelling

## 2.1 Mathematical models

If the dynamic behaviour of a physical system can be represented by an equation, or a set of equations, this is referred to as the mathematical model of the system. Such models can be constructed from knowledge of the physical characteristics of the system, i.e. mass for a mechanical system or resistance for an electrical system. Alternatively, a mathematical model may be determined by experimentation, by measuring how the system output responds to known inputs.

## 2.2 Simple mathematical model of a motor vehicle

Assume that a mathematical model for a motor vehicle is required, relating the accelerator pedal angle $\theta$ to the forward speed $u$, a simple mathematical model might be

$$u(t) = a\theta(t) \tag{2.1}$$

Since $u$ and $\theta$ are functions of time, they are written $u(t)$ and $\theta(t)$. The constant $a$ could be calculated if the following vehicle data for engine torque $T$, wheel traction force $F$, aerodynamic drag $D$ were available

$$T = b\theta(t) \tag{2.2}$$

$$F = cT \tag{2.3}$$

$$D = du(t) \tag{2.4}$$

Now aerodynamic drag $D$ must equal traction force $F$

$$D = F$$
$$du(t) = cT$$

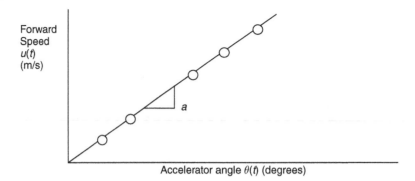

**Fig. 2.1** Vehicle forward speed plotted against accelerator angle.

from (2.2)

$$du(t) = cb\theta(t)$$

giving

$$u(t) = \left(\frac{cb}{d}\right)\theta(t) \tag{2.5}$$

Hence the constant for the vehicle is

$$a = \left(\frac{cb}{d}\right) \tag{2.6}$$

If the constants $b$, $c$ and $d$ were not available, then the vehicle model could be obtained by measuring the forward speed $u(t)$ for a number of different accelerator angles $\theta(t)$ and plotting the results, as shown in Figure 2.1.

Since Figure 2.1 shows a linear relationship, the value of the vehicle constant $a$ is the slope of the line.

## 2.3  More complex mathematical models

Equation (2.1) for the motor vehicle implies that when there is a change in accelerator angle, there is an instantaneous change in vehicle forward speed. As all car drivers know, it takes time to build up to the new forward speed, so to model the dynamic characteristics of the vehicle accurately, this needs to be taken into account.

Mathematical models that represent the dynamic behaviour of physical systems are constructed using differential equations. A more accurate representation of the motor vehicle would be

$$e\frac{du}{dt} + fu = g\theta(t) \tag{2.7}$$

Here, $du/dt$ is the acceleration of the vehicle. When it travels at constant velocity, this term becomes zero. So then

$$fu(t) = g\theta(t)$$
$$u(t) = \left(\frac{g}{f}\right)\theta(t) \tag{2.8}$$

Hence $(g/f)$ is again the vehicle constant, or parameter $a$ in equation (2.1)

### 2.3.1 Differential equations with constant coefficients

In general, consider a system whose output is $x(t)$, whose input is $y(t)$ and contains constant coefficients of values $a, b, c, \ldots, z$. If the dynamics of the system produce a first-order differential equation, it would be represented as

$$a\frac{dx}{dt} + bx = cy(t) \tag{2.9}$$

If the system dynamics produced a second-order differential equation, it would be represented by

$$a\frac{d^2x}{dt^2} + b\frac{dx}{dt} + cx = ey(t) \tag{2.10}$$

If the dynamics produce a third-order differential equation, its representation would be

$$a\frac{d^3x}{dt^3} + b\frac{d^2x}{dt^2} + c\frac{dx}{dt} + ex = fy(t) \tag{2.11}$$

Equations (2.9), (2.10) and (2.11) are linear differential equations with constant coefficients. Note that the order of the differential equation is the order of the highest derivative. Systems described by such equations are called linear systems of the same order as the differential equation. For example, equation (2.9) describes a first-order linear system, equation (2.10) a second-order linear system and equation (2.11) a third-order linear system.

## 2.4 Mathematical models of mechanical systems

Mechanical systems are usually considered to comprise of the linear lumped parameter elements of stiffness, damping and mass.

### 2.4.1 Stiffness in mechanical systems

An elastic element is assumed to produce an extension proportional to the force (or torque) applied to it.

For the translational spring

$$\text{Force} \propto \text{Extension}$$

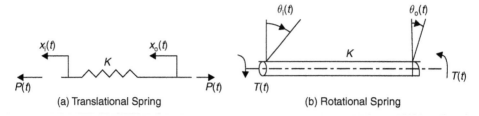

(a) Translational Spring  (b) Rotational Spring

**Fig. 2.2** Linear elastic elements.

If $x_i(t) > x_o(t)$, then

$$P(t) = K(x_i(t) - x_o(t)) \tag{2.12}$$

And for the rotational spring

$$\text{Torque} \propto \text{Twist}$$

If $\theta_i(t) > \theta_o(t)$, then

$$T(t) = K(\theta_i(t) - \theta_o(t)) \tag{2.13}$$

Note that $K$, the spring stiffness, has units of (N/m) in equation (2.12) and (Nm/rad) in equation (2.13).

### 2.4.2  Damping in mechanical systems

A damping element (sometimes called a dashpot) is assumed to produce a velocity proportional to the force (or torque) applied to it.

For the translational damper

$$\text{Force} \propto \text{Velocity}$$

$$P(t) = Cv(t) = C\frac{dx_o}{dt} \tag{2.14}$$

And for the rotational damper

$$\text{Torque} \propto \text{Angular velocity}$$

$$T(t) = C\omega(t) = C\frac{d\theta_o}{dt} \tag{2.15}$$

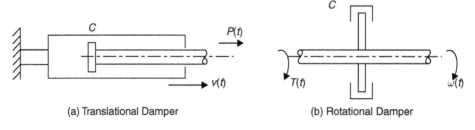

(a) Translational Damper  (b) Rotational Damper

**Fig. 2.3** Linear damping elements.

Note that $C$, the damping coefficient, has units of (Ns/m) in equation (2.14) and (Nm s/rad) in equation (2.15).

### 2.4.3 Mass in mechanical systems

The force to accelerate a body is the product of its mass and acceleration (Newton's second law).

For the translational system

$$\text{Force} \propto \text{Acceleration}$$

$$P(t) = ma(t) = m\frac{dv}{dt} = m\frac{d^2x_o}{dt^2} \tag{2.16}$$

For the rotational system

$$\text{Torque} \propto \text{Angular acceleration}$$

$$T(t) = I\alpha(t) = I\frac{d\omega_\omega}{dt} = I\frac{d^2\theta_o}{dt^2} \tag{2.17}$$

In equation (2.17) $I$ is the moment of inertia about the rotational axis.

When analysing mechanical systems, it is usual to identify all external forces by the use of a 'Free-body diagram', and then apply Newton's second law of motion in the form:

$$\sum F = ma \quad \text{for translational systems}$$

or

$$\sum M = I\alpha \quad \text{for rotational systems} \tag{2.18}$$

*Example 2.1*
Find the differential equation relating the displacements $x_i(t)$ and $x_o(t)$ for the spring–mass–damper system shown in Figure 2.5. What would be the effect of neglecting the mass?

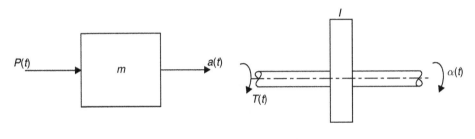

(a) Translational Acceleration    (b) Angular Acceleration

**Fig. 2.4** Linear mass elements.

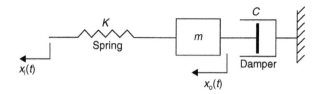

**Fig. 2.5** Spring–mass–damper system.

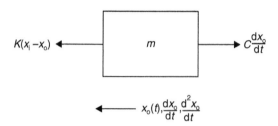

**Fig. 2.6** Free-body diagram for spring–mass–damper system.

*Solution*
Using equations (2.12) and (2.14) the free-body diagram is shown in Figure 2.6.
From equation (2.18), the equation of motion is

$$\sum F_x = ma_x$$

$$K(x_i - x_o) - C\frac{dx_o}{dt} = m\frac{d^2x_o}{dt^2}$$

$$Kx_i - Kx_o = m\frac{d^2x_o}{dt^2} + C\frac{dx_o}{dt}$$

Putting in the form of equation (2.10)

$$m\frac{d^2x_o}{dt^2} + C\frac{dx_o}{dt} + Kx_o = Kx_i(t) \qquad (2.19)$$

Hence a spring–mass–damper system is a second-order system.
  If the mass is zero then

$$\sum F_x = 0$$

$$K(x_i - x_o) - C\frac{dx_o}{dt} = 0$$

$$Kx_i - Kx_o = C\frac{dx_o}{dt}$$

Hence

$$C\frac{dx_o}{dt} + Kx_o = Kx_i(t) \qquad (2.20)$$

Thus if the mass is neglected, the system becomes a first-order system.

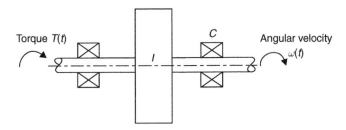

**Fig. 2.7** Flywheel in bearings.

*Example 2.2*
A flywheel of moment of inertia $I$ sits in bearings that produce a frictional moment of $C$ times the angular velocity $\omega(t)$ of the shaft as shown in Figure 2.7. Find the differential equation relating the applied torque $T(t)$ and the angular velocity $\omega(t)$.

*Solution*
From equation (2.18), the equation of motion is

$$\sum M = I\alpha$$

$$T(t) - C\omega = I\frac{d\omega}{dt}$$

$$I\frac{d\omega}{dt} + C\omega = T(t) \tag{2.21}$$

*Example 2.3*
Figure 2.8 shows a reduction gearbox being driven by a motor that develops a torque $T_m(t)$. It has a gear reduction ratio of '$n$' and the moments of inertia on the motor and output shafts are $I_m$ and $I_o$, and the respective damping coefficients $C_m$ and $C_o$. Find the differential equation relating the motor torque $T_m(t)$ and the output angular position $\theta_o(t)$.

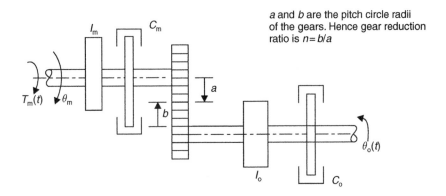

a and b are the pitch circle radii of the gears. Hence gear reduction ratio is $n = b/a$

**Fig. 2.8** Reduction gearbox.

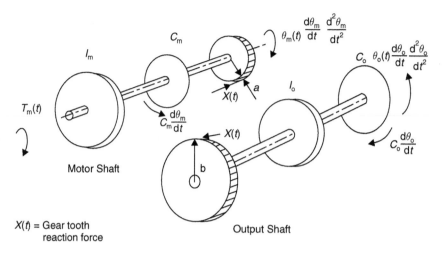

X(t) = Gear tooth reaction force

**Fig. 2.9** Free-body diagrams for reduction gearbox.

Gearbox parameters

$$I_{\mathrm{m}} = 5 \times 10^{-6}\,\mathrm{kg\,m^2}$$
$$I_{\mathrm{o}} = 0.01\,\mathrm{kg\,m^2}$$
$$C_{\mathrm{m}} = 60 \times 10^{-6}\,\mathrm{Nm\,s/rad}$$
$$C_{\mathrm{o}} = 0.15\,\mathrm{Nm\,s/rad}$$
$$n = 50{:}1$$

*Solution*
The free-body diagrams for the motor shaft and output shaft are shown in Figure 2.9.
　Equations of Motion are

(1) Motor shaft

$$\sum M = I_{\mathrm{m}}\frac{d^2\theta_{\mathrm{m}}}{dt^2}$$

$$T_{\mathrm{m}}(t) - C_{\mathrm{m}}\frac{d\theta_{\mathrm{m}}}{dt} - aX(t) = I_{\mathrm{m}}\frac{d^2\theta_{\mathrm{m}}}{dt^2}$$

re-arranging the above equation,

$$X(t) = \frac{1}{a}\left(T_{\mathrm{m}}(t) - I_{\mathrm{m}}\frac{d^2\theta_{\mathrm{m}}}{dt^2} - C_{\mathrm{m}}\frac{d\theta_{\mathrm{m}}}{dt}\right) \qquad (2.22)$$

(2) Output shaft

$$\sum M = I_{\mathrm{o}}\frac{d^2\theta_{\mathrm{m}}}{dt^2}$$

$$bX(t) - C_{\mathrm{o}}\frac{d\theta_{\mathrm{o}}}{dt} = I_{\mathrm{o}}\frac{d^2\theta_{\mathrm{o}}}{dt^2}$$

re-arranging the above equation,

$$X(t) = \frac{1}{b}\left(I_o\frac{d^2\theta_o}{dt^2} + C_o\frac{d\theta_o}{dt}\right)$$
(2.23)

Equating equations (2.22) and (2.23)

$$\frac{b}{a}\left(T_m(t) - I_m\frac{d^2\theta_m}{dt^2} - C_m\frac{d\theta_m}{dt}\right) = \left(I_o\frac{d^2\theta_o}{dt^2} + C_o\frac{d\theta_o}{dt}\right)$$

Kinematic relationships

$$\frac{b}{a} = n \quad \theta_m(t) = n\theta_o(t)$$

$$\frac{d\theta_m}{dt} = n\frac{d\theta_o}{dt}$$

$$\frac{d^2\theta_m}{dt^2} = n\frac{d^2\theta_o}{dt^2}$$

Hence

$$n\left(T_m(t) - nI_m\frac{d^2\theta_o}{dt^2} - nC_m\frac{d\theta_o}{dt}\right) = \left(I_o\frac{d^2\theta_o}{dt^2} + C_o\frac{d\theta_o}{dt}\right)$$

giving the differential equation

$$(I_o + n^2I_m)\frac{d^2\theta_o}{dt^2} + (C_o + n^2C_m)\frac{d\theta_o}{dt} = nT_m(t)$$
(2.24)

The terms $(I_o + n^2I_m)$ and $(C_o + n^2C_m)$ are called the equivalent moment of inertia $I_e$ and equivalent damping coefficient $C_e$ referred to the output shaft.

Substituting values gives

$$I_e = (0.01 + 50^2 \times 5 \times 10^{-6}) = 0.0225\,\text{kg m}^2$$
$$C_e = (0.15 + 50^2 \times 60 \times 10^{-6}) = 0.3\,\text{Nm s/rad}$$

From equation (2.24)

$$0.0225\frac{d^2\theta_o}{dt^2} + 0.3\frac{d\theta_o}{dt} = 50T_m(t)$$
(2.25)

## 2.5 Mathematical models of electrical systems

The basic passive elements of electrical systems are resistance, inductance and capacitance as shown in Figure 2.10.

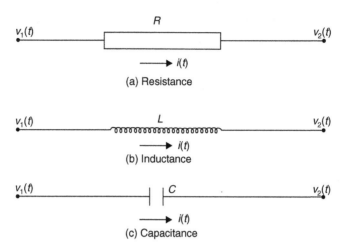

**Fig. 2.10** Passive elements of an electrical system.

For a resistive element, Ohm's Law can be written

$$(v_1(t) - v_2(t)) = Ri(t) \tag{2.26}$$

For an inductive element, the relationship between voltage and current is

$$(v_1(t) - v_2(t)) = L\frac{di}{dt} \tag{2.27}$$

For a capacitive element, the electrostatic equation is

$$Q(t) = C(v_1(t) - v_2(t))$$

Differentiating both sides with respect to $t$

$$\frac{dQ}{dt} = i(t) = C\frac{d}{dt}(v_1(t) - v_2(t)) \tag{2.28}$$

Note that if both sides of equation (2.28) are integrated then

$$(v_1(t) - v_2(t)) = \frac{1}{C}\int i dt \tag{2.29}$$

*Example 2.4*
Find the differential equation relating $v_1(t)$ and $v_2(t)$ for the $RC$ network shown in
Figure 2.11.

*Solution*
From equations (2.26) and (2.29)

$$v_1(t) - v_2(t) = Ri(t)$$

$$v_2(t) = \frac{1}{C}\int i dt \tag{2.30}$$

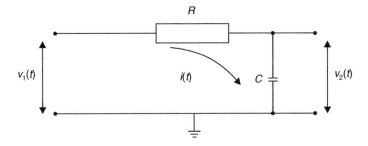

**Fig. 2.11** *RC* network.

or

$$C\frac{\mathrm{d}v_2}{\mathrm{d}t} = i(t) \tag{2.31}$$

substituting (2.31) into (2.30)

$$v_1(t) - v_2(t) = RC\frac{\mathrm{d}v_2}{\mathrm{d}t} \tag{2.32}$$

Equation (2.32) can be expressed as a first-order differential equation

$$RC\frac{\mathrm{d}v_2}{\mathrm{d}t} + v_2 = v_1(t) \tag{2.33}$$

*Example 2.5*
Find the differential equations relating $v_1(t)$ and $v_2(t)$ for the networks shown in Figure 2.12.

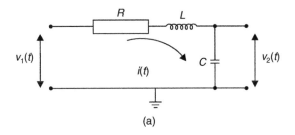

(a)

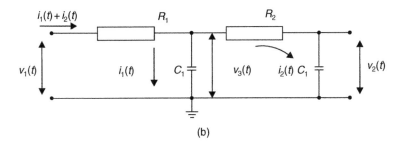

(b)

**Fig. 2.12** Electrical networks.

*Solution for Network (a) Figure 2.12*
From equations (2.26), (2.27) and (2.29)

$$v_1(t) - v_2(t) = Ri(t) + L\frac{di}{dt}$$

$$v_2(t) = \frac{1}{C}\int i\, dt$$

(2.34)

or

$$C\frac{dv_2}{dt} = i(t)$$

(2.35)

substituting (2.35) into (2.34)

$$v_1(t) - v_2(t) = RC\frac{dv_2}{dt} + L\frac{d}{dt}\left(C\frac{dv_2}{dt}\right)$$

or

$$v_1(t) - v_2(t) = RC\frac{dv_2}{dt} + LC\frac{d^2v_2}{dt^2}$$

(2.36)

Equation (2.36) can be expressed as a second-order differential equation

$$LC\frac{d^2v_2}{dt^2} + RC\frac{dv_2}{dt} + v_2 = v_1(t)$$

(2.37)

*Solution for Network (b) Figure 2.12*
System equations

$$v_1(t) - v_3(t) = R_1(i_1(t) + i_2(t))$$

(2.38)

$$v_3(t) = \frac{1}{C_1}\int i_1\, dt \quad \text{or} \quad C_1\frac{dv_3}{dt} = i_1(t)$$

(2.39)

$$v_3(t) - v_2(t) = R_2 i_2(t)$$

(2.40)

$$v_2(t) = \frac{1}{C_2}\int i_2\, dt \quad \text{or} \quad C_2\frac{dv_2}{dt} = i_2(t)$$

(2.41)

From equation (2.40)

$$v_3(t) = R_2 i_2(t) + v_2(t)$$

Substituting for $i_2(t)$ using equation (2.41)

$$v_3(t) = R_2 C_2\frac{dv_2}{dt} + v_2(t)$$

(2.42)

Hence from equations (2.42) and (2.39)

$$i_1(t) = C_1 \frac{d}{dt}\left\{ R_2 C_2 \frac{dv_2}{dt} + v_2(t) \right\}$$

$$= R_2 C_1 C_2 \frac{d^2 v_2}{dt^2} + C_1 \frac{dv_2}{dt} \qquad (2.43)$$

Substituting equations (2.41), (2.42) and (2.43) into equation (2.38)

$$v_1(t) - \left\{ R_2 C_2 \frac{dv_2}{dt} + v_2(t) \right\} = R_1 \left\{ R_2 C_1 C_2 \frac{d^2 v_2}{dt^2} + C_1 \frac{dv_2}{dt} + C_2 \frac{dv_2}{dt} \right\}$$

which produces the second-order differential equation

$$R_1 R_2 C_1 C_2 \frac{d^2 v_2}{dt^2} + (R_1 C_1 + R_1 C_2 + R_2 C_2)\frac{dv_2}{dt} + v_2 = v_1(t) \qquad (2.44)$$

## 2.6 Mathematical models of thermal systems

It is convenient to consider thermal systems as being analogous to electrical systems so that they contain both resistive and capacitive elements.

### 2.6.1 Thermal resistance $R_T$

Heat flow by conduction is given by Fourier's Law

$$Q_T = \frac{KA(\theta_1 - \theta_2)}{\ell} \qquad (2.45)$$

The parameters in equation (2.45) are shown in Figure 2.13. They are

$$(\theta_1 - \theta_2) = \text{Temperature differential (K)}$$
$$A = \text{Normal cross sectional area (m}^2)$$
$$\ell = \text{Thickness (m)}$$
$$K = \text{Thermal conductivity (W/mK)}$$
$$Q_T = \text{Heat flow (J/s} = \text{W)}$$

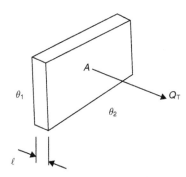

**Fig. 2.13** Heat flow through a flat plate.

Equation (2.45) can be written in the same form as Ohm's Law (equation (2.26))

$$(\theta_1(t) - \theta_2(t)) = R_T Q_T(t) \qquad (2.46)$$

where $R_T$ is the thermal resistance and is

$$R_T = \frac{\ell}{KA} \qquad (2.47)$$

## 2.6.2   Thermal capacitance $C_T$

The heat stored by a body is

$$H(t) = mC_p\theta(t) \qquad (2.48)$$

where

$H =$ Heat (J)
$m =$ Mass (kg)
$C_p =$ Specific heat at constant
pressure (J/kg K)
$\theta =$ Temperature rise (K)

If equation (2.48) is compared with the electrostatic equation

$$Q(t) = Cv(t) \qquad (2.49)$$

then the thermal capacitance $C_T$ is

$$C_T = mC_p \qquad (2.50)$$

To obtain the heat flow $Q_T$, equation (2.48) is differentiated with respect to time

$$\frac{dH}{dt} = mC_p\frac{d\theta}{dt} \qquad (2.51)$$

or

$$Q_T(t) = C_T\frac{d\theta}{dt} \qquad (2.52)$$

*Example 2.6*
Heat flows from a heat source at temperature $\theta_1(t)$ through a wall having ideal thermal resistance $R_T$ to a heat sink at temperature $\theta_2(t)$ having ideal thermal capacitance $C_T$ as shown in Figure 2.14. Find the differential equation relating $\theta_1(t)$ and $\theta_2(t)$.

*Solution*
(1) Wall: From equation (2.46)

$$Q_T(t) = \frac{(\theta_1(t) - \theta_2(t))}{R_T} \qquad (2.53)$$

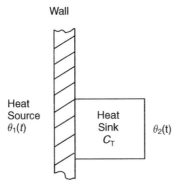

**Fig. 2.14** Heat transfer system.

(2) Heat sink: From equation (2.52)

$$Q_T(t) = C_T \frac{\mathrm{d}\theta_2}{\mathrm{d}t} \qquad (2.54)$$

Equating equations (2.53) and (2.54)

$$\frac{(\theta_1(t) - \theta_2(t))}{R_T} = C_T \frac{\mathrm{d}\theta_2}{\mathrm{d}t}$$

Re-arranging to give the first-order differential equation

$$R_T C_T \frac{\mathrm{d}\theta_2}{\mathrm{d}t} + \theta_2 = \theta_1(t) \qquad (2.55)$$

## 2.7 Mathematical models of fluid systems

Like thermal systems, it is convenient to consider fluid systems as being analogous to electrical systems. There is one important difference however, and this is that the relationship between pressure and flow-rate for a liquid under turbulent flow conditions is nonlinear. In order to represent such systems using linear differential equations it becomes necessary to linearize the system equations.

### 2.7.1 Linearization of nonlinear functions for small perturbations

Consider a nonlinear function $Y = f(x)$ as shown in Figure 2.15. Assume that it is necessary to operate in the vicinity of point $a$ on the curve (the operating point) whose co-ordinates are $X_a, Y_a$.

For the small perturbations $\Delta X$ and $\Delta Y$ about the operating point $a$ let

$$\Delta X = x$$
$$\Delta Y = y \qquad (2.56)$$

If the slope at the operating point is

$$\left.\frac{\mathrm{d}Y}{\mathrm{d}X}\right|_a$$

then the approximate linear relationship becomes

$$y = \left.\frac{\mathrm{d}Y}{\mathrm{d}X}\right|_a x \qquad (2.57)$$

*Example 2.7*
The free-body diagram of a ship is shown in Figure 2.16. It has a mass of $15 \times 10^6$ kg and the propeller produces a thrust of $K_n$ times the angular velocity $n$ of the propeller, $K_n$ having a value of $110 \times 10^3$ Ns/rad. The hydrodynamic resistance is given by the relationship $R = C_v V^2$, where $C_v$ has a value of $10{,}000$ Ns$^2$/m$^2$. Determine, using small perturbation theory, the linear differential equation relating the forward speed $v(t)$ and propeller angular velocity $n(t)$ when the forward speed is 7.5 m/s.

*Solution*
Linearize hydrodynamic resistance equation for an operating speed $V_a$ of 7.5 m/s.

$$R = C_v V^2$$

$$\frac{\mathrm{d}R}{\mathrm{d}V} = 2C_v V$$

$$\left.\frac{\mathrm{d}R}{\mathrm{d}V}\right|_a = 2C_v V_a$$

$$= 2 \times 10\,000 \times 7.5$$

$$\left.\frac{\mathrm{d}R}{\mathrm{d}V}\right|_a = C = 150\,000 \text{ Ns/m}$$

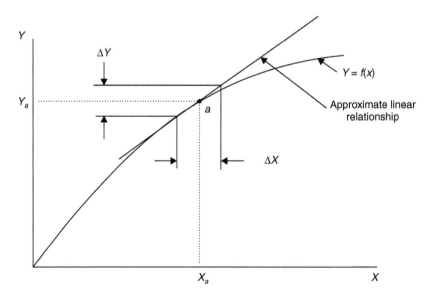

**Fig. 2.15** Linearization of a nonlinear function.

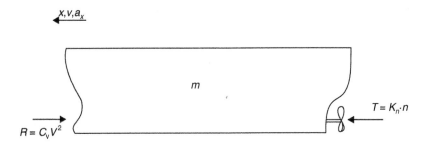

**Fig. 2.16** Free-body diagram of ship.

Hence the linear relationship is

$$R = Cv \tag{2.58}$$

Using Newton's second law of motion

$$\sum F_x = ma_x$$

$$T - R = m\frac{\mathrm{d}v}{\mathrm{d}t}$$

$$K_n n - Cv = m\frac{\mathrm{d}v}{\mathrm{d}t}$$

$$m\frac{\mathrm{d}v}{\mathrm{d}t} + Cv = K_n n \tag{2.59}$$

Substituting values gives

$$(15 \times 10^6)\frac{\mathrm{d}v}{\mathrm{d}t} + (150 \times 10^3)v = (110 \times 10^3)n(t) \tag{2.60}$$

*Example 2.8*
In Figure 2.17 the tank of water has a cross-sectional area $A$, and under steady conditions both the outflow and inflow is $V_a$ and the head is $H_a$.

(a) Under these conditions find an expression for the linearized valve resistance $R_f$ given that flow through the valve is

$$V = A_v C_d \sqrt{2gH},$$

where

$V$ = volumetric flow-rate (m³/s)

$A_v$ = valve flow area (m²)

$C_d$ = coefficient of discharge

$g$ = acceleration due to gravity (m/s²)

$H$ = head across the valve (m)

(b) If the steady value of the head $H_a$ is 1.5 m, what is the valve resistance $R_f$ when

$$A_v = 15 \times 10^{-3}\,\mathrm{m}^2$$
$$g = 9.81\,\mathrm{m/s}^2$$
$$C_d = 0.6$$

(c) If the inflow now increases an amount $v_1$ producing an increase in head $h$ and an increase in outflow $v_2$, find the differential equation relating $v_1$ and $v_2$ when the tank cross-sectional area $A$ is $0.75\,\mathrm{m}^2$.

*Solution*

(a) Flow through the valve is given by

$$V = A_v C_d \sqrt{2gH}$$

now

$$\left.\frac{dV}{dH}\right|_a = A_v C_d (2g)^{1/2} \times 0.5 H_a^{-1/2}$$

$$\left.\frac{dV}{dH}\right|_a = A_v C_d \sqrt{\frac{g}{2H_a}} = \frac{v_2}{h}$$

The linearized relationship is

$$h = R_f v_2$$

hence

$$R_f = \frac{1}{A_v C_d}\sqrt{\frac{2H_a}{g}} \tag{2.61}$$

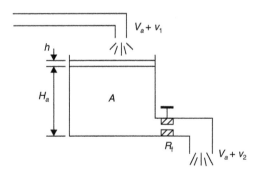

**Fig. 2.17** Tank and valve system.

(b) Inserting values gives

$$R_f = \frac{1}{15 \times 10^{-3} \times 0.6} \sqrt{\frac{2 \times 1.5}{9.81}}$$

$$R_f = 61.45 \, \text{s/m}^2 \tag{2.62}$$

(c) Tank (Continuity Equation)

$$\text{Inflow} - \text{Outflow} = A\frac{dh}{dt}$$

$$(V_a + v_1) - (V_a + v_2) = A\frac{dh}{dt}$$

$$v_1 - v_2 = A\frac{dh}{dt} \tag{2.63}$$

Valve (Linearized Equation)

$$h = R_f v_2$$

and

$$\frac{dh}{dt} = R_f \frac{dv_2}{dt} \tag{2.64}$$

Substituting equation (2.64) into equation (2.63)

$$v_1 - v_2 = AR_f \frac{dv_2}{dt}$$

giving

$$AR_f \frac{dv_2}{dt} + v_2 = v_1(t) \tag{2.65}$$

Inserting values gives

$$46.09\frac{dv_2}{dt} + v_2 = v_1(t) \tag{2.66}$$

## 2.8 Further problems

*Example 2.9*
A solenoid valve is shown in Figure 2.18. The coil has an electrical resistance of $4\,\Omega$, an inductance of $0.6\,\text{H}$ and produces an electromagnetic force $F_c(t)$ of $K_c$ times the current $i(t)$. The valve has a mass of $0.125\,\text{kg}$ and the linear bearings produce a resistive force of $C$ times the velocity $u(t)$. The values of $K_c$ and $C$ are $0.4\,\text{N/A}$ and $0.25\,\text{Ns/m}$ respectively. Develop the differential equations relating the voltage $v(t)$ and current $i(t)$ for the electrical circuit, and also for the current $i(t)$ and velocity $u(t)$ for the mechanical elements. Hence deduce the overall differential equation relating the input voltage $v(t)$ to the output velocity $u(t)$.

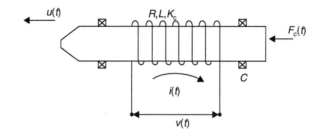

**Fig. 2.18** Solenoid valve.

*Solution*

$$L\frac{di}{dt} + Ri = v(t)$$

$$m\frac{du}{dt} + Cu = K_c i(t)$$

$$0.075\frac{d^2u}{dt^2} + 0.65\frac{du}{dt} + u = 0.4v(t)$$

*Example 2.10*

The laser-guided missile shown in Figure 2.19 has a pitch moment of inertia of $90\,\text{kg}\,\text{m}^2$. The control fins produce a moment about the pitch mass centre of $360\,\text{Nm}$ per radian of fin angle $\beta(t)$. The fin positional control system is described by the differential equation

$$0.2\frac{d\beta}{dt} + \beta(t) = u(t)$$

where $u(t)$ is the control signal. Determine the differential equation relating the control signal $u(t)$ and the pitch angle $\theta(t)$.

*Solution*

$$\frac{d^3\theta}{dt^3} + 5\frac{d^2\theta}{dt^2} = 20u(t)$$

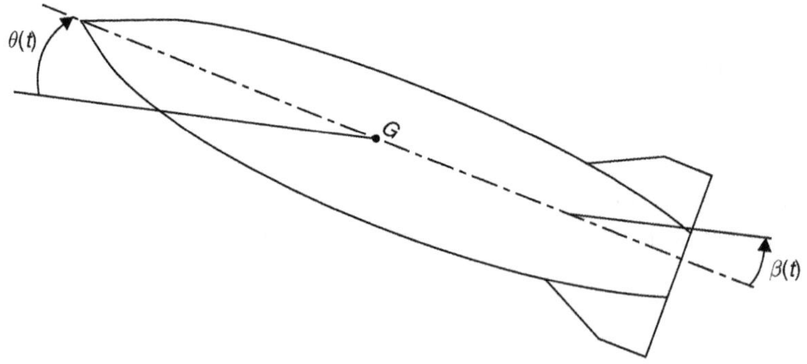

**Fig. 2.19** Laser-guided missile.

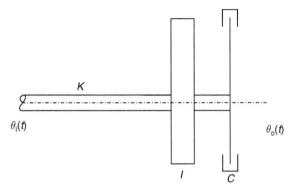

**Fig. 2.20** Torsional spring–mass–damper system.

*Example 2.11*
A torsional spring of stiffness $K$, a mass of moment of inertia $I$ and a fluid damper with damping coefficient $C$ are connected together as shown in Figure 2.20. If the angular displacement of the free end of the spring is $\theta_i(t)$ and the angular displacement of the mass and damper is $\theta_o(t)$, find the differential equation relating $\theta_i(t)$ and $\theta_o(t)$ given that

$$I = 2.5 \, \text{kg m}^2$$
$$C = 12.5 \, \text{Nm s/rad}$$
$$K = 250 \, \text{Nm/rad}$$

*Solution*

$$2.5\frac{\text{d}^2\theta_o}{\text{d}t^2} + 12.5\frac{\text{d}\theta_o}{\text{d}t} + 250\theta_o = 250\theta_i(t)$$

*Example 2.12*
A field controlled d.c. motor develops a torque $T_m(t)$ proportional to the field current $i_f(t)$. The rotating parts have a moment of inertia $I$ of $1.5 \, \text{kg m}^2$ and a viscous damping coefficient $C$ of $0.5 \, \text{Nm s/rad}$.

When a current of $1.0 \, \text{A}$ is passed through the field coil, the shaft finally settles down to a steady speed $\omega_o(t)$ of $5 \, \text{rad/s}$.

(a) Determine the differential equations relating $i_f(t)$ and $\omega_o(t)$.
(b) What is the value of the coil constant $K_c$, and hence what is the torque developed by the motor when a current of $0.5 \, \text{A}$ flows through the field coil?

*Solution*
(a) $I\dfrac{\text{d}\omega_o}{\text{d}t} + C\omega_o = K_c i_f(t)$

(b) $K_c = 2.5 \, \text{Nm/A}$. $T_m = 1.25 \, \text{Nm}$

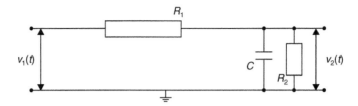

**Fig. 2.21** Passive *RC* network.

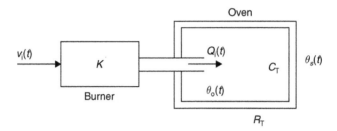

**Fig. 2.22** Drying oven.

*Example 2.13*
Figure 2.21 shows a passive electrical network. Determine the differential equation relating $v_1(t)$ and $v_2(t)$.

*Solution*

$$R_1 C \frac{dv_2}{dt} + \left( \frac{R_1 + R_2}{R_2} \right) v_2 = v_1(t)$$

*Example 2.14*
A drying oven which is constructed of firebrick walls is heated by an electrically operated gas burner as shown in Figure 2.22. The system variables and constants are

$v_1(t)$ = burner operating voltage (V)
$Q_i(t)$ = heat input to oven (W)
$\theta_o(t)$ = internal oven temperature (K)
$\theta_s(t)$ = temperature of surroundings (K)
$K$ = burner constant = 2000 W/V
$R_T$ = thermal resistance of walls = $0.5 \times 10^{-3}$ min K/J
$C_T$ = oven thermal capacitance = $1 \times 10^4$ J/K

Find the differential equation relating $v_i(t)$, $\theta_o(t)$ and $\theta_s(t)$.

*Solution*

$$5 \frac{d\theta_o}{dt} + \theta_o = v_1(t) + \theta_s(t)$$

# 3

# Time domain analysis

## 3.1 Introduction

The manner in which a dynamic system responds to an input, expressed as a function of time, is called the time response. The theoretical evaluation of this response is said to be undertaken in the time domain, and is referred to as time domain analysis. It is possible to compute the time response of a system if the following is known:

- the nature of the input(s), expressed as a function of time
- the mathematical model of the system.

The time response of any system has two components:

(a) *Transient response*: This component of the response will (for a stable system) decay, usually exponentially, to zero as time increases. It is a function only of the system dynamics, and is independent of the input quantity.
(b) *Steady-state response*: This is the response of the system after the transient component has decayed and is a function of both the system dynamics and the input quantity.

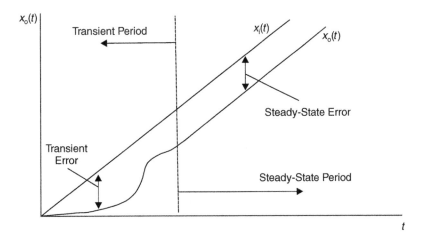

**Fig. 3.1** Transient and steady-state periods of time response.

The total response of the system is always the sum of the transient and steady-state components. Figure 3.1 shows the transient and steady-state periods of time response. Differences between the input function $x_i(t)$ (in this case a ramp function) and system response $x_o(t)$ are called transient errors during the transient period, and steady-state errors during the steady-state period. One of the major objectives of control system design is to minimize these errors.

## 3.2  Laplace transforms

In order to compute the time response of a dynamic system, it is necessary to solve the differential equations (system mathematical model) for given inputs. There are a number of analytical and numerical techniques available to do this, but the one favoured by control engineers is the use of the Laplace transform.

This technique transforms the problem from the time (or $t$) domain to the Laplace (or $s$) domain. The advantage in doing this is that complex time domain differential equations become relatively simple $s$ domain algebraic equations. When a suitable solution is arrived at, it is inverse transformed back to the time domain. The process is shown in Figure 3.2.

The Laplace transform of a function of time $f(t)$ is given by the integral

$$\mathscr{L}[f(t)] = \int_0^\infty f(t)e^{-st}dt = F(s) \tag{3.1}$$

where $s$ is a complex variable $\sigma \pm j\omega$ and is called the Laplace operator.

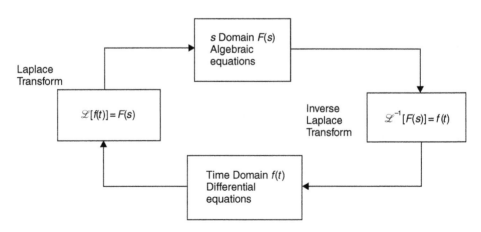

**Fig. 3.2** The Laplace transform process.

## 3.2.1   Laplace transforms of common functions

*Example 3.1*
$f(t) = 1$ (called a unit step function).

*Solution*
From equation (3.1)

$$\mathscr{L}[f(t)] = F(s) = \int_0^\infty 1e^{-st}dt$$

$$= \left[ -\frac{1}{s}(e^{-st}) \right]_0^\infty$$

$$= \left[ -\frac{1}{s}(0-1) \right] = \frac{1}{s} \qquad (3.2)$$

*Example 3.2*

$$f(t) = e^{-at}$$

$$\mathscr{L}[f(t)] = F(s) = \int_0^\infty e^{-at}e^{-st}dt$$

$$= \int_0^\infty e^{-(s+a)t}dt$$

$$= \left[ -\frac{1}{s+a}(e^{-(s+a)t}) \right]_0^\infty$$

$$= \left[ -\frac{1}{s+a}(0-1) \right]$$

$$= \frac{1}{s+a} \qquad (3.3)$$

Table 3.1 gives further Laplace transforms of common functions (called Laplace transform pairs).

## 3.2.2   Properties of the Laplace transform

(a) Derivatives: The Laplace transform of a time derivative is

$$\frac{d^n}{dt^n}f(t) = s^n F(s) - f(0)s^{n-1} - f'(0)s^{n-2} - \cdots \qquad (3.4)$$

where $f(0), f'(0)$ are the initial conditions, or the values of $f(t)$, $d/dt\, f(t)$ etc. at $t = 0$
(b) Linearity

$$\mathscr{L}[f_1(t) \pm f_2(t)] = F_1(s) \pm F_2(s) \qquad (3.5)$$

**Table 3.1** Common Laplace transform pairs

| Time function $f(t)$ | Laplace transform $\mathscr{L}[f(t)] = F(s)$ |
|---|---|
| 1 unit impulse $\delta(t)$ | 1 |
| 2 unit step 1 | $1/s$ |
| 3 unit ramp $t$ | $1/s^2$ |
| 4 $t^n$ | $\dfrac{n!}{s^{n+1}}$ |
| 5 $e^{-at}$ | $\dfrac{1}{(s+a)}$ |
| 6 $1 - e^{-at}$ | $\dfrac{a}{s(s+a)}$ |
| 7 $\sin \omega t$ | $\dfrac{\omega}{s^2 + \omega^2}$ |
| 8 $\cos \omega t$ | $\dfrac{s}{s^2 + \omega^2}$ |
| 9 $e^{-at} \sin \omega t$ | $\dfrac{\omega}{(s+a)^2 + \omega^2}$ |
| 10 $e^{-at}(\cos \omega t - \dfrac{a}{\omega} \sin \omega t)$ | $\dfrac{s}{(s+a)^2 + \omega^2}$ |

(c) Constant multiplication

$$\mathscr{L}[af(t)] = aF(s) \tag{3.6}$$

(d) Real shift theorem

$$\mathscr{L}[f(t - T)] = e^{-Ts}F(s) \quad \text{for} \quad T \geq 0 \tag{3.7}$$

(e) Convolution integral

$$\int_0^t f_1(\tau)f_2(t - \tau)d\tau = F_1(s)F_2(s) \tag{3.8}$$

(f) Initial value theorem

$$f(0) = \lim_{t \to 0}[f(t)] = \lim_{s \to \infty}[sF(s)] \tag{3.9}$$

(g) Final value theorem

$$f(\infty) = \lim_{t \to \infty}[f(t)] = \lim_{s \to 0}[sF(s)] \tag{3.10}$$

## 3.2.3 Inverse transformation

The inverse transform of a function of $s$ is given by the integral

$$f(t) = \mathscr{L}^{-1}[F(s)] = \frac{1}{2\pi j} \int_{\sigma - j\omega}^{\sigma + j\omega} F(s)e^{st}ds \tag{3.11}$$

In practice, inverse transformation is most easily achieved by using partial fractions to break down solutions into standard components, and then use tables of Laplace transform pairs, as given in Table 3.1.

## 3.2.4 Common partial fraction expansions

(i) Factored roots

$$\frac{K}{s(s+a)} = \frac{A}{s} + \frac{B}{(s+a)} \tag{3.12}$$

(ii) Repeated roots

$$\frac{K}{s^2(s+a)} = \frac{A}{s} + \frac{B}{s^2} + \frac{C}{(s+a)} \tag{3.13}$$

(iii) Second-order real roots ($b^2 > 4ac$)

$$\frac{K}{s(as^2 + bs + c)} = \frac{K}{s(s+d)(s+e)} = \frac{A}{s} + \frac{B}{(s+d)} + \frac{C}{(s+e)}$$

(iv) Second-order complex roots ($b^2 < 4ac$)

$$\frac{K}{s(as^2 + bs + c)} = \frac{A}{s} + \frac{Bs + C}{as^2 + bs + c}$$

Completing the square gives

$$\frac{A}{s} + \frac{Bs + C}{(s+\alpha)^2 + \omega^2} \tag{3.14}$$

*Note*: In (iii) and (iv) the coefficient $a$ is usually factored to a unity value.

## 3.3 Transfer functions

A transfer function is the Laplace transform of a differential equation with zero initial conditions. It is a very easy way to transform from the time to the $s$ domain, and a powerful tool for the control engineer.

*Example 3.3*
Find the Laplace transform of the following differential equation given:

(a) initial conditions $x_o = 4$, $dx_o/dt = 3$
(b) zero initial conditions

$$\frac{d^2 x_o}{dt^2} + 3\frac{dx_o}{dt} + 2x_o = 5$$

**Fig. 3.3** The transfer function approach.

*Solution*
(a) *Including initial conditions*: Take Laplace transforms (equation (3.4), Table 3.1).

$$(s^2 X_o(s) - 4s - 3) + 3(sX_o(s) - 4) + 2X_o(s) = \frac{5}{s}$$

$$s^2 X_o(s) + 3sX_o(s) + 2X_o(s) = \frac{5}{s} + 4s + 3 + 12$$

$$(s^2 + 3s + 2)X_o(s) = \frac{5 + 4s^2 + 15s}{s}$$

$$X_o(s) = \frac{4s^2 + 15s + 5}{s(s^2 + 3s + 2)} \qquad (3.15)$$

(b) *Zero initial conditions*
At $t = 0$, $x_o = 0$, $dx_o/dt = 0$.
   Take Laplace transforms

$$s^2 X_o(s) + 3sX_o(s) + 2X_o(s) = \frac{5}{s}$$

$$X_o(s) = \frac{5}{s(s^2 + 3s + 2)} \qquad (3.16)$$

Example 3.3(b) is easily solved using transfer functions. Figure 3.3 shows the general approach. In Figure 3.3

- $X_i(s)$ is the Laplace transform of the input function.
- $X_o(s)$ is the Laplace transform of the output function, or system response.
- $G(s)$ is the transfer function, i.e. the Laplace transform of the differential equation for zero initial conditions.

The solution is therefore given by

$$X_o(s) = G(s)X_i(s) \qquad (3.17)$$

Thus, for a general second-order transfer function

$$a\frac{d^2 x_o}{dt^2} + b\frac{dx_o}{dt} + cx_o = Kx_i(t)$$

$$(as^2 + bs + c)X_o(s) = KX_i(s)$$

Hence

$$X_o(s) = \left\{\frac{K}{as^2 + bs + c}\right\}X_i(s) \qquad (3.18)$$

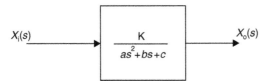

**Fig. 3.4** General second-order transfer function.

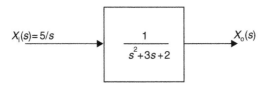

**Fig. 3.5** Example 3.3(b) expressed as a transfer function.

Comparing equations (3.17) and (3.18), the transfer function $G(s)$ is

$$G(s) = \frac{K}{as^2 + bs + c} \qquad (3.19)$$

which, using the form shown in Figure 3.3, can be expressed as shown in Figure 3.4.

Returning to Example 3.3(b), the solution, using the transfer function approach is shown in Figure 3.5. From Figure 3.5

$$X_o(s) = \frac{5}{s(s^2 + 3s + 2)} \qquad (3.20)$$

which is the same as equation (3.16).

## 3.4 Common time domain input functions

### 3.4.1 The impulse function

An impulse is a pulse with a width $\Delta t \rightarrow 0$ as shown in Figure 3.6. The strength of an impulse is its area $A$, where

$$A = \text{height } h \times \Delta t. \qquad (3.21)$$

The Laplace transform of an impulse function is equal to the area of the function. The impulse function whose area is unity is called a unit impulse $\delta(t)$.

### 3.4.2 The step function

A step function is described as $x_i(t) = B$; $X_i(s) = B/s$ for $t > 0$ (Figure 3.7). For a unit step function $x_i(t) = 1$; $X_i(s) = 1/s$. This is sometimes referred to as a 'constant position' input.

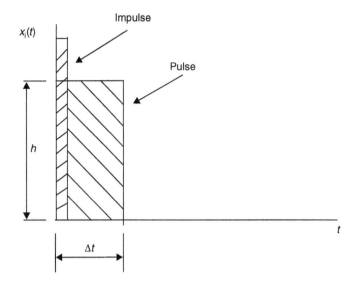

**Fig. 3.6** The impulse function.

**Fig. 3.7** The step function.

### 3.4.3   The ramp function

A ramp function is described as $x_i(t) = Qt$; $X_i(s) = Q/s^2$ for $t > 0$ (Figure 3.8). For a unit ramp function $x_i(t) = t$; $X_i(s) = 1/s^2$. This is sometimes referred to as a 'constant velocity' input.

### 3.4.4   The parabolic function

A parabolic function is described as $x_i(t) = Kt^2$; $X_i(s) = 2K/s^3$ for $t > 0$ (Figure 3.9). For a unit parabolic function $x_i(t) = t^2$; $X_i(s) = 2/s^3$. This is sometimes referred to as a 'constant acceleration' input.

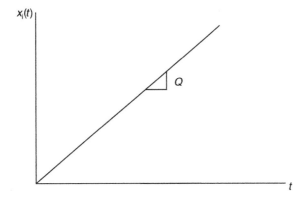

**Fig. 3.8** The ramp function.

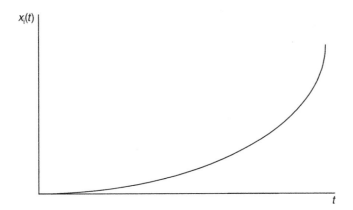

**Fig. 3.9** The parabolic function.

## 3.5 Time domain response of first-order systems

### 3.5.1 Standard form

Consider a first-order differential equation

$$a\frac{\mathrm{d}x_\mathrm{o}}{\mathrm{d}t} + bx_\mathrm{o} = cx_\mathrm{i}(t) \tag{3.22}$$

Take Laplace transforms, zero initial conditions

$$asX_\mathrm{o}(s) + bX_\mathrm{o}(s) = cX_\mathrm{i}(s)$$
$$(as + b)X_\mathrm{o}(s) = cX_\mathrm{i}(s)$$

The transfer function is

$$G(s) = \frac{X_o}{X_i}(s) = \frac{c}{as + b}$$

To obtain the standard form, divide by $b$

$$G(s) = \frac{\frac{c}{b}}{1 + \frac{a}{b}s}$$

which is written

$$G(s) = \frac{K}{1 + Ts} \tag{3.23}$$

Equation (3.23) is the standard form of transfer function for a first-order system, where $K$ = steady-state gain constant and $T$ = time constant (seconds).

## 3.5.2 Impulse response of first-order systems

*Example 3.4* (See also Appendix 1, *examp34.m*)
Find an expression for the response of a first-order system to an impulse function of area $A$.

*Solution*
From Figure 3.10

$$X_o(s) = \frac{AK}{1 + Ts} = \frac{AK/T}{(s + 1/T)} \tag{3.24}$$

or

$$X_o(s) = \frac{AK}{T}\left(\frac{1}{(s + a)}\right) \tag{3.25}$$

Equation (3.25) is in the form given in Laplace transform pair 5, Table 3.1, so the inverse transform becomes

$$x_o(t) = \frac{AK}{T}e^{-at} = \frac{AK}{T}e^{-t/T} \tag{3.26}$$

The impulse response function, equation (3.26) is shown in Figure 3.11.

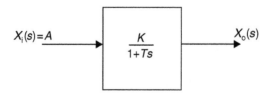

**Fig. 3.10** Impulse response of a first-order system.

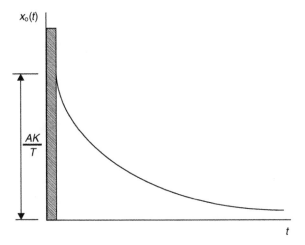

**Fig. 3.11** Response of a first-order system to an impulse function of area $A$.

### 3.5.3 Step response of first-order systems

*Example 3.5* (See also Appendix 1, *examp35.m*)
Find an expression for the response of a first-order system to a step function of height $B$.

*Solution*
From Figure 3.12

$$X_o(s) = \frac{BK}{s(1 + Ts)} = BK\left(\frac{1/T}{s(s + 1/T)}\right) \tag{3.27}$$

Equation (3.27) is in the form given in Laplace transform pair 6 Table 3.1, so the inverse transform becomes

$$x_o(t) = BK\left(1 - e^{-t/T}\right) \tag{3.28}$$

If $B = 1$ (unit step) and $K = 1$ (unity gain) then

$$x_o(t) = \left(1 - e^{-t/T}\right) \tag{3.29}$$

When time $t$ is expressed as a ratio of time constant $T$, then Table 3.2 and Figure 3.13 can be constructed.

**Table 3.2** Unit step response of a first-order system

| $t/T$ | 0 | 0.25 | 0.5 | 0.75 | 1 | 1.5 | 2 | 2.5 | 3 | 4 |
|-------|---|------|-----|------|---|-----|---|-----|---|---|
| $x_o(t)$ | 0 | 0.221 | 0.393 | 0.527 | 0.632 | 0.770 | 0.865 | 0.920 | 0.950 | 0.980 |

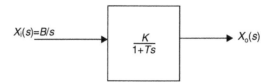

**Fig. 3.12** Step response of a first-order system.

### 3.5.4 Experimental determination of system time constant using step response

*Method one*: The system time constant is the time the system takes to reach 63.2% of its final value (see Table 3.2).

*Method two*: The system time constant is the intersection of the slope at $t = 0$ with the final value line (see Figure 3.13) since

$$x_o(t) = 1 - e^{-t/T}$$

$$\frac{dx_o}{dt} = 0 - \left(-\frac{1}{T}\right)e^{-t/T} = \frac{1}{T}e^{-t/T} \tag{3.30}$$

$$\frac{dx_o}{dt}\bigg|_{t=0} = \frac{1}{T} \quad \text{at } t = 0 \tag{3.31}$$

This also applies to any other tangent, see Figure 3.13.

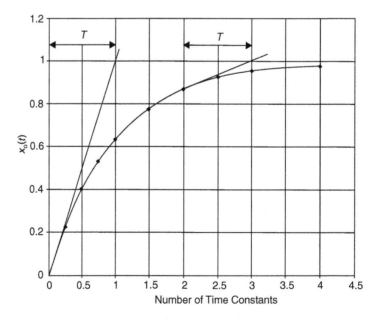

**Fig. 3.13** Unit step response of a first-order system.

### 3.5.5 Ramp response of first-order systems

*Example 3.6*
Find an expression for the response of a first-order system to a ramp function of slope Q.

*Solution*
From Figure 3.14

$$X_o(s) = \frac{QK}{s^2(1 + Ts)} = \frac{QK/T}{s^2(s + 1/T)} = \frac{A}{s} + \frac{B}{s^2} + \frac{C}{(s + 1/T)} \tag{3.32}$$

(See partial fraction expansion equation (3.13)). Multiplying both sides by $s^2(s + 1/T)$, we get

$$\frac{QK}{T} = As\left(s + \frac{1}{T}\right) + B\left(s + \frac{1}{T}\right) + Cs^2$$

i.e. $$\frac{QK}{T} = As^2 + \frac{A}{T}s + Bs + \frac{B}{T} + Cs^2 \tag{3.33}$$

Equating coefficients on both sides of equation (3.33)

$$(s^2): \quad 0 = A + C \tag{3.34}$$

$$(s^1): \quad 0 = \frac{A}{T} + B \tag{3.35}$$

$$(s^0): \quad \frac{QK}{T} = \frac{B}{T} \tag{3.36}$$

From (3.34)
$$C = -A$$

From (3.36)
$$B = QK$$

Substituting into (3.35)
$$A = -QKT$$

Hence from (3.34)
$$C = QKT$$

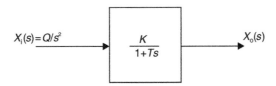

$X_i(s) = Q/s^2 \longrightarrow \boxed{\dfrac{K}{1+Ts}} \longrightarrow X_o(s)$

**Fig. 3.14** Ramp response of a first-order system (see also Figure A1.1).

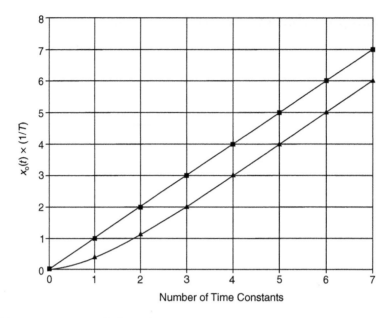

**Fig. 3.15** Unit ramp response of a first-order system.

Inserting values of $A$, $B$ and $C$ into (3.32)

$$X_o(s) = -\frac{QKT}{s} + \frac{QK}{s^2} + \frac{QKT}{(s + 1/T)}$$  (3.37)

Inverse transform, and factor out $KQ$

$$x_o(t) = KQ\left(t - T + Te^{-t/T}\right)$$  (3.38)

If $Q = 1$ (unit ramp) and $K = 1$ (unity gain) then

$$x_o(t) = t - T + Te^{-t/T}$$  (3.39)

The first term in equation (3.39) represents the input quantity, the second is the steady-state error and the third is the transient component. When time $t$ is expressed as a ratio of time constant $T$, then Table 3.3 and Figure 3.15 can be constructed. In Figure 3.15 the distance along the time axis between the input and output, in the steady-state, is the time constant.

**Table 3.3** Unit ramp response of a first-order system

| $t/T$ | 0 | 1 | 2 | 3 | 4 | 5 | 6 | 7 |
|---|---|---|---|---|---|---|---|---|
| $x_i(t)/T$ | 0 | 1 | 2 | 3 | 4 | 5 | 6 | 7 |
| $x_o(t)/T$ | 0 | 0.368 | 1.135 | 2.05 | 3.018 | 4.007 | 5 | 6 |

## 3.6 Time domain response of second-order systems

### 3.6.1 Standard form

Consider a second-order differential equation

$$a\frac{d^2 x_o}{dt^2} + b\frac{dx_o}{dt} + cx_o = ex_i(t) \tag{3.40}$$

Take Laplace transforms, zero initial conditions

$$as^2 X_o(s) + bsX_o(s) + cX_o(s) = eX_i(s)$$

$$(as^2 + bs + c)X_o(s) = eX_i(s) \tag{3.41}$$

The transfer function is

$$G(s) = \frac{X_o}{X_i}(s) = \frac{e}{as^2 + bs + c}$$

To obtain the standard form, divide by $c$

$$G(s) = \frac{\frac{e}{c}}{\frac{a}{c}s^2 + \frac{b}{c}s + 1}$$

which is written as

$$G(s) = \frac{K}{\frac{1}{\omega_n^2}s^2 + \frac{2\zeta}{\omega_n}s + 1} \tag{3.42}$$

This can also be normalized to make the $s^2$ coefficient unity, i.e.

$$G(s) = \frac{K\omega_n^2}{s^2 + 2\zeta\omega_n s + \omega_n^2} \tag{3.43}$$

Equations (3.42) and (3.43) are the standard forms of transfer functions for a second-order system, where $K$ = steady-state gain constant, $\omega_n$ = undamped natural frequency (rad/s) and $\zeta$ = damping ratio. The meaning of the parameters $\omega_n$ and $\zeta$ are explained in sections 3.6.4 and 3.6.3.

### 3.6.2 Roots of the characteristic equation and their relationship to damping in second-order systems

As discussed in Section 3.1, the transient response of a system is independent of the input. Thus for transient response analysis, the system input can be considered to be zero, and equation (3.41) can be written as

$$(as^2 + bs + c)X_o(s) = 0$$

If $X_o(s) \neq 0$, then

$$as^2 + bs + c = 0 \tag{3.44}$$

**Table 3.4** Transient behaviour of a second-order system

| Discriminant | Roots | Transient response type |
|---|---|---|
| $b^2 > 4ac$ | $s_1$ and $s_2$ real and unequal (−ve) | Overdamped Transient Response |
| $b^2 = 4ac$ | $s_1$ and $s_2$ real and equal (−ve) | Critically Damped Transient Response |
| $b^2 < 4ac$ | $s_1$ and $s_2$ complex conjugate of the form: $s_1, s_2 = -\sigma \pm j\omega$ | Underdamped Transient Response |

This polynomial in $s$ is called the *Characteristic Equation* and its roots will determine the system transient response. Their values are

$$s_1, \ s_2 = \frac{-b \pm \sqrt{b^2 - 4ac}}{2a} \tag{3.45}$$

The term $(b^2 - 4ac)$, called the discriminant, may be positive, zero or negative which will make the roots real and unequal, real and equal or complex. This gives rise to the three different types of transient response described in Table 3.4.

The transient response of a second-order system is given by the general solution

$$x_o(t) = Ae^{s_1 t} + Be^{s_2 t} \tag{3.46}$$

This gives a step response function of the form shown in Figure 3.16.

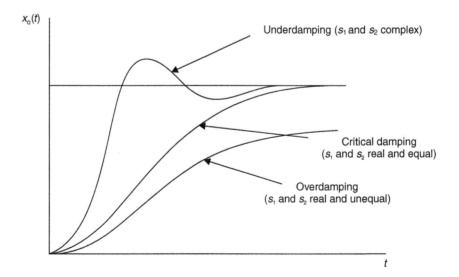

**Fig. 3.16** Effect that roots of the characteristic equation have on the damping of a second-order system.

### 3.6.3 Critical damping and damping ratio

#### Critical damping
When the damping coefficient $C$ of a second-order system has its critical value $C_c$, the system, when disturbed, will reach its steady-state value in the minimum time without overshoot. As indicated in Table 3.4, this is when the roots of the Characteristic Equation have equal negative real roots.

#### Damping ratio $\zeta$
The ratio of the damping coefficient $C$ in a second-order system compared with the value of the damping coefficient $C_c$ required for critical damping is called the Damping Ratio $\zeta$ (Zeta). Hence

$$\zeta = \frac{C}{C_c} \tag{3.47}$$

Thus

$$\zeta = 0 \quad \text{No damping}$$
$$\zeta < 1 \quad \text{Underdamping}$$
$$\zeta = 1 \quad \text{Critical damping}$$
$$\zeta > 1 \quad \text{Overdamping}$$

*Example 3.7*
Find the value of the critical damping coefficient $C_c$ in terms of $K$ and $m$ for the spring–mass–damper system shown in Figure 3.17.

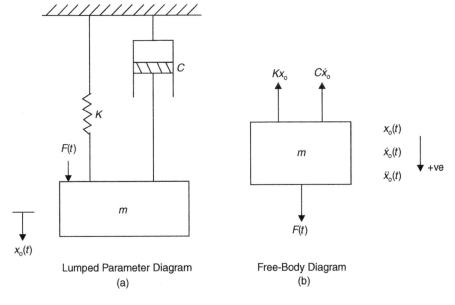

Lumped Parameter Diagram
(a)

Free-Body Diagram
(b)

**Fig. 3.17** Spring–mass–damper system.

*Solution*

From Newton's second law

$$\sum Fx = m\ddot{x}_\mathrm{o}$$

From the free-body diagram

$$F(t) - Kx_\mathrm{o}(t) - C\dot{x}_\mathrm{o}(t) = m\ddot{x}_\mathrm{o}(t) \tag{3.48}$$

Taking Laplace transforms, zero initial conditions

$$F(s) - KX_\mathrm{o}(s) - CsX_\mathrm{o}(s) = ms^2 X_\mathrm{o}(s)$$

or

$$(ms^2 + Cs + K)X_\mathrm{o}(s) = F(s) \tag{3.49}$$

Characteristic Equation is

$$ms^2 + Cs + K = 0$$

$$\text{i.e.} \quad s^2 + \frac{C}{m} + \frac{K}{m} = 0$$

and the roots are

$$s_1, s_2 = \frac{1}{2}\left\{ \frac{C}{m} \pm \sqrt{\left(\frac{C}{m}\right)^2 - 4\frac{K}{m}} \right\} \tag{3.50}$$

For critical damping, the discriminant is zero, hence the roots become

$$s_1 = s_2 = -\frac{C_\mathrm{c}}{2m}$$

Also, for critical damping

$$\frac{C_\mathrm{c}^2}{m^2} = \frac{4K}{m}$$

$$C_\mathrm{c}^2 = \frac{4Km^2}{m}$$

giving

$$C_\mathrm{c} = 2\sqrt{Km} \tag{3.51}$$

## 3.6.4   Generalized second-order system response to a unit step input

Consider a second-order system whose steady-state gain is $K$, undamped natural frequency is $\omega_\mathrm{n}$ and whose damping ratio is $\zeta$, where $\zeta < 1$. For a unit step input, the block diagram is as shown in Figure 3.18. From Figure 3.18

$$X_\mathrm{o}(s) = \frac{K\omega_\mathrm{n}^2}{s(s^2 + 2\zeta\omega_\mathrm{n}s + \omega_\mathrm{n}^2)} \tag{3.52}$$

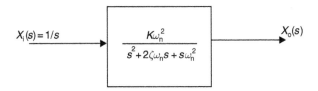

**Fig. 3.18** Step response of a generalized second-order system for $\zeta < 1$.

Expanding equation (3.52) using partial fractions

$$X_o(s) = \frac{A}{s} + \frac{Bs + C}{(s^2 + 2\zeta\omega_n s + \omega_n^2)} \qquad (3.53)$$

Equating (3.52) and (3.53) and multiply by $s(s^2 + 2\zeta\omega_n s + \omega_n^2)$

$$K\omega_n^2 = A(s^2 + 2\zeta\omega_n s + \omega_n^2) + Bs^2 + Cs$$

Equating coefficients

$$(s^2): \quad 0 = A + B$$
$$(s^1): \quad 0 = 2\zeta\omega_n A + C$$
$$(s^0): \quad K\omega_n^2 = \omega_n^2 A$$

giving

$$A = K, \quad B = -K \quad \text{and} \quad C = -2\zeta\omega_n K$$

Substituting back into equation (3.53)

$$X_o(s) = K\left[\frac{1}{s} - \left\{\frac{s + 2\zeta\omega_n}{s^2 + 2\zeta\omega_n s + \omega_n^2}\right\}\right]$$

Completing the square

$$X_o(s) = K\left[\frac{1}{s} - \left\{\frac{s + 2\zeta\omega_n}{(s + \zeta\omega_n)^2 + \omega_n^2 - \zeta^2\omega_n^2}\right\}\right]$$

$$= K\left[\frac{1}{s} - \left\{\frac{s + 2\zeta\omega_n}{(s + \zeta\omega_n)^2 + \left(\omega_n\sqrt{1 - \zeta^2}\right)^2}\right\}\right] \qquad (3.54)$$

The terms in the brackets { } can be written in the standard forms 10 and 9 in Table 3.1.

$$\text{Term (1)} = \frac{-s}{(s + \zeta\omega_n)^2 + \left(\omega_n\sqrt{1 - \zeta^2}\right)^2}$$

$$\text{Term (2)} = -\left\{\frac{2\zeta\omega_n}{\omega_n\sqrt{1 - \zeta^2}}\right\}\left\{\frac{\omega_n\sqrt{1 - \zeta^2}}{(s^2 + \zeta\omega_n)^2 + \left(\omega_n\sqrt{1 - \zeta^2}\right)^2}\right\}$$

Inverse transform

$$x_o(t) = K\left[1 - \left\{e^{-\zeta\omega_n t}\left(\cos\left(\omega_n\sqrt{1-\zeta^2}\right)t - \frac{\zeta\omega_n}{\omega_n\sqrt{1-\zeta^2}}\sin\left(\omega_n\sqrt{1-\zeta^2}\right)t\right)\right\}\right.$$
$$\left. - \left\{\frac{2\zeta}{\sqrt{1-\zeta^2}}\right\}\left\{e^{-\zeta\omega_n t}\left(\sin\left(\omega_n\sqrt{1-\zeta^2}\right)t\right)\right\}\right] \qquad (3.55)$$

Equation (3.55) can be simplified to give

$$x_o(t) = K\left[1 - e^{-\zeta\omega_n t}\left\{\cos\left(\omega_n\sqrt{1-\zeta^2}\right)t + \left(\frac{\zeta}{\sqrt{1-\zeta^2}}\right)\sin\left(\omega_n\sqrt{1-\zeta^2}\right)t\right\}\right] \quad (3.56)$$

When $\zeta = 0$

$$x_o(t) = K[1 - e^0\{\cos\omega_n t + 0\}]$$
$$= K[1 - \cos\omega_n t] \qquad (3.57)$$

From equation (3.57) it can be seen that when there is no damping, a step input will cause the system to oscillate continuously at $\omega_n$ (rad/s).

### Damped natural frequency $\omega_d$

From equation (3.56), when $0 < \zeta > 1$, the frequency of transient oscillation is given by

$$\omega_d = \omega_n\sqrt{1-\zeta^2} \qquad (3.58)$$

where $\omega_d$ is called the damped natural frequency. Hence equation (3.56) can be written as

$$x_o(t) = K\left[1 - e^{-\zeta\omega_n t}\left\{\cos\omega_d t + \left(\frac{\zeta}{\sqrt{1-\zeta^2}}\right)\sin\omega_d t\right\}\right] \qquad (3.59)$$

$$= K\left[1 - \frac{e^{-\zeta\omega_n t}}{\sqrt{1-\zeta^2}}\sin(\omega_d t + \phi)\right] \qquad (3.60)$$

where

$$\tan\phi = \frac{\sqrt{1-\zeta^2}}{\zeta} \qquad (3.61)$$

When $\zeta = 1$, the unit step response is

$$x_o(t) = K[1 - e^{-\omega_n t}(1 + \omega_n t)] \qquad (3.62)$$

and when $\zeta > 1$, the unit step response from equation (3.46) is given by

$$x_o(t) = K\left[1 - \left\{\left(\frac{1}{2} + \frac{\zeta}{2\sqrt{\zeta^2 - 1}}\right)e^{\left(-\zeta + \sqrt{\zeta^2 - 1}\right)\omega_n t}\right.\right.$$
$$\left.\left. + \left(\frac{1}{2} - \frac{\zeta}{2\sqrt{\zeta^2 - 1}}\right)e^{\left(-\zeta - \sqrt{\zeta^2 - 1}\right)\omega_n t}\right\}\right] \qquad (3.63)$$

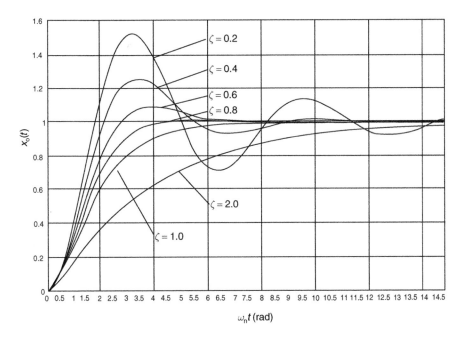

**Fig. 3.19** Unit step response of a second-order system.

The generalized second-order system response to a unit step input is shown in Figure 3.19 for the condition $K = 1$ (see also Appendix 1, *sec_ord.m*).

## 3.7 Step response analysis and performance specification

### 3.7.1 Step response analysis

It is possible to identify the mathematical model of an underdamped second-order system from its step response function.

Consider a unity-gain ($K = 1$) second-order underdamped system responding to an input of the form

$$x_i(t) = B \tag{3.64}$$

The resulting output $x_o(t)$ would be as shown in Figure 3.20. There are two methods for calculating the damping ratio.

*Method (a)*: Percentage Overshoot of first peak

$$\%\text{Overshoot} = \frac{a_1}{B} \times 100 \tag{3.65}$$

Now

$$a_1 = B e^{-\zeta \omega_n (\tau/2)}$$

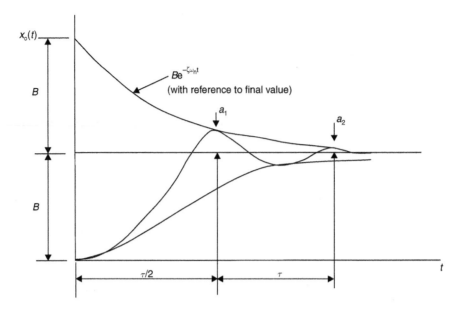

**Fig. 3.20** Step response analysis.

Thus,

$$\% \text{Overshoot} = \frac{Be^{-\zeta\omega_n(\tau/2)}}{B} \times 100 \tag{3.66}$$

Since the frequency of transient oscillation is $\omega_d$, then,

$$\tau = \frac{2\pi}{\omega_d}$$

$$= \frac{2\pi}{\omega_n\sqrt{1-\zeta^2}} \tag{3.67}$$

Substituting (3.67) into (3.66)

$$\% \text{Overshoot} = e^{-2\pi\zeta\omega_n/2\omega_n\sqrt{1-\zeta^2}} \times 100$$

$$\% \text{Overshoot} = e^{-\zeta\pi/\sqrt{1-\zeta^2}} \tag{3.68}$$

*Method (b)*: Logarithmic decrement. Consider the ratio of successive peaks $a_1$ and $a_2$

$$a_1 = Be^{-\zeta\omega_n(\tau/2)} \tag{3.69}$$

$$a_2 = Be^{-\zeta\omega_n(3\tau/2)} \tag{3.70}$$

Hence

$$\frac{a_1}{a_2} = \frac{e^{-\zeta\omega_n(\tau/2)}}{e^{-\zeta\omega_n(3\tau/2)}} = e^{\{-\zeta\omega_n(\tau/2)+\zeta\omega_n(3\tau/2)\}}$$

$$= e^{\zeta\omega_n\tau} = e^{2\zeta\pi/\sqrt{1-\zeta^2}} \tag{3.71}$$

Equation (3.71) can only be used if the damping is light and there is more than one overshoot. Equation (3.67) can now be employed to calculate the undamped natural frequency

$$\omega_n = \frac{2\pi}{\tau\sqrt{1-\zeta^2}} \qquad (3.72)$$

## 3.7.2 Step response performance specification

The three parameters shown in Figure 3.21 are used to specify performance in the time domain.

(a) *Rise time $t_r$*: The shortest time to achieve the final or steady-state value, for the first time. This can be 100% rise time as shown, or the time taken for example from 10% to 90% of the final value, thus allowing for non-overshoot response.
(b) *Overshoot*: The relationship between the percentage overshoot and damping ratio is given in equation (3.68). For a control system an overshoot of between 0 and 10% ($1 < \zeta > 0.6$) is generally acceptable.
(c) *Settling time $t_s$*: This is the time for the system output to settle down to within a tolerance band of the final value, normally between $\pm 2$ or 5%.

Using 2% value, from Figure 3.21

$$0.02B = Be^{-\zeta\omega_n t_s}$$

Invert

$$50 = e^{\zeta\omega_n t_s}$$

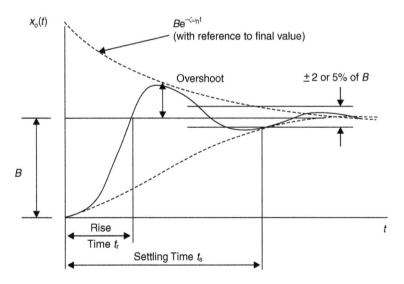

**Fig. 3.21** Step response performance specification.

Take natural logs

$$\ln 50 = \zeta \omega_n t_s$$

giving

$$t_s = \left(\frac{1}{\zeta \omega_n}\right) \ln 50 \qquad (3.73)$$

The term $(1/\zeta \omega_n)$ is sometimes called the equivalent time constant $T_c$ for a second-order system. Note that $\ln 50$ (2% tolerance) is 3.9, and $\ln 20$ (5% tolerance) is 3.0. Thus the transient period for both first and second-order systems is three times the time constant to within a 5% tolerance band, or four times the time constant to within a 2% tolerance band, a useful rule-of-thumb.

## 3.8  Response of higher-order systems

Transfer function techniques can be used to calculate the time response of higher-order systems.

*Example 3.8* (See also Appendix 1, *examp38.m*)
Figure 3.22 shows, in block diagram form, the transfer functions for a resistance thermometer and a valve connected together. The input $x_i(t)$ is temperature and the output $x_o(t)$ is valve position. Find an expression for the unit step response function when there are zero initial conditions.

*Solution*
From Figure 3.22

$$X_o(s) = \frac{25}{s(1 + 2s)(s^2 + s + 25)} \qquad (3.74)$$

$$= \frac{12.5}{s(s + 0.5)(s^2 + s + 25)} \qquad (3.75)$$

$$= \frac{A}{s} + \frac{B}{(s + 0.5)} + \frac{Cs + D}{(s + 0.5)^2 + (4.97)^2} \qquad (3.76)$$

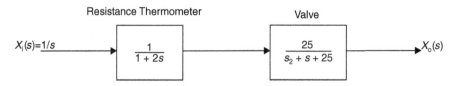

**Fig. 3.22** Block diagram representation of a resistance thermometer and valve.

Note that the second-order term in equation (3.76) has had the 'square' completed since its roots are complex ($b^2 < 4ac$). Equate equations (3.75) and (3.76) and multiply both sides by $s(s + 0.5)(s^2 + s + 25)$.

$$12.5 = (s^3 + 1.5s^2 + 25.5s + 12.5)A + (s^3 + s^2 + 25s)B$$
$$+ (s^3 + 0.5s^2)C + (s^2 + 0.5s)D \tag{3.77}$$

Equating coefficients

$$(s^3): \quad 0 = A + B + C$$
$$(s^2): \quad 0 = 1.5A + B + 0.5C + D$$
$$(s^1): \quad 0 = 25.5A + 25B + 0.5D$$
$$(s^0): \quad 12.5 = 12.5A$$

Solving the four simultaneous equations

$$A = 1, \quad B = -1.01, \quad C = 0.01, \quad D = -0.5$$

Substituting back into equation (3.76) gives

$$X_o(s) = \frac{1}{s} - \frac{1.01}{(s + 0.5)} + \frac{0.01s - 0.5}{(s + 0.5)^2 + (4.97)^2} \tag{3.78}$$

Inverse transform

$$x_o(t) = 1 - 1.01e^{-0.5t} - 0.01e^{-0.5t}(10.16 \sin 4.97t - \cos 4.97t) \tag{3.79}$$

Equation (3.79) shows that the third-order transient response contains both first-order and second-order elements whose time constants and equivalent time constants are 2 seconds, i.e. a transient period of about 8 seconds. The second-order element has a predominate negative sine term, and a damped natural frequency of 4.97 rad/s. The time response is shown in Figure 3.23.

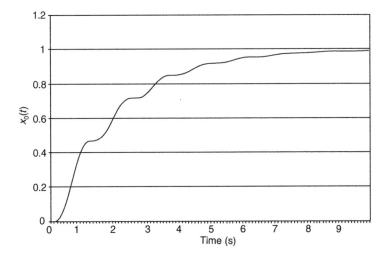

**Fig. 3.23** Time response of third-order system.

## 3.9  Further problems

*Example 3.9*
A ship has a mass $m$ and a resistance $C$ times the forward velocity $u(t)$. If the thrust from the propeller is $K$ times its angular velocity $\omega(t)$, determine:

(a)  The first-order differential equation and hence the transfer function relating $U(s)$ and $\omega(s)$.

When the vessel has the parameters: $m = 18\,000 \times 10^3$ kg, $C = 150\,000$ Ns/m, and $K = 96\,000$ Ns/rad, find,

(b)  the time constant.
(c)  an expression for the time response of the ship when there is a step change of $\omega(t)$ from 0 to 12.5 rad/s. Assume that the vessel is initially at rest.
(d)  What is the forward velocity after
    (i)   one minute
    (ii)  ten minutes.

*Solution*

(a)  $m(du/dt) + Cu = K\omega(t)$

$$\frac{U}{\omega}(s) = \frac{K/C}{1 + (m/C)s}$$

(b)  120 seconds
(c)  $u(t) = 8(1 - e^{-0.00833t})$
(d)  (i)   3.148 m/s
    (ii)  7.946 m/s

*Example 3.10*

(a)  Determine the transfer function relating $V_2(s)$ and $V_1(s)$ for the passive electrical network shown in Figure 3.24.
(b)  When $C = 2\,\mu$F and $R_1 = R_2 = 1\,M\Omega$, determine the steady-state gain $K$ and time constant $T$.
(c)  Find an expression for the unit step response.

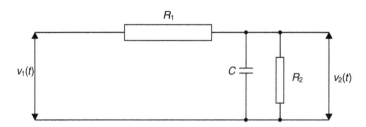

**Fig. 3.24**  Passive electrical network.

*Solution*

(a) $\dfrac{V_2}{V_1}(s) = \dfrac{R_2/R_1 + R_2}{1 + (R_1 R_2 C/R_1 + R_2)s}$

(b) 0.5

   1.0 seconds

(c) $v_o(t) = 0.5(1 - e^{-t})$

*Example 3.11*

Determine the values of $\omega_n$ and $\zeta$ and also expressions for the unit step response for the systems represented by the following second-order transfer functions

(i) $\dfrac{X_o}{X_i}(s) = \dfrac{1}{0.25s^2 + s + 1}$

(ii) $\dfrac{X_o}{X_i}(s) = \dfrac{10}{s^2 + 6s + 5}$

(iii) $\dfrac{X_o}{X_i}(s) = \dfrac{1}{s^2 + s + 1}$

*Solution*

(i)  2.0

   1.0 (Critical damping)

   $x_o(t) = 1 - e^{-2t}(1 + 2t)$

(ii) 2.236

   1.342 (Overdamped)

   $x_o(t) = 2 - 2.5e^{-t} + 0.5e^{-5t}$

(iii) 1.0

   0.5 (Underdamped)

   $x_o(t) = 1 - e^{-0.5t}(\cos 0.866t + 0.577 \sin 0.866t)$

*Example 3.12*

A torsional spring of stiffness $K$, a mass of moment of inertia $I$ and a fluid damper with damping coefficient $C$ are connected together as shown in Figure 3.25. The angular displacement of the free end of the spring is $\theta_i(t)$ and the angular displacement of the mass and damper is $\theta_o(t)$.

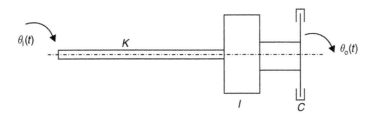

**Fig. 3.25** Torsional system.

(a) Develop the transfer function relating $\theta_i(s)$ and $\theta_o(s)$.
(b) If the time relationship for $\theta_i(t)$ is given by $\theta_i(t) = 4t$ then find an expression for the time response of $\theta_o(t)$. Assume zero initial conditions. What is the steady-state error between $\theta_i(t)$ and $\theta_o(t)$?

*Solution*

(a) $\dfrac{\theta_o}{\theta_i}(s) = \dfrac{1}{\left(\frac{I}{K}\right)s^2 + \left(\frac{C}{K}\right)s + 1}$

(b) $\theta_o(t) = 4t - 0.2 + e^{-2.5t}(0.2\cos 9.682t - 0.361\sin 9.682t)$

0.2 radians.

*Example 3.13*
When a unity gain second-order system is subject to a unit step input, its transient response contains a first overshoot of 77%, occurring after 32.5 ms has elapsed. Find

(a) the damped natural frequency
(b) the damping ratio
(c) the undamped natural frequency
(d) the system transfer function
(e) the time to settle down to within $\pm 2\%$ of the final value

*Solution*
(a) 96.66 rad/s
(b) 0.083
(c) 96.99 rad/s

(d) $G(s) = \dfrac{1}{0.106 \times 10^{-3}s^2 + 1.712 \times 10^{-3}s + 1}$

(e) 0.486 seconds

*Example 3.14*
A system consists of a first-order element linked to a second-order element without interaction. The first-order element has a time constant of 5 seconds and a steady-state gain constant of 0.2. The second-order element has an undamped natural frequency of 4 rad/s, a damping ratio of 0.25 and a steady-state gain constant of unity.

If a step input function of 10 units is applied to the system, find an expression for the time response. Assume zero initial conditions.

*Solution*

$$x_o(t) = 2.0 - 2.046e^{-0.2t} + e^{-t}\left(0.046\cos\sqrt{15}t - 0.094\sin\sqrt{15}t\right)$$

# 4

# Closed-loop control systems

Any system in which the output quantity is monitored and compared with the input, any difference being used to actuate the system until the output equals the input is called a *closed-loop* or *feedback* control system.

The elements of a closed-loop control system are represented in block diagram form using the transfer function approach. The general form of such a system is shown in Figure 4.1.

The transfer function relating $R(s)$ and $C(s)$ is termed the closed-loop transfer function.

From Figure 4.1

$$C(s) = G(s)E(s) \tag{4.1}$$

$$B(s) = H(s)C(s) \tag{4.2}$$

$$E(s) = R(s) - B(s) \tag{4.3}$$

Substituting (4.2) and (4.3) into (4.1)

$$C(s) = G(s)\{R(s) - H(s)C(s)\}$$

$$C(s) = G(s)R(s) - G(s)H(s)C(s)$$

$$C(s)\{1 + G(s)H(s)\} = G(s)R(s)$$

$$\frac{C}{R}(s) = \frac{G(s)}{1 + G(s)H(s)} \tag{4.4}$$

The closed-loop transfer function is the forward-path transfer function divided by one plus the open-loop transfer function.

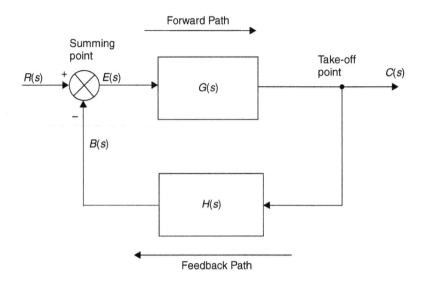

**Fig. 4.1** Block diagram of a closed-loop control system. $R(s)$ = Laplace transform of reference input $r(t)$; $C(s)$ = Laplace transform of controlled output $c(t)$; $B(s)$ = Primary feedback signal, of value $H(s)C(s)$; $E(s)$ = Actuating or error signal, of value $R(s) - B(s)$; $G(s)$ = Product of all transfer functions along the forward path; $H(s)$ = Product of all transfer functions along the feedback path; $G(s)H(s)$ = Open-loop transfer function; $\otimes$ = summing point symbol, used to denote algebraic summation; $\bullet$ = Signal take-off point; $\rightarrow$ = Direction of information flow.

## 4.2  Block diagram reduction

### 4.2.1  Control systems with multiple loops

A control system may have several feedback control loops. For example, with a ship autopilot, the rudder-angle control loop is termed the minor loop, whereas the heading control loop is referred to as the major loop. When analysing multiple loop systems, the minor loops are considered first, until the system is reduced to a single overall closed-loop transfer function.

To reduce complexity, in the following examples the function of $s$ notation $(s)$ used for transfer functions is only included in the final solution.

*Example 4.1*
Find the closed-loop transfer function for the system shown in Figure 4.2.

*Solution*
In Figure 4.2, the first minor loop to be considered is $G_3H_3$. Using equation (4.4), this may be replaced by

$$G_{m1} = \frac{G_3}{1 + G_3H_3} \tag{4.5}$$

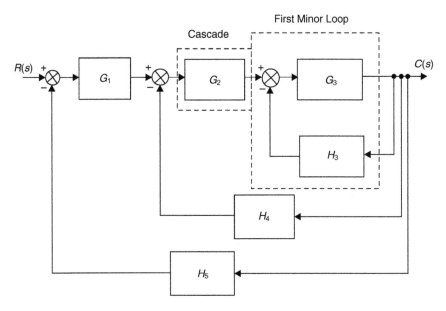

**Fig. 4.2** Multiple loop control system.

Now $G_{m1}$ is multiplied by, or in cascade with $G_2$. Hence the combined transfer function is

$$G_2 G_{m1} = \frac{G_2 G_3}{1 + G_3 H_3} \qquad (4.6)$$

The reduced block diagram is shown in Figure 4.3.

Following a similar process, the second minor loop $G_{m2}$ may be written

$$G_{m2} = \frac{\frac{G_2 G_3}{1 + G_3 H_3}}{1 + \frac{G_2 G_3 H_2}{1 + G_3 H_3}}$$

Multiplying numerator and denominator by $1 + G_3 H_3$

$$G_{m2} = \frac{G_2 G_3}{1 + G_3 H_3 + G_2 G_3 H_2}$$

But $G_{m2}$ is in cascade with $G_1$, hence

$$G_1 G_{m2} = \frac{G_1 G_2 G_3}{1 + G_3 H_3 + G_2 G_3 H_2} \qquad (4.7)$$

Transfer function (4.7) now becomes the complete forward-path transfer function as shown in Figure 4.4.

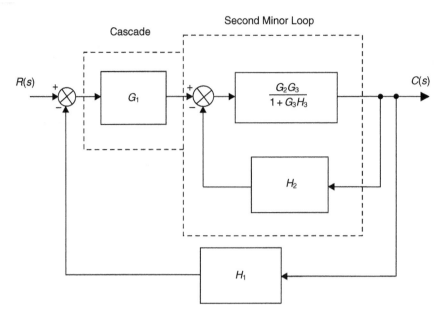

**Fig. 4.3** First stage of block diagram reduction.

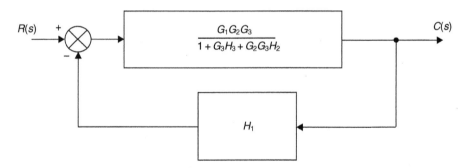

**Fig. 4.4** Second stage of block diagram reduction.

The complete, or overall closed-loop transfer function can now be evaluated

$$\frac{C}{R}(s) = \frac{\frac{G_1 G_2 G_3}{1+G_3 H_3+G_2 G_3 H_2}}{1+\frac{G_1 G_2 G_3 H_1}{1+G_3 H_3+G_2 G_3 H_2}}$$

Multiplying numerator and denominator by $1 + G_3 H_3 + G_2 G_3 H_2$

$$\frac{C}{R}(s) = \frac{G_1(s)G_2(s)G_3(s)}{1 + G_3(s)H_3(s) + G_2(s)G_3(s)H_2(s) + G_1(s)G_2(s)G_3(s)H_1(s)} \qquad (4.8)$$

## 4.2.2 Block diagram manipulation

There are occasions when there is interaction between the control loops and, for the purpose of analysis, it becomes necessary to re-arrange the block diagram configuration. This can be undertaken using Block Diagram Transformation Theorems.

**Table 4.1** Block Diagram Transformation Theorems

| Transformation | Equation | Block diagram | Equivalent block diagram |
|---|---|---|---|
| 1. Combining blocks in cascade | $Y=(G_1G_2)X$ | | |
| 2. Combining blocks in parallel; or eliminating a forward loop | $Y=G_1X\pm G_2X$ | | |
| 3. Removing a block from a forward path | $Y=G_1X\pm G_2X$ | | |
| 4. Eliminating a feedback loop | $Y=G_1(X\pm G_2Y)$ | | |
| 5. Removing a block from a feedback loop | $Y=G_1(X\pm G_2Y)$ | | |
| 6. Rearranging summing points | $Z=W\pm X\pm Y$ | | |
| 7. Moving a summing point ahead of a block | $Z=GX\pm Y$ | | |
| 8. Moving a summing point beyond a block | $Z=G(X\pm Y)$ | | |
| 9. Moving a take-off point ahead of a block | $Y=GX$ | | |
| 10. Moving a take-off point beyond a block | $Y=GX$ | | |

*Example 4.2*
Moving a summing point ahead of a block.

|  | Equation | Equation |  |
|---|---|---|---|
|  | $Z = GX \pm Y$ | $Z = \{X \pm (1/G)Y\}G$ | (4.9) |
|  |  | $Z = GX \pm Y$ |  |

A complete set of Block Diagram Transformation Theorems is given in Table 4.1.

*Example 4.3*
Find the overall closed-loop transfer function for the system shown in Figure 4.6.

*Solution*
Moving the first summing point ahead of $G_1$, and the final take-off point beyond $G_4$ gives the modified block diagram shown in Figure 4.7. The block diagram shown in Figure 4.7 is then reduced to the form given in Figure 4.8. The overall closed-loop transfer function is then

$$
\frac{C}{R}(s) = \frac{\dfrac{G_1G_2G_3G_4}{(1+G_1G_2H_1)(1+G_3G_4H_2)}}{1 + \dfrac{G_1G_2G_3G_4H_3}{(G_1G_4)(1+G_1G_2H_1)(1+G_3G_4H_2)}}
$$

$$
= \frac{G_1(s)G_2(s)G_3(s)G_4(s)}{(1 + G_1(s)G_2(s)H_1(s))(1 + G_3(s)G_4(s)H_2(s)) + G_2(s)G_3(s)H_3(s)} \tag{4.10}
$$

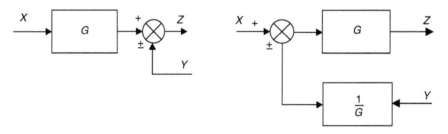

**Fig. 4.5** Moving a summing point ahead of a block.

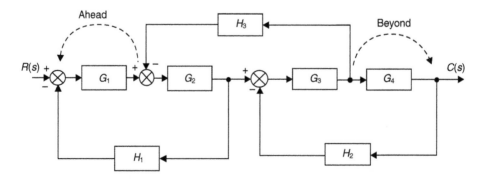

**Fig. 4.6** Block diagram with interaction.

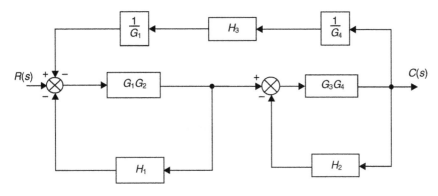

**Fig. 4.7** Modified block diagram with no interaction.

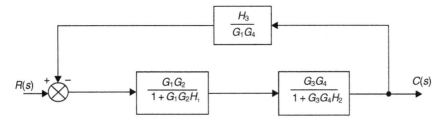

**Fig. 4.8** Reduced block diagram.

## 4.3 Systems with multiple inputs

### 4.3.1 Principle of superposition

A dynamic system is linear if the Principle of Superposition can be applied. This states that 'The response $y(t)$ of a linear system due to several inputs $x_1(t)$, $x_2(t), \ldots, x_n(t)$, acting simultaneously is equal to the sum of the responses of each input acting alone'.

*Example 4.4*
Find the complete output for the system shown in Figure 4.9 when both inputs act simultaneously.

*Solution*
The block diagram shown in Figure 4.9 can be reduced and simplified to the form given in Figure 4.10. Putting $R_2(s) = 0$ and replacing the summing point by $+1$ gives the block diagram shown in Figure 4.11. In Figure 4.11 note that $C^{\mathrm{I}}(s)$ is response to $R_1(s)$ acting alone. The closed-loop transfer function is therefore

$$\frac{C^{\mathrm{I}}}{R_1}(s) = \frac{\frac{G_1 G_2}{1 + G_2 H_2}}{1 + \frac{G_1 G_2 H_1}{1 + G_2 H_2}}$$

or

$$C^{\mathrm{I}}(s) = \frac{G_1(s)G_2(s)R_1(s)}{1 + G_2(s)H_2(s) + G_1(s)G_2(s)H_1(s)} \tag{4.11}$$

Now if $R_1(s) = 0$ and the summing point is replaced by $-1$, then the response $C^{\mathrm{II}}(s)$ to input $R_2(s)$ acting alone is given by Figure 4.12. The choice as to whether the summing point is replaced by $+1$ or $-1$ depends upon the sign at the summing point.

Note that in Figure 4.12 there is a positive feedback loop. Hence the closed-loop transfer function relating $R_2(s)$ and $C^{\mathrm{II}}(s)$ is

$$\frac{C^{\mathrm{II}}}{R_2}(s) = \frac{\frac{-G_1 G_2 H_1}{1 + G_2 H_2}}{1 - \left(\frac{-G_1 G_2 H_1}{1 + G_2 H_2}\right)}$$

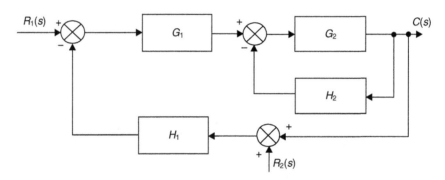

**Fig. 4.9** System with multiple inputs.

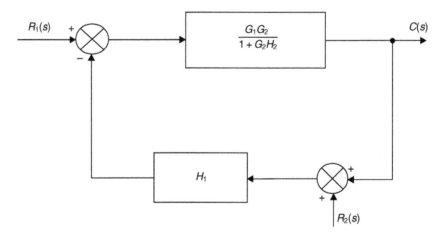

**Fig. 4.10** Reduced and simplified block diagram.

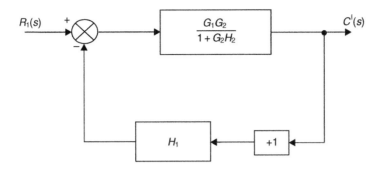

**Fig. 4.11** Block diagram for $R_1(s)$ acting alone.

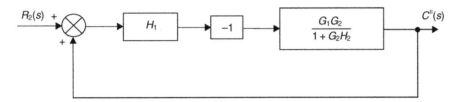

**Fig. 4.12** Block diagram for $R_2(s)$ acting alone.

or

$$C^{\mathrm{II}}(s) = \frac{-G_1(s)G_2(s)H_1(s)R_2(s)}{1 + G_2(s)H_2(s) + G_1(s)G_2(s)H_1(s)} \qquad (4.12)$$

It should be noticed that the denominators for equations (4.11) and (4.12) are identical. Using the Principle of Superposition, the complete response is given by

$$C(s) = C^{\mathrm{I}}(s) + C^{\mathrm{II}}(s) \qquad (4.13)$$

or

$$C(s) = \frac{(G_1(s)G_2(s))R_1(s) - (G_1(s)G_2(s)H_1(s))R_2(s)}{1 + G_2(s)H_2(s) + G_1(s)G_2(s)H_1(s)} \qquad (4.14)$$

## 4.4 Transfer functions for system elements

### 4.4.1 DC servo-motors

One of the most common devices for actuating a control system is the DC servo-motor shown in Figure 4.13, and can operate under either armature or field control.

(a) *Armature control*: This arrangement is shown in schematic form in Figure 4.14. Now air gap flux $\Phi$ is proportional to $i_f$, or

$$\Phi = K_{\mathrm{fd}} i_f \qquad (4.15)$$

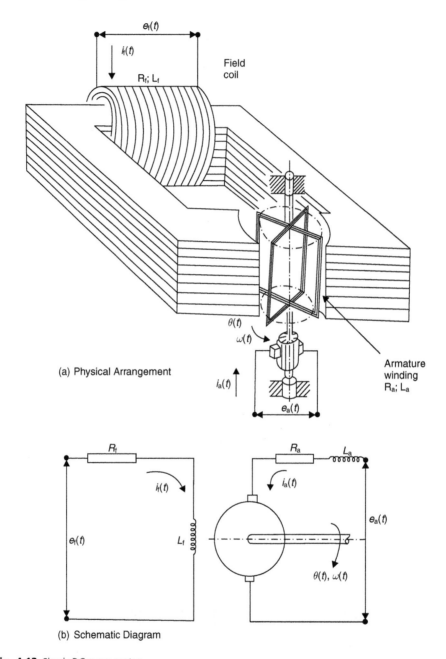

(a) Physical Arrangement

(b) Schematic Diagram

**Fig. 4.13** Simple DC servo-motor.

where $K_{fd}$ is the field coil constant.

Also, torque developed $T_m$ is proportional to the product of the air gap flux and the armature current

$$T_m(t) = \Phi K_{am} i_a(t) \tag{4.16}$$

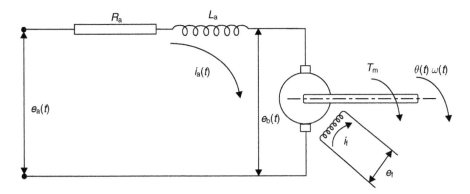

**Fig. 4.14** DC servo-motor under armature control. $e_a(t)$ = Armature excitation voltage; $e_b(t)$ = Back emf; $i_a(t)$ = Armature current; $R_a$ = Armature resistance; $L_a$ = Armature inductance; $e_f$ = Constant field voltage; $i_f$ = Constant field current; $T_m$ = Torque developed by motor; $\theta(t)$ = Shaft angular displacement; $w(t)$ = Shaft angular velocity = $d\theta/dt$.

where $K_{am}$ is the armature coil constant.

Substituting (4.15) into (4.16) gives

$$T_m(t) = (K_{fd}K_{am}i_f)i_a(t) \tag{4.17}$$

Since $i_f$ is constant

$$T_m(t) = K_a i_a(t) \tag{4.18}$$

where the overall armature constant $K_a$ is

$$K_a = K_{fd}K_{am}i_f \tag{4.19}$$

When the armature rotates, it behaves like a generator, producing a back emf $e_b(t)$ proportional to the shaft angular velocity

$$e_b(t) = K_b\frac{d\theta}{dt} = K_b w(t) \tag{4.20}$$

where $K_b$ is the back emf constant.

The potential difference across the armature winding is therefore

$$e_a(t) - e_b(t) = L_a\frac{di_a}{dt} + R_a i_a \tag{4.21}$$

Taking Laplace transforms of equation (4.21) with zero initial conditions

$$E_a(s) - E_b(s) = (L_a s + R_a)I_a(s) \tag{4.22}$$

Figure 4.15 combines equations (4.18), (4.20) and (4.22) in block diagram form.

Under steady-state conditions, the torque developed by the DC servo-motor is

$$T_m(t) = \{e_a(t) - K_b w(t)\}\frac{K_a}{R_a}$$

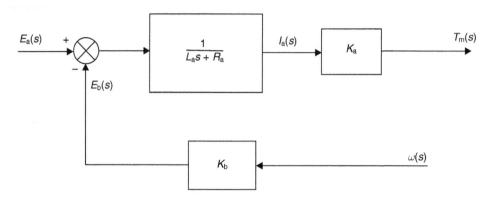

**Fig. 4.15** Block diagram representation of armature controlled DC servo-motor.

or

$$T_m(t) = \left(\frac{K_a}{R_a}\right)e_a(t) - \left(\frac{K_a K_b}{R_a}\right)\omega(t) \tag{4.23}$$

From equation (4.23), the relationship between $T_m(t)$, $\omega(t)$ and $E_a(t)$ under steady-state conditions is shown in Figure 4.16.

(b) *Field control*: This arrangement is shown in schematic form in Figure 4.13, with the exception that the armature current $i_a$ is held at a constant value. Equation (4.17) may now be written as

$$T_m(t) = (K_{fd} K_{am} i_a)i_f(t) \tag{4.24}$$

and since $i_a$ is a constant, then

$$T_m(t) = K_f i_f(t) \tag{4.25}$$

where the overall field constant $K_f$ is

$$K_f = K_{fd} K_{am} i_a \tag{4.26}$$

In this instance, the back emf $e_b$ does not play a part in the torque equation, but it can produce difficulties in maintaining a constant armature current $i_a$.

The potential difference across the field coil is

$$e_f(t) = L_f \frac{di_f}{dt} + R_f i_f \tag{4.27}$$

Taking Laplace transforms of equation (4.27) with zero initial conditions

$$E_f(s) = (L_f s + R_f)I_f(s) \tag{4.28}$$

Figure 4.17 combines equations (4.25) and (4.28) in block diagram form.

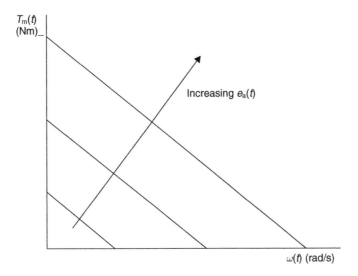

**Fig. 4.16** Steady-state relationship between $T_m(t)$, $\omega(t)$ and $e_a(t)$ for an armature controlled DC servo-motor.

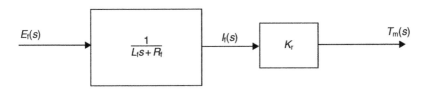

**Fig. 4.17** Block diagram representation of field controlled DC servo-motor.

Under steady-state conditions, the torque developed by the DC servo-motor is

$$T_m(t) = \left(\frac{K_f}{R_f}\right) e_f(t) \tag{4.29}$$

The relationship between $T_m(t)$, $e_f(t)$ and $\omega(t)$ under steady-state conditions is shown in Figure 4.18.

## 4.4.2 Linear hydraulic actuators

Hydraulic actuators are employed in such areas as the aerospace industry because they possess a good power to weight ratio and have a fast response.

Figure 4.19 shows a spool-valve controlled linear actuator. When the spool-valve is moved to the right, pressurized hydraulic oil flows into chamber (1) causing the piston to move to the left, and in so doing forces oil in chamber (2) to be expelled to the exhaust port.

The following analysis will be linearized for small perturbations of the spool-valve and actuator.

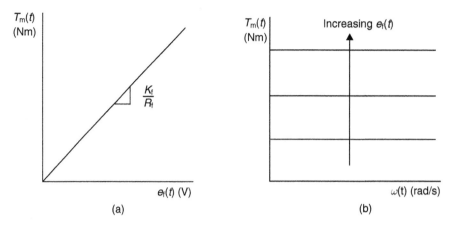

**Fig. 4.18** Steady-state relationship between $T_m(t)$, $e_f(t)$ and $\omega(t)$ for a field controlled DC servo-motor.

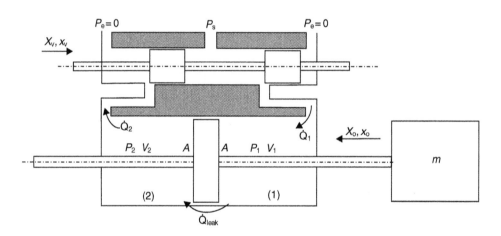

**Fig. 4.19** Spool-valve controlled linear actuator.

It is assumed that:

- the supply pressure $P_s$ is constant
- the exhaust pressure $P_e$ is atmospheric
- the actuator is in mid-position so that $V_1 = V_2 = V_o$ which is half the total volume of hydraulic fluid $V_t$
- the hydraulic oil is compressible
- the piston faces have equal areas $A$
- $\dot{Q}_1$ and $\dot{Q}_2$ are the volumetric flow-rates into chamber (1) and out of chamber (2)
- the average, or load flow-rate $\dot{Q}_L$ has a value $(\dot{Q}_1 + \dot{Q}_2)/2$
- $P_1$ and $P_2$ are the fluid pressures in chamber (1) and chamber (2)
- the load pressure $P_L$ has a value $(P_1 - P_2)$

(a) *Actuator analysis*: The continuity equation for the chambers may be written

$$\sum \dot{Q}_{in} - \sum \dot{Q}_{out} = \text{(rate of change of chamber volume)}$$
$$+ \text{(rate of change of oil volume)} \tag{4.30}$$

In equation (4.30), the rate of change of chamber volume is due to the piston movement, i.e. $dV/dt$. The rate of change of oil volume is due to compressibility effects, i.e.:

Bulk Modulus of oil, $\beta$ = Volumetric stress/Volumetric strain

$$\beta = \frac{dP}{-dV/V} \tag{4.31}$$

Note that in equation (4.31) the denominator is negative since an increase in pressure causes a reduction in oil volume.

Hence

$$\frac{dV}{V} = \frac{dP}{-\beta}$$

Giving, when differentiated with respect to time

$$\frac{dV}{dt} = -\left(\frac{V}{\beta}\right)\frac{dP}{dt} \tag{4.32}$$

For chamber (1), equation (4.30) may be expressed as

$$\dot{Q}_1 - \dot{Q}_{leak} = \frac{dV_1}{dt} + \left(\frac{V_1}{\beta}\right)\frac{dP_1}{dt} \tag{4.33}$$

and for chamber (2)

$$\dot{Q}_{leak} - \dot{Q}_2 = \frac{dV_2}{dt} + \left(\frac{V_2}{\beta}\right)\frac{dP_2}{dt} \tag{4.34}$$

Now

$$\dot{Q}_L = \frac{\dot{Q}_1 + \dot{Q}_2}{2}$$

Thus, from eqations (4.33) and (4.34)

$$\dot{Q}_L = \dot{Q}_{leak} + \frac{1}{2}\left(\frac{dV_1}{dt} - \frac{dV_2}{dt}\right) + \frac{1}{2\beta}\left(V_1\frac{dP_1}{dt} - V_2\frac{dP_2}{dt}\right) \tag{4.35}$$

If leakage flow-rate $\dot{Q}_{leak}$ is laminar, then

$$\dot{Q}_{leak} = C_p P_L \tag{4.36}$$

where $C_p$ is the leakage coefficient. Also, if $V_1 = V_2 = V_o$, then

$$-\frac{dV_2}{dt} = \frac{dV_1}{dt} = \frac{dV_o}{dt} \tag{4.37}$$

Hence equation (4.35) can be written

$$\dot{Q}_L = C_P P_L + \frac{dV_o}{dt} + \frac{V_o}{2\beta}\frac{d}{dt}(P_1 - P_2) \tag{4.38}$$

or

$$\dot{Q}_L = C_P P_L + A\frac{dX_o}{dt} + \frac{V_t}{4\beta}\frac{dP_L}{dt} \tag{4.39}$$

where

$$\frac{dV_o}{dt} = A\frac{dX_o}{dt}$$

and

$$V_o = \frac{V_t}{2}$$

(b) *Linearized spool-valve analysis*: Assume that the spool-valve ports are rectangular in form, and have area

$$A_v = WX_v \tag{4.40}$$

where $W$ is the width of the port.
   From orifice theory

$$\dot{Q}_1 = C_d WX_v\sqrt{\frac{2}{\rho}(P_s - P_1)} \tag{4.41}$$

and

$$\dot{Q}_2 = C_d WX_v\sqrt{\frac{2}{\rho}(P_2 - 0)} \tag{4.42}$$

whereo $C_d$ is a coefficient of discharge and $\rho$ is the fluid density.
   Equating (4.41) and (4.42)

$$P_s - P_1 = P_2 \tag{4.43}$$

since

$$P_L = P_1 - P_2$$

Equation (4.43) may be re-arranged to give

$$P_1 = \frac{P_s + P_L}{2} \tag{4.44}$$

From equations (4.41) and (4.42) the load flow-rate may be written as

$$\dot{Q}_L = C_d W X_v \sqrt{\frac{2}{\rho}\left(\frac{P_s - P_L}{2}\right)} \tag{4.45}$$

Hence

$$\dot{Q}_L = F(X_V, P_L) \tag{4.46}$$

Equation (4.45) can be linearized using the technique described in section 2.7.1. If $q_L$, $x_v$ and $p_L$ are small perturbations of parameters $Q_L$, $X_V$ and $P_L$ about some operating point 'a', then from equation (4.46)

$$\dot{q}_L = \left.\frac{\partial \dot{Q}_L}{\partial X_v}\right|_a x_v + \left.\frac{\partial \dot{Q}_L}{\partial P_L}\right|_a p_L \tag{4.47}$$

or

$$\dot{q}_L = K_q x_v - K_c p_L \tag{4.48}$$

where

$$K_q \text{ (flow gain)} = C_d W \sqrt{\frac{1}{\rho}(P_s - P_{La})} \tag{4.49}$$

and

(flow-pressure coefficient which has a negative value)

$$K_c = \frac{-C_d W X_{va}}{2}\sqrt{\frac{1}{\rho(P_s - P_{La})}} \tag{4.50}$$

Note that $P_{La}$ and $X_{va}$ are the values of $P_L$ and $X_v$ at the operating point 'a'.

The relationship between $\dot{Q}_L$, $P_L$ and $X_v$, from equation (4.45), together with the linearized relations $\dot{q}_L$, $P_L$ and $x_v$ are shown in Figure 4.20.

Equation (4.39) is true for both large and small perturbations, and so can be written

$$\dot{q}_L = A\frac{dx_o}{dt} + C_P p_L + \frac{V_t}{4\beta}\frac{dp_L}{dt} \tag{4.51}$$

Equating (4.48) and (4.51) gives

$$K_q x_v = A\frac{dx_o}{dt} + (C_P + K_c)p_L + \frac{V_t}{4\beta}\frac{dp_L}{dt} \tag{4.52}$$

Taking Laplace transforms (zero initial conditions), but retaining the lower-case small perturbation notation gives

$$K_q x_v(s) = A s x_o(s) + \left\{(C_P + K_c) + \frac{V_t}{4\beta}s\right\}p_L(s) \tag{4.53}$$

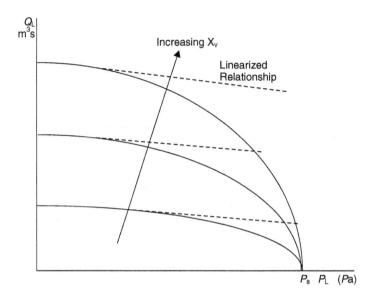

**Fig. 4.20** Pressure–Flow-rate characteristics for a spool-valve.

The force to accelerate the mass $m$ is shown in Figure 4.21. From Figure 4.21

$$\sum F_x = m\ddot{x}_o$$
$$AP_L = m\ddot{x}_o \qquad (4.54)$$

Take Laplace transforms with zero initial conditions and using lower-case notation

$$p_L(s) = \frac{m}{A}s^2 x_o(s) \qquad (4.55)$$

Inserting equation (4.55) into (4.53) gives

$$K_q x_v(s) = A s x_o(s) + \left\{ (C_P + K_c) + \frac{V_t}{4\beta}s \right\} \left( \frac{m}{A}s^2 \right) x_o(s) \qquad (4.56)$$

Equation (4.56) may be re-arranged to give the transfer function relating $x_o(s)$ and $x_v(s)$

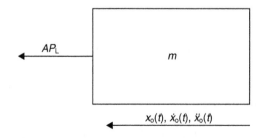

**Fig. 4.21** Free-body diagram of load on hydraulic actuator.

$$\frac{x_o}{x_v}(s) = \frac{\frac{K_q}{A}}{s\left\{\left(\frac{mV_t}{4\beta A^2}\right)s^2 + \left(\frac{C_p + K_c}{A^2}\right)s + 1\right\}} \quad (4.57)$$

Equation (4.57) can be written in the standard form

$$\frac{x_o}{x_v}(s) = \frac{K_h}{s\left(\frac{1}{\omega_{nh}^2}s^2 + \frac{2\zeta_h}{\omega_{nh}}s + 1\right)} \quad (4.58)$$

where

$$K_h \text{ (hydraulic gain)} = \frac{K_q}{A}$$

$$\omega_{nh} \text{ (hydraulic natural frequency)} = \sqrt{\frac{4\beta A^2}{mV_t}}$$

$$\zeta_h \text{ (hydraulic damping ratio)} = \left(\frac{C_p + K_c}{2}\right)\sqrt{\frac{4\beta}{mV_t A^2}}$$

Since the Bulk Modulus of hydraulic oil is in the order of 1.4 GPa, if $m$ and $V_t$ are small, a large hydraulic natural frequency is possible, resulting in a rapid response. Note that the hydraulic damping ratio is governed by $C_P$ and $K_c$. To control the level of damping, it is sometimes necessary to drill small holes through the piston.

## 4.5 Controllers for closed-loop systems

### 4.5.1 The generalized control problem

A generalized closed-loop control system is shown in Figure 4.22. The control problem can be stated as: 'The control action $u(t)$ will be such that the controlled output $c(t)$ will be equal to the reference input $r_1(t)$ for all values of time, irrespective of the value of the disturbance input $r_2(t)$'.

In practice, there will always be transient errors, but the transient period should be kept as small as possible. It is usually possible to design the controller so that steady-state errors are minimized, or ideally, eliminated.

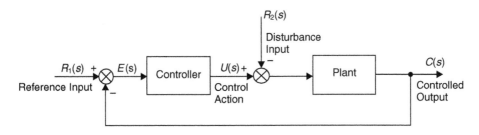

**Fig. 4.22** Generalized closed-loop control system.

### 4.5.2 Proportional control

In this case, the control action, or signal is proportional to the error $e(t)$

$$u(t) = K_1 e(t) \tag{4.59}$$

where $K_1$ is the proportional gain constant.

If the plant dynamics are first-order, then Figure 4.22 can be described as shown in Figure 4.23. The plant transfer function is

$$(U(s) - R_2(s))\left(\frac{K}{1 + Ts}\right) = C(s) \tag{4.60}$$

And the proportional control law, from equation (4.59) becomes

$$U(s) = K_1(R_1(s) - C(s)) \tag{4.61}$$

Inserting equation (4.61) into equation (4.60) gives

$$C(s) = \frac{\{K_1(R_1(s) - C(s)) - R_2(s)\}K}{(1 + Ts)} \tag{4.62}$$

which can be written as

$$\{(1 + K_1 K) + Ts\}C(s) = K_1 K R_1(s) - K R_2(s) \tag{4.63}$$

Re-arranging equation (4.63) gives

$$C(s) = \frac{\left(\frac{K_1 K}{1 + K_1 K}\right)R_1(s) - \left(\frac{K}{1 + K_1 K}\right)R_2(s)}{\left\{1 + \left(\frac{T}{1 + K_1 K}\right)s\right\}} \tag{4.64}$$

When $r_1(t)$ is a unit step, and $r_2(t)$ is zero, the final value theorem (equation (3.10)) gives the steady-state response

$$c(t) = \left(\frac{K_1 K}{1 + K_1 K}\right) \quad \text{as } t \to \infty.$$

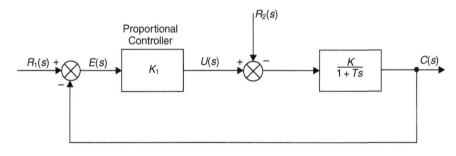

**Fig. 4.23** Proportional control of a first-order plant.

When $r_2(t)$ is a unit step, and $r_1(t)$ is zero, the final value theorem (equation (3.10)) gives the steady-state response

$$c(t) = -\left(\frac{K}{1 + K_1 K}\right) \quad \text{as } t \to \infty.$$

Hence, for the system to have zero steady-state error, the terms in equation (4.64) should be

$$\left(\frac{K_1 K}{1 + K_1 K}\right) = 1$$
$$\left(\frac{K}{1 + K_1 K}\right) = 0 \tag{4.65}$$

This can only happen if the open-loop gain constant $K_1 K$ is infinite. In practice this is not possible and therefore the proportional control system proposed in Figure 4.23 will always produce steady-state errors. These can be minimized by keeping the open-loop gain constant $K_1 K$ as high as possible.

Since the closed-loop time-constant form equation (4.64) is

$$T_\text{c} = \left(\frac{T}{1 + K_1 K}\right) \tag{4.66}$$

Then maintaining $K_1 K$ at a high value will reduce the closed-loop time constant and therefore improve the system transient response.

This is illustrated in Figure 4.24 which shows a step change in $r_1(t)$ followed by a step change in $r_2(t)$.

### Summary
For a first-order plant, proportional control will always produce steady-state errors. This is discussed in more detail in Chapter 6 under 'system type classification' where equations (6.63)–(6.65) define a set of error coefficients. Increasing the open-loop

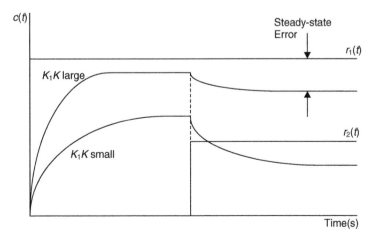

**Fig. 4.24** Step response of a first-order plant using proportional control.

gain constant (which is usually achieved by increasing the controller gain $K_1$) will reduce, but not eliminate them. A high controller gain will also reduce the transient period. However, as will be shown in Chapters 5 and 6, high open-loop gain constants can result in the instability of higher-order plant transfer functions.

## 4.5.3 Proportional plus Integral (PI) control

Including a term that is a function of the integral of the error can, with the type of plant shown in Figure 4.23, eliminate steady-state errors.

Consider a control law of the form

$$u(t) = K_1 e(t) + K_2 \int e \, dt \tag{4.67}$$

Taking Laplace transforms

$$U(s) = \left( K_1 + \frac{K_2}{s} \right) E(s)$$

$$= K_1 \left( 1 + \frac{K_2}{K_1 s} \right) E(s)$$

$$= K_1 \left( 1 + \frac{1}{T_i s} \right) E(s) \tag{4.68}$$

In equation (4.68), $T_i$ is called the integral action time, and is formally defined as: 'The time interval in which the part of the control signal due to integral action increases by an amount equal to the part of the control signal due to proportional action when the error is unchanging'. (BS 1523).

Inserting the PI control law given in equation (4.68) into the first-order plant transfer function shown in equation (4.60) gives

$$C(s) = \frac{(K_1(1 + 1/T_i s)(R_1(s) - C(s)) - R_2(s))K}{(1 + Ts)} \tag{4.69}$$

which can be written as

$$\{ T_i T s^2 + T_i (1 + K_1 K)_s + K_1 K \} C(s) = K_1 K (1 + T_i s) R_1(s) - K_1 K T_i s R_2(s) \tag{4.70}$$

Re-arranging gives

$$C(s) = \frac{(1 + T_i s) R_1(s) - T_i s R_2(s)}{\left( \frac{T_i T}{K_1 K} \right) s^2 + T_i \left( 1 + \frac{1}{K_1 K} \right) s + 1} \tag{4.71}$$

The denominator is now in the standard second-order system form of equation (3.42). The steady-state response may be obtained using the final value theorem given in equation (3.10).

$$c(t) = (1 + 0) r_1(t) - (0) r_2(t) \quad \text{as } t \to \infty \tag{4.72}$$

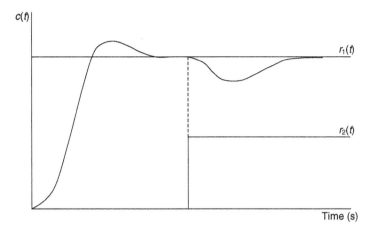

**Fig. 4.25** Step response of a first-order plant using PI control.

When there are step changes in $r_1(t)$ and $r_2(t)$:

$$C(s) = \frac{(1+0)sR_1(s)}{s} - (0)\frac{sR_2(s)}{s}$$

$$= R_1(s)$$

$$c(t) = r_1(t) \qquad\qquad (4.73)$$

Thus, when $r_1(t)$ and $r_2(t)$ are unchanging, or have step changes, there are no steady-state errors as can be seen in Figure 4.25. The second-order dynamics of the closed-loop system depend upon the values of $T_i$, $T$, $K_1$ and $K$. Again, a high value of $K_1$ will provide a fast transient response since it increases the undamped natural frequency, but with higher order plant transfer functions can give rise to instability.

### Summary
For a first-order plant, PI control will produce a second-order response. There will be zero steady-state errors if the reference and disturbance inputs $r_1(t)$ and $r_2(t)$ are either unchanging or have step changes. The process of including an integrator within the control loop to reduce or eliminate steady-state errors is discussed in more detail in Chapter 6 under 'system type classification'.

*Example 4.5* (See also Appendix 1, *examp45.m*)
A liquid-level process control system is shown in Figure 4.26. The system parameters are

$$A = 2\,\text{m}^2 \qquad R_f = 15\,\text{s/m}^2$$

$$H_1 = 1\,\text{V/m} \quad K_v = 0.1\,\text{m}^3/\text{sV} \quad K_1 = 1 \quad \text{(controller again)}$$

(a) What are the values of $T_i$ and $\zeta$ when the undamped natural frequency $\omega_n$ is 0.1 rad/s?
(b) Find an expression for the time response of the system when there is a step change of $h_d(t)$ from 0 to 4 m. Assume zero initial conditions.

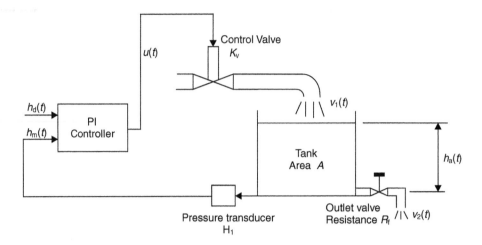

**Fig. 4.26** Liquid-level process control system.

The controller is given in equation (4.68). The inflow to the tank is

$$v_1(t) = K_v u(t) \tag{4.74}$$

The tank dynamics are expressed, using equation (2.63) as

$$v_1(t) - v_2(t) = A\frac{dh_a}{dt} \tag{4.75}$$

and the linearized outflow is

$$v_2(t) = \frac{h_a(t)}{R_f} \tag{4.76}$$

The measured head $h_m(t)$ is obtained from the pressure transducer

$$h_m(t) = H_1 h_a(t) \tag{4.77}$$

From equations (4.75) and (4.76), the tank and outflow valve transfer function is

$$\frac{H_a}{V_1}(s) = \frac{R_f}{1 + AR_f s} \tag{4.78}$$

The block diagram for the control system is shown in Figure 4.27. From the block diagram, the forward-path transfer function $G(s)$ is

$$
\begin{aligned}
G(s) &= \frac{K_1 K_v R_f \left(1 + \frac{1}{T_i s}\right)}{(1 + AR_f s)} \\
&= \frac{K_1 K_v R_f (1 + T_i s)}{T_i s (1 + AR_f)}
\end{aligned}
\tag{4.79}
$$

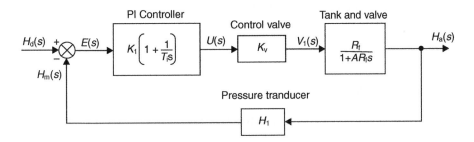

**Fig. 4.27** Block diagram for liquid-level process control system.

Using equation (4.4), the closed-loop transfer function becomes

$$\frac{H_a}{H_d}(s) = \frac{\frac{K_1 K_v R_f(1+T_i s)}{(AR_f T_i s^2 + T_i s)}}{1 + \frac{K_1 K_v R_f H_1(1+T_i s)}{(AR_f T_i s^2 + T_i s)}} \tag{4.80}$$

which simplifies to

$$\frac{H_a}{H_d}(s) = \frac{K_1 K_v R_f(1 + T_i s)}{(AR_f T_i)s^2 + T_i(1 + K_1 K_v R_f H_1)s + K_1 K_v R_f H_1} \tag{4.81}$$

Equation (4.81) can be expressed in the standard form of equation (3.42) for a second-order system.

Putting $H_1 = 1$, then

$$\frac{H_a}{H_d}(s) = \frac{(1 + T_i s)}{\left(\frac{AT_i}{K_1 K_v}\right)s^2 + T_i\left(\frac{1}{K_1 K_v R_f} + 1\right)s + 1} \tag{4.82}$$

(a) Comparing the denominator terms with the standard form given in equation (3.42)

$$\left(\frac{AT_i}{K_1 K_v}\right) = \frac{1}{\omega_n^2} \tag{4.83}$$

$$T_i\left(\frac{1}{K_1 K_v R_f} + 1\right) = \frac{2\zeta}{\omega_n} \tag{4.84}$$

From equation (4.83)

$$T_i = \frac{K_1 K_v}{\omega_n^2 A} = \frac{1 \times 0.1}{0.1^2 \times 2} = 5 \text{ seconds}$$

From equation (4.84)

$$\zeta = \frac{\omega_n T_i}{2}\left(\frac{1}{K_1 K_v R_f} + 1\right)$$

$$= \frac{0.1 \times 5}{2}\left(\frac{1}{1 \times 0.1 \times 15} + 1\right) = 0.417$$

(b)  Inserting values into equation (4.82)

$$\frac{H_a}{H_d}(s) = \frac{(1 + 5s)}{100s^2 + 8.34s + 1} \tag{4.85}$$

For a step input of height 4 m

$$H_a(s) = \left[\frac{0.01(1 + 5s)}{s^2 + 0.0834s + 0.01}\right]\frac{4}{s}$$

Expanding by partial fractions using 3.2.4 (iv)

$$H_a(s) = \frac{0.04 + 0.2s}{s(s^2 + 0.0834s + 0.01)} = \frac{A}{s} + \frac{Bs + C}{s^2 + 0.0834s + 0.01} \tag{4.86}$$

Multiplying through by $s(s^2 + 0.0834s + 0.01)$

$$0.04 + 0.2s = A(s^2 + 0.0834s + 0.01) + Bs^2 + Cs$$

Equating coefficients

$$(s^2): \quad 0 = A + B$$
$$(s^1): \quad 0.2 = 0.0834A + C$$
$$(s^0): \quad 0.04 = 0.01A$$

giving

$$A = 4 \quad B = -4 \quad C = -0.1336$$

Substituting values back into (4.86) and complete the square to give

$$H_a(s) = \frac{4}{s} + \frac{-4s - 0.1336}{(s + 0.0417)^2 + 0.0909^2} \tag{4.87}$$

Inverse transform using Laplace transform pairs (9) and (10) in Table 3.1.

$$H_a(s) = \frac{4}{s} - \left\{\frac{4s}{(s + 0.0417)^2 + 0.0909^2}\right\} - 1.4697\left\{\frac{0.0909}{(s + 0.0417)^2 + 0.0909^2}\right\}$$

$$h_a(t) = 4 - 4e^{-0.0417t}\left(\cos 0.0909t - \frac{0.0417}{0.0909}\sin 0.0909t\right)$$
$$- 1.4697e^{-0.0417t}\sin 0.0909t$$

which simplifies to give

$$h_a(t) = 4[1 - e^{-0.0417t}(\cos 0.0909t - 0.0913\sin 0.0909t)] \tag{4.88}$$

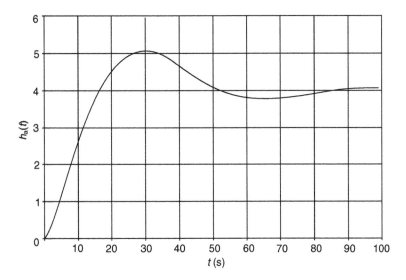

**Fig. 4.28** Response of the PI controlled liquid-level system shown in Figure 4.26 to a step change in $h_d(t)$ from 0 to 4 m.

In equation (4.88) the amplitude of the sine term is small, compared with the cosine term, and can be ignored. Hence

$$h_a(t) = 4(1 - e^{-0.0417t} \cos 0.0909t) \tag{4.89}$$

The time response depicted by equation (4.89) is shown in Figure 4.28.

## 4.5.4 Proportional plus Integral plus Derivative (PID) control

Most commercial controllers provide full PID (also called three-term) control action. Including a term that is a function of the derivative of the error can, with high-order plants, provide a stable control solution.

Proportional plus Integral plus Derivative control action is expressed as

$$u(t) = K_1 e(t) + K_2 \int e \, dt + K_3 \frac{de}{dt} \tag{4.90}$$

Taking Laplace transforms

$$U(s) = \left( K_1 + \frac{K_2}{s} + K_3 s \right) E(s)$$

$$= K_1 \left( 1 + \frac{K_2}{K_1 s} + \frac{K_3}{K_1} s \right) E(s)$$

$$= K_1 \left( 1 + \frac{1}{T_i s} + T_d s \right) E(s) \tag{4.91}$$

In equation (4.91), $T_d$ is called the derivative action time, and is formally defined as: 'The time interval in which the part of the control signal due to proportional action increases by an amount equal to the part of the control signal due to derivative action when the error is changing at a constant rate' (BS 1523).

Equation (4.91) can also be expressed as

$$U(s) = \frac{K_1(T_i T_d s^2 + T_i s + 1)}{T_i s} E(s) \tag{4.92}$$

## 4.5.5 The Ziegler–Nichols methods for tuning PID controllers

The selection of the PID controller parameters $K_1$, $T_i$ and $T_d$ can be obtained using the classical control system design techniques described in Chapters 5 and 6. In the 1940s, when such tools were just being developed, Ziegler and Nichols (1942) devised two empirical methods for obtaining the controller parameters. These methods are still in use.

(a) *The Process Reaction Method:* This is based on the assumption that the open-loop step response of most process control systems has an S-shape, called the process reaction curve, as shown in Figure 4.29. The process reaction curve may be approximated to a time delay $D$ (also called a transportation lag) and a first-order system of maximum tangential slope $R$ as shown in Figure 4.29 (see also Figure 3.13).

The Process Reaction Method assumes that the optimum response for the closed-loop system occurs when the ratio of successive peaks, as defined by equation (3.71), is 4:1. From equation (3.71) it can be seen that this occurs when the closed-loop damping ratio has a value of 0.21. The controller parameters, as a function of $R$ and $D$, to produce this response, are given in Table 4.2.

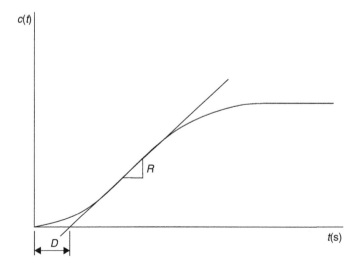

**Fig. 4.29** Process reaction curve.

**Table 4.2** Ziegler–Nichols PID parameters using the Process Reaction Method

| Controller type | $K_1$ | $T_i$ | $T_d$ |
|---|---|---|---|
| P | $1/RD$ | – | – |
| PI | $0.9/RD$ | $D/0.3$ | – |
| PID | $1.2/RD$ | $2D$ | $0.5D$ |

**Table 4.3** Ziegler–Nichols PID parameters using the Continuous Cycling Method

| Controller type | $K_1$ | $T_i$ | $T_d$ |
|---|---|---|---|
| P | $K_u/2$ | – | – |
| PI | $K_u/2.2$ | $T_u/1.2$ | – |
| PID | $K_u/1.7$ | $T_u/2$ | $T_u/8$ |

Note that the Process Reaction Method cannot be used if the open-loop step response has an overshoot, or contains a pure integrator(s).

(b) *The Continuous Cycling Method*: This is a closed-loop technique whereby, using proportional control only, the controller gain $K_1$ is increased until the system controlled output $c(t)$ oscillates continually at constant amplitude, like a second-order system with no damping. This condition is referred to as marginal stability and is discussed further in Chapters 5 and 6. This value of controller gain is called the ultimate gain $K_u$, and the time period for one oscillation of $c(t)$ is called the ultimate period $T_u$. The controller parameters, as a function of $K_u$ and $T_u$, to provide a similar closed-loop response to the Process Reaction Method, are given in Table 4.3.

The two Ziegler–Nichols PID tuning methods provide a useful 'rule of thumb' empirical approach. The control system design techniques discussed in Chapters 5 and 6 however will generally yield better design solutions.

Of the two techniques, the Process Reaction Method is the easiest and least disruptive to implement. In practice, the measurement of $R$ and $D$ is very subjective, and can lead to errors.

The Continuous Cycling Method, although more disruptive, has the potential to give better results. There is the risk however, particularly with high performance servo-mechanisms, that if $K_u$ is increased by accident to slightly above the marginal stability value, then full instability can occur, resulting in damage to the system.

### Actuator saturation and integral wind-up

One of the practical problems of implementing PID control is that of actuator saturation and integral wind-up. Since the range of movement in say, a control valve, has physical limits, once it has saturated, increasing the magnitude of the control signal further has no effect. However, if there is a difference between desired and measured values, the resulting error will cause a continuing increase in the integral term, referred to as integral wind-up. When the error term changes its sign, the integral term starts to 'unwind,' and this can cause long time delays and possible instability. The solution is to limit the maximum value that the integral term can have.

### 4.5.6  Proportional plus Derivative (PD) control

Proportional plus Derivative control action is expressed as

$$u(t) = K_1 e(t) + K_3 \frac{de}{dt} \qquad (4.93)$$

Taking Laplace transforms

$$U(s) = K_1 \left( 1 + \frac{K_3}{K_1} \right) E(s)$$

$$= K_1 (1 + T_d s) E(s) \qquad (4.94)$$

The inclusion of a derivative term in the controller generally gives improved damping and stability. This is discussed in more detail in Chapters 5 and 6.

## 4.6  Case study examples

*Example 4.6.1*  CNC Machine-Tool Positional Control (See also Appendix 1, *examp461.m*)
The physical configuration and block diagram representation of a CNC machine-tool is shown in Figures 1.10 and 1.11. The fundamental control problem here is that, by design, the lead-screw (by the use of re-circulating ball-bearings) is friction-free. This means that the positional control system will have no damping, and will oscillate continuously at the undamped natural frequency of the closed-loop system.
    Damping can be introduced in a number of ways:

(a) *A dashpot attached to the lead-screw*: This is wasteful on energy and defeats the objective of a friction-free system.
(b) *Velocity feedback*: A signal from a sensor that is the first derivative of the output (i.e. velocity) will produce a damping term in the closed-loop transfer function.
(c) *PD control*: A PD controller will also provide a damping term. However, the practical realization will require an additional filter to remove unwanted high frequency noise (see Chapter 6 for further details on lead-lag compensation).

Most machine-tool manufacturers employ velocity feedback to obtain the necessary damping. Since overshoot in a cutting operation usually cannot be tolerated, the damping coefficient for the system must be unity, or greater.
    For this study, the machine-tool configuration will be essentially the same as shown in Figure 1.10, with the exception that:

(i)   A gearbox will be placed between the servo-motor and the lead-screw to provide additional torque.
(ii)  The machine table movement will be measured by a linear displacement trans-ducer attached to the table. This has the advantage of bringing the table 'within the control-loop' and hence providing more accurate control.

*System element dynamic equations*: With reference to Figures 1.11 and 4.31

1. *Controller*

$$\text{Proportional control, gain } K_1\text{(V/m)}$$
$$\text{Control signal } U_1(s) = K_1(X_d(s) - X_m(s)) \tag{4.95}$$

2. *Power amplifier*

$$\text{Gain } K_2\text{(V/V)}$$
$$\text{Control signal } U_2(s) = K_2(U_1(s) - B_2(s)) \tag{4.96}$$

3. *DC servo-motor*: Field controlled, with transfer function as shown in Figure 4.17. It will be assumed that the field time constant $L_f/R_f$ is small compared with the dynamics of the machine table, and therefore can be ignored. Hence, DC servo-motor gain $K_3$ (Nm/V).

$$\text{Motor Torque } T_m(s) = K_3 U_2(s) \tag{4.97}$$

4. *Gearbox, lead-screw and machine-table*: With reference to Figure 2.9 (free-body diagram of a gearbox), the motor-shaft will have zero viscous friction $C_m$, hence equation (2.22), using Laplace notation, becomes

$$X(s) = \frac{1}{a}(T_m(s) - I_m s^2 \theta_m(s)) \tag{4.98}$$

The output shaft in this case is the lead screw, which is assumed to have zero moment of inertia $I_o$ and viscous friction $C_o$. The free-body diagrams of the machine-table and lead-screw are shown in Figure 4.30.

For lead-screw

$$\text{Work in} = \text{Work out}$$
$$bX(t)\theta_o(t) = F(t)x_o(t)$$

or

$$F(t) = bX(t)\frac{\theta_o(t)}{x_o(t)} \tag{4.99}$$

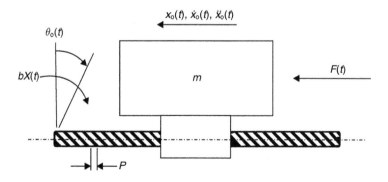

**Fig. 4.30** Free-body diagrams of lead-screw and machine-table.

Now the pitch $p$ of the lead-screw is

$$p = \frac{x_o(t)}{\theta_o(t)} \tag{4.100}$$

Substituting (4.100) into (4.99)

$$F(t) = \frac{bX(t)}{p} \tag{4.101}$$

The equation of motion for the machine-table is

$$F(t) = m\ddot{x}_o \tag{4.102}$$

Equating (4.101) and (4.102) gives

$$X(t) = \frac{1}{b}(pm\ddot{x}_o)$$

Taking Laplace transforms

$$X(s) = \frac{1}{b}(pms^2 X_o(s)) \tag{4.103}$$

Equating (4.98) and (4.103) gives

$$pms^2 X_o(s) = \frac{b}{a}\left(T_m(s) - I_m s^2 \theta_m(s)\right) \tag{4.104}$$

Now

$$b/a = \text{gear ratio } n$$
$$\theta_m(s) = n\theta_o(s)$$

Hence

$$s^2 \theta_m(s) = ns^2 \theta_o(s)$$

and

$$\theta_o(s) = \frac{X_o(s)}{p} \tag{4.105}$$

Equation (4.105) can be substituted into (4.104)

$$pms^2 X_o(s) = nT_m(s) - nI_m \frac{n}{p}s^2 X_o(s)$$

or

$$nT_m(s) = \left(pm + \frac{n^2 I_m}{p}\right)s^2 X_o(s) \tag{4.106}$$

giving the transfer function for the gearbox, lead-screw and machine-table as

$$\frac{X_o}{T_m}(s) = \frac{n}{(pm + n^2 I_m/p)s^2} \tag{4.107}$$

where the term $n^2 I_m/p$ may be considered to be equivalent mass of $I_m$ referred to the machine-table.

5. *Tachogenerator*

$$\text{Gain } H_2 \text{ (V s/rad)}$$

$$\text{Feedback signal } B_2(s) = H_2 s \theta_o(s) \tag{4.108}$$

or, from equation (4.105)

$$B_2(s) = \frac{H_2}{p} s X_o(s) \tag{4.109}$$

6. *Position transducer*

$$\text{Gain } H_1 \text{(V/m)}$$

$$\text{Feedback signal } X_m(s) = H_1 X_o(s) \tag{4.110}$$

The system element dynamic equations can now be combined in the block diagram shown in Figure 4.31. Using equation (4.4), the inner-loop transfer function is

$$G(s) = \frac{K_2 K_3 n p}{(p^2 m + n^2 I_m)s + K_2 K_3 n H_2} \tag{4.111}$$

Again, using equation (4.4), the overall closed-loop transfer function becomes

$$\frac{X_o}{X_d}(s) = \frac{K_1 K_2 K_3 n p}{(p^2 m + n^2 I_m)s^2 + K_2 K_3 n H_2 s + K_1 K_2 K_3 n p H_1} \tag{4.112}$$

which can be written in standard form

$$\frac{X_o}{X_d}(s) = \frac{\frac{1}{H_1}}{\left(\frac{p^2 m + n^2 I_m}{K_1 K_2 K_3 n p H_1}\right)s^2 + \left(\frac{H_2}{K_1 p H_1}\right)s + 1} \tag{4.113}$$

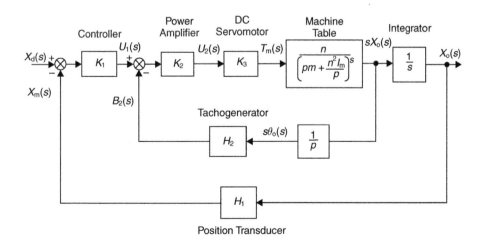

**Fig. 4.31** Block diagram of CNC machine-tool control system.

*Specification*: The CNC machine-table control system is to be critically damped with a settling time of 0.1 seconds.

*Control problem*: To select the controller gain $K_1$ to achieve the settling time and tachogenerator constant to provide critical damping.

*System parameters*

$$K_2 = 2\,\text{V/V} \quad K_3 = 4\,\text{Nm/V}$$
$$n = 10:1 \quad p = 5 \times 10^{-3}\,\text{m}$$
$$m = 50\,\text{kg} \quad I_m = 10 \times 10^{-6}\,\text{kgm}^2$$
$$H_1 = 60\,\text{V/m}$$

*Calculation of $K_1$*: In general, the settling time of a system with critical damping is equal to the periodic time of the undamped system, as can be seen in Figure 3.19.

This can be demonstrated using equation (3.62) for critical damping

$$x_o(t) = [1 - e^{-\omega_n t}(1 + \omega_n t)]$$

when

$$t = 2\pi/\omega_n$$
$$x_o(t) = \left[1 - e^{-2\pi}(1 + 2\pi)\right]$$
$$= 0.986 \tag{4.114}$$

Thus, for a settling time of 0.1 seconds for a system that is critically damped, the undamped natural frequency is

$$\omega_n = \frac{2\pi}{0.1} = 62.84\,\text{rad/s} \tag{4.115}$$

Comparing the closed-loop transfer function given in equation (4.113) with the standard form given in (3.42)

$$\omega_n^2 = \left(\frac{K_1 K_2 K_3 n p H_1}{p^2 m + n^2 I_m}\right) \tag{4.116}$$

Hence

$$K_1 = \frac{(p^2 m + n^2 I_m)\omega_n^2}{K_2 K_3 n p H_1}$$
$$= \left[\frac{\{(5 \times 10^{-3})^2 \times 50\} + (10^2 \times 10 \times 10^{-6})}{(2 \times 4 \times 10 \times 5 \times 10^{-3} \times 60)}\right] \times 62.84^2$$
$$= 0.365\,\text{V/V} \tag{4.117}$$

Again, comparing equation (4.113) with the standard form (3.42)

$$\frac{2\zeta}{\omega_n} = \frac{H_2}{K_1 p H_1} \tag{4.118}$$

Hence

$$H_2 = \frac{2\zeta K_1 p H_1}{\omega_n}$$

$$= \frac{2 \times 1 \times 0.365 \times 5 \times 10^{-3} \times 60}{62.84}$$

$$= 3.485 \times 10^{-3} \, \text{V s/rad} \tag{4.119}$$

*Example 4.6.2* Temperature control system (See also Appendix 1, *examp462.m*)
The general form of a temperature control system is shown in Figure 1.6 with the corresponding block diagram given in Figure 1.7.

The system variables are:

$\theta_d(t) = $ Desired temperature (°C)

$\theta_m(t) = $ Measured temperature (V)

$\theta_o(t) = $ Actual temperature (°C)

$\theta_s(t) = $ Temperature of surroundings (°C)

$u(t) = $ Control signal (v)

$\dot{v}(t) = $ Gas flow-rate (m$^3$/s)

$Q_i(t) = $ Heat flow into room (J/s = W)

$Q_o(t) = $ Heat flow though walls (W)

*System equations*

1. *Controller*: The control action is PID of the form given in equation (4.91)

$$U(s) = K_1 \left( 1 + \frac{1}{T_i s} + T_d s \right) (\theta_d(s) - \theta_m(s)) \tag{4.120}$$

2. *Gas solenoid valve*: This is assumed to have first-order dynamics of the form

$$\frac{\dot{V}}{U}(s) = \frac{K_2}{1 + T_1 s} \tag{4.121}$$

where $K_2$ is the valve constant (m$^3$/s V).

3. *Gas burner*: This converts gas flow-rate $\dot{v}(t)$ into heat flow $Q_i(t)$ i.e.:

$$Q_i(s) = K_3 \dot{V}(s) \tag{4.122}$$

where $K_3$ is the burner constant (Ws/m$^3$).

4. *Room dynamics*: The thermal dynamics of the room are

$$Q_i(t) - Q_o(t) = C_T \frac{d\theta_o}{dt} \tag{4.123}$$

Equation (4.123) is similar to equation (2.54), where $C_T$ is the thermal capacitance of the air in the room.

The heat flow through the walls of the building is as given in equation (2.53), i.e.

$$Q_o(t) = \frac{(\theta_o(t) - \theta_s(t))}{R_T} \tag{4.124}$$

where $R_T$ is the thermal resistance of the walls, see equation (2.47).
Substituting equation (4.124) into (4.123) gives

$$Q_i(t) - \left( \frac{\theta_o(t) - \theta_s(t)}{R_T} \right) = C_T \frac{d\theta_o}{dt} \tag{4.125}$$

Multiplying through by $R_T$

$$R_T Q_i(t) + \theta_s(t) = \theta_o(t) + R_T C_T \frac{d\theta_o}{dt} \tag{4.126}$$

Taking Laplace transforms

$$R_T Q_i(s) + \theta_s(s) = (1 + R_T C_T s)\theta_o(s) \tag{4.127}$$

Equation (4.127) can be represented in block diagram form as shown in Figure 4.32.

5. *Thermometer*: The thermometer equation is

$$\theta_m(s) = H_1 \theta_o(s) \tag{4.128}$$

The complete block diagram of the temperature control system is shown in Figure 4.33.
From Figure 4.33

$$\frac{K_1 K_2 K_3 R_T (T_i T_d s^2 + T_i s + 1)(\theta_d(s) - H_1 \theta_o(s))}{T_i s(1 + T_1 s)} + \theta_s(s) = (1 + R_T C_T s)\theta_o(s) \tag{4.129}$$

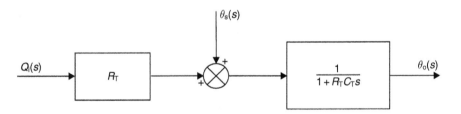

**Fig. 4.32** Block diagram representation of the thermal dynamics of the room.

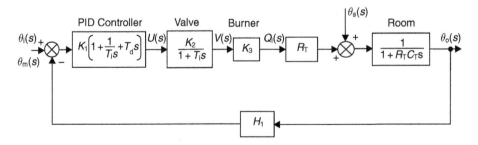

**Fig. 4.33** Block diagram of temperature control system.

Equation (4.129) can be re-arranged to give

$$\theta_o(s) = \frac{\frac{1}{H_1}(T_i T_d s^2 + T_i s + 1)\theta_d(s) + \frac{T_i s(1 + T_1 s)}{K_F H_1}\theta_s(s)}{\left(\frac{T_i T_1 T_2}{K_F H_1}\right)s^3 + \left(\frac{T_i(T_1 + T_2)}{K_F H_1} + T_i T_d\right)s^2 + T_i\left(\frac{1}{K_F H_1} + 1\right)s + 1} \qquad (4.130)$$

where the forward-path gain $K_F$ is

$$K_F = K_1 K_2 K_3 R_T \qquad (4.131)$$

*Control problem*: Given the system parameters, the control problem is to determine the controller settings for $K_1$, $T_i$ and $T_d$. This will be undertaken using the Zeigler–Nichols process reaction method described in Section 4.5.5(a).

*System parameters:*

$$K_2 K_3 = 5\,\text{W/V} \quad R_T = 0.1\,\text{Ks/J}$$
$$C_T = 80\,\text{J/K} \qquad H_1 = 1.0\,\text{V/K}$$
$$T_1 = 4\,\text{seconds}$$

*Process reaction curve*: This can be obtained from the forward-path transfer function

$$\frac{\theta_o}{U}(s) = \frac{K_2 K_3 R_T}{(1 + T_1 s)(1 + R_T C_T s)} \qquad (4.132)$$

Inserting values into equation (4.132) gives

$$\frac{\theta_o}{U}(s) = \frac{0.5}{(1 + 4s)(1 + 8s)} \qquad (4.133)$$

Figure 4.34 shows the response to a unit step, or the process reaction curve.

From the $R$ and $D$ values obtained from the process reaction curve, using the Zeigler–Nichols PID controller settings given in Table 4.2

$$K_1 = 1.2/RD = 26.144$$
$$T_i = 2D = 3.0\,\text{seconds}$$
$$T_d = 0.5D = 0.75\,\text{seconds}$$

Assuming that the temperature of the surroundings $\theta_s(t)$ remains constant, the closed-loop transfer function (using equation (4.130)) for the temperature control system, is

$$\frac{\theta_o}{\theta_d}(s) = \frac{(2.25s^2 + 3s + 1)}{7.344s^3 + 5.004s^2 + 3.229s + 1} \qquad (4.134)$$

The response to a step change in the desired temperature of 0–20 °C for the closed-loop transfer function given by equation (4.134) is shown in Figure 4.35.

From Figure 4.35, the ratio of successive peaks is

$$\frac{a_1}{a_2} = \frac{8.92}{1.98} = 4.5 \qquad (4.135)$$

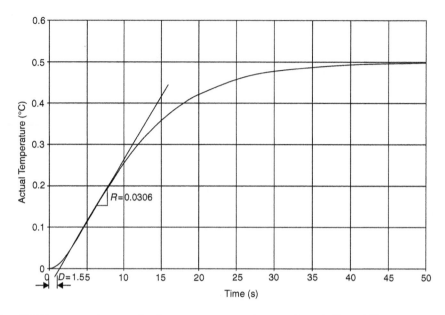

**Fig. 4.34** Process reaction curve for the temperature control system shown in Figure 4.33.

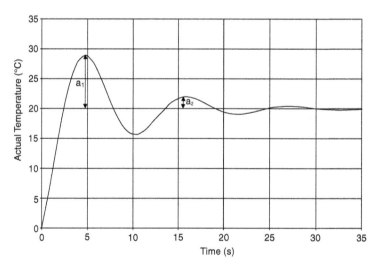

**Fig. 4.35** Closed-loop step response of temperature control system using PID controller tuned using Zeigler–Nichols process reaction method.

This corresponds to a damping ratio of 0.23. These values are very close to the Zeigler–Nichols optimum values of 4.0 and 0.21 respectively.

*Example 4.6.3*  Ship Autopilot (See also Appendix 1, *examp463.m*)
A ship has six degrees-of-freedom, i.e. it is free to move in six directions simultaneously, namely three linear motions – surge (forward), sway (lateral) and heave

(vertical) together with three rotational motions – roll, pitch and yaw. This analysis considers rotation about the yaw axis (i.e. heading control) only.

Figure 1.12 shows a typical ship autopilot system and Figure 1.13 shows the corresponding block diagram. Rotation about the yaw axis is in effect rotation about the $z$, or vertical axis of the vessel, called the '$r$' direction since $r$ is the symbol for yaw-rate. Hence hydrodynamic coefficients for the yaw axis are therefore given the subscript '$r$'. Yaw hydrodynamic coefficients are given the symbol '$N$'. In this analysis the dynamics of the steering gear are neglected.

The system variables are

$\Psi_d(t)$ = Desired heading (radians)

$\Psi_a(t)$ = Actual heading (radians)

$\delta(t)$ = Rudder angle (radians)

Figure 4.36 shows the hull free-body diagram. Disturbance effects (wind, waves and current) are not included.

*System equations*
1. *Hull dynamics*: In Figure 4.36, $X_o Y_o$ is the earth co-ordinate system where $X_o$ is aligned to north. All angles are measured with respect to $X_o$. A consistent right-hand system of co-ordinates is used, with the exception of the rudder-angle, which has been selected to be left-hand to avoid negative coefficients in the hull transfer function.

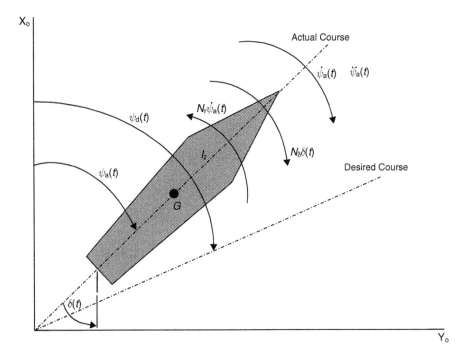

**Fig. 4.36** Free-body diagram of ship hull dynamics.

From Figure 4.36 the equation of motion for the yaw axis is

$$\sum M_G = I_z\ddot{\psi}_a(t)$$
$$(N_\delta\delta(t)) - (N_r\dot{\psi}_a(t)) = I_z\ddot{\psi}_a(t)$$

(4.136)

Taking Laplace transforms

$$N_\delta\delta(s) = (I_zs^2 + N_rs)\psi_a(s)$$

(4.137)

Hence the hull transfer function becomes

$$\frac{\psi_a}{\delta}(s) = \frac{N_\delta}{s(I_zs + N_r)}$$

(4.138)

2. *Control action*: In this case, the autopilot (controller) is considered to provide proportional control only.

$$\delta(s) = K_1(\psi_d(s) - \psi_a(s))$$

(4.139)

3. *Gyro-compass*: This provides a measured heading proportional to the actual heading

$$\psi_m(s) = H_1\psi_a(s)$$

(4.140)

Combining equations (4.138), (4.139) and (4.140) produces the block diagram shown in Figure 4.37.
Using equation (4.4), the closed-loop transfer function is

$$\frac{\psi_a}{\psi_d}(s) = \frac{\frac{K_1K_2N_\delta}{s(I_zs+N_r)}}{1 + \frac{K_1K_2N_\delta H_1}{s(I_zs+N_r)}}$$

(4.141)

Equation (4.141) simplifies to give

$$\frac{\psi_a}{\psi_d}(s) = \frac{\frac{1}{H_1}}{\left(\frac{I_z}{K_1K_2N_\delta H_1}\right)s^2 + \left(\frac{N_r}{K_1K_2N_\delta H_1}\right)s + 1}$$

(4.142)

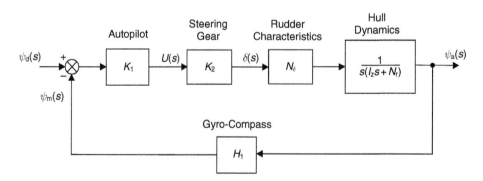

**Fig. 4.37** Block diagram of ship autopilot control system.

Equation (4.142) is in the standard form given in equation (3.42).

*Control problem*: For a specific hull, the control problem is to determine the autopilot setting ($K_1$) to provide a satisfactory transient response. In this case, this will be when the damping ratio has a value of 0.5. Also to be determined are the rise time, settling time and percentage overshoot.

*System parameters*: The ship to be controlled is a cargo vessel of length 161 m with a MARINER type hull of total displacement 17 000 tonnes.

$$K_2 = 1.0 \, \text{rad/V} \qquad N_\delta = 80 \times 10^6 \, \text{Nm/rad}$$
$$N_r = 2 \times 10^9 \, \text{Nms/rad} \quad I_z = 20 \times 10^9 \, \text{kg m}^2$$
$$H_1 = 1.0 \, \text{V/rad}$$

Inserting values into equation (4.142) gives

$$\frac{\psi_a}{\psi_d}(s) = \frac{1}{\left(\frac{20 \times 10^9}{K_1 \times 80 \times 10^6}\right)s^2 + \left(\frac{2 \times 10^9}{K_1 \times 80 \times 10^6}\right)s + 1} \tag{4.143}$$

which simplifies to

$$\frac{\psi_a}{\psi_d}(s) = \frac{1}{\left(\frac{250}{K_1}\right)s^2 + \left(\frac{25}{K_1}\right)s + 1} \tag{4.144}$$

Comparing equation (4.144) with the standard form given in equation (3.42)

$$\frac{1}{\omega_n^2} = \frac{250}{K_1} \tag{4.145}$$

and

$$\frac{2\zeta}{\omega_n} = \frac{25}{K_1} \tag{4.146}$$

Given that $\zeta = 0.5$, then from equation (4.146)

$$\omega_n = \frac{K_1}{25} \tag{4.147}$$

Substituting (4.147) into (4.145) gives

$$\frac{25^2}{K_1^2} = \frac{250}{K_1}$$

Hence

$$K_1 = \frac{625}{250} = 2.5 \tag{4.148}$$

and from equation (4.147),

$$\omega_n = \frac{2.5}{25} = 0.1 \, \text{rad/s} \tag{4.149}$$

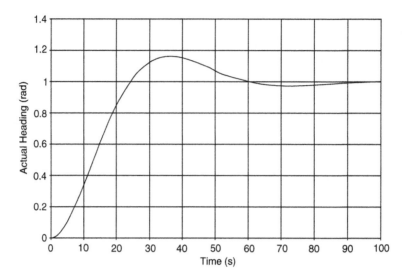

**Fig. 4.38** Unit step response of ship autopilot control system. RiseTime (to 95%) = 23 seconds;
Percentage Overshoot = 16.3%;
Settling time (to ±2%) = 81 seconds.

From equations (3.58) and (3.59) the unit step response for the ship autopilot control system is given by the expression

$$\psi_a(t) = 1 - e^{-0.05t}(\cos 0.0866t + 0.577 \sin 0.0866t) \qquad (4.150)$$

Figure 4.38 shows the system unit step response. From Figure 4.38

## 4.7 Further problems

*Example 4.7*
For the block diagrams shown in Figure 4.39 find an expression for the complete output when all inputs act simultaneously.

*Solutions*

(a) $C(s) = \dfrac{(G_1(s)G_2(s)G_3(s))R_1(s) + G_3(s)(1 + G_2(s)H_3(s))R_2(s)}{1 + G_3(s)H_2(s) + G_2(s)H_3(s) + G_1(s)G_2(s)G_3(s)H_1(s)}$

(b) $C(s) = \dfrac{(G_1(s)G_2(s)G_3(s)G_4(s))R_1(s) - (G_1(s)G_2(s)G_3(s)G_4(s)H_1(s))R_2(s) - (G_3(s)G_4(s))R_3(s)}{1 + G_3(s)H_2(s) + G_1(s)G_2(s)G_3(s)G_4(s)H_1(s)}$

*Example 4.8*
The speed control system shown in Figure 4.40 consists of an amplifier, a field-controlled DC servomotor and a tachogenerator. The load consists of a mass of moment of inertia $I$ and a fluid damper $C$. The system parameters are:

$$I = 0.75 \, \text{kg m}^2 \quad C = 0.5 \, \text{Nms/rad}$$
$$K_2 = 5 \, \text{Nm/A} \quad H_1 = 0.1 \, \text{Vs/rad}$$

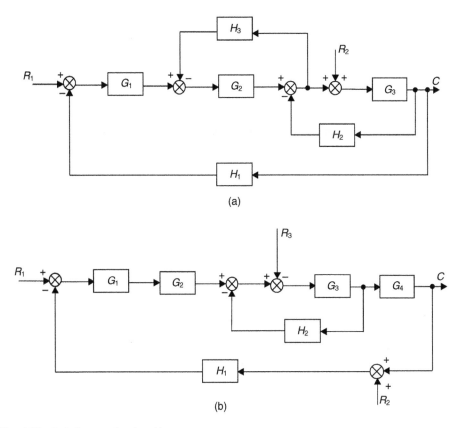

(a)

(b)

**Fig. 4.39** Block diagrams for closed-loop systems.

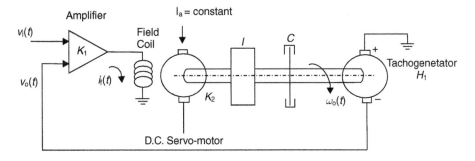

**Fig. 4.40** Speed control system.

Find the value of $K_1$ to give the system a closed-loop time constant of one second. What is the steady-state value of $\omega_o(t)$ when $v_i(t)$ has a value of 10 V.

*Solution*
0.5 A/V
33.3 rad/s

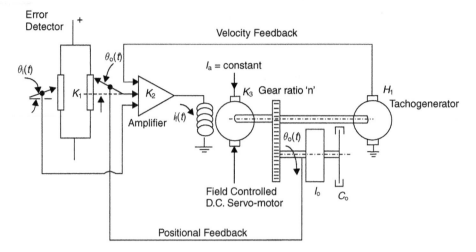

**Fig. 4.41** Angular positional control system. $K_1$ = Error detector gain (V/rad); $K_2$ = Amplifier gain (A/V); $K_3$ = Motor constant (Nm/A); $n$ = Gear ratio; $H_1$ = Tachogenerator constant (V s/rad); $I_o$ = Load moment of inertia (kg m$^2$); $C_o$ = Load damping coefficient (Nms/rad).

*Example 4.9*
Find an expression for the closed-loop transfer function relating $\theta_i(s)$ and $\theta_o(s)$ for the angular positional control system shown in Figure 4.41.

*Solution*

$$\frac{\theta_o}{\theta_i}(s) = \frac{1}{\left(\frac{I_o}{K_1 K_2 K_3 n}\right)s^2 + \left(\frac{C_0 + K_2 K_3 n^2 H_1}{K_1 K_2 K_3 n}\right)s + 1}$$

*Example 4.10*
A hydraulic servomechanism consists of the spool-valve/actuator arrangement shown in Figure 4.19 together with a 'walking beam' feedback linkage as shown in Figure 4.42. The spool-valve displacement $x_v(t)$ is given by the relationship

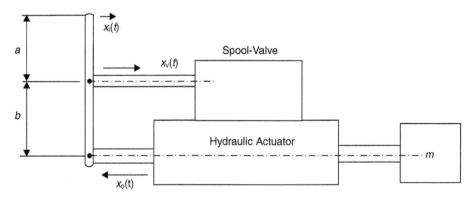

**Fig. 4.42** Hydraulic servomechanism with 'walking beam' feedback linkage.

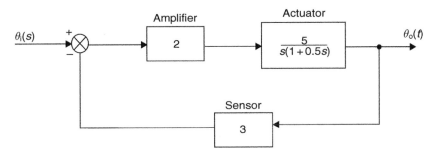

**Fig. 4.43** Block diagram of a servomechanism.

$$x_v(t) = \left(\frac{b}{a+b}\right) x_i(t) - \left(\frac{a}{a+b}\right) x_o(t)$$

If the forward-path transfer function is given by equation (4.57), find an expression for the closed-loop transfer function relating $X_i(s)$ and $X_o(s)$. The system parameters are

$$m = 50\,\text{kg} \qquad V_t = 4 \times 10^{-3}\,\text{m}^3 \quad \beta = 1.4\,\text{GPa}$$
$$A = 0.01\,\text{m}^2 \qquad K_q = 10.0\,\text{m}^2/\text{s} \quad K_c = 6 \times 10^{-9}\,\text{m}^5/\text{Ns}$$
$$C_p = 6 \times 10^{-9}\,\text{m}^5/\text{Ns} \quad a = b = 0.15\,\text{m}$$

*Solution*

$$\frac{X_o}{X_i}(s) = \frac{500}{0.357 \times 10^{-6}s^3 + 0.12 \times 10^{-3}s^2 + s + 500}$$

*Example 4.11*
A servomechanism consists of an amplifier, actuator and sensor as shown in block diagram form in Figure 4.43. If the input to the system is a constant velocity of the form

$$\theta_i(t) = 2t$$

find an expression for the time response of the system.

*Solution*

$$\theta_o(t) = 0.667t - 0.0222 + e^{-t}(0.0222\cos 7.68t - 0.083\sin 7.68t)$$

*Example 4.12*
Figure 4.44 shows the elements of a closed-loop temperature control system. A proportional controller compares the desired value $\theta_i(t)$ with the measured value $v_o(t)$ and provides a control signal $u(t)$ of $K_1$ times their difference to actuate the valve and burner unit. The heat input to the oven $Q_i(t)$ is $K_2$ times the control signal.

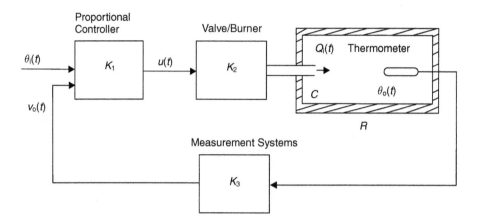

**Fig. 4.44** Closed-loop temperature control system.

The walls of the oven have a thermal resistance $R_T$ and the oven has a thermal capacitance $C_T$ and operating temperature $\theta_o(t)$. The heat transfer equation for the oven may be written

$$Q_i(t) - \frac{\theta_o(t)}{R_T} = C_T \frac{d\theta_o}{dt}$$

The thermometer and measurement system feed a measured value of $H_1$ times $\theta_o(t)$ to the controller. The system parameters are

$$K_1 = 5 \qquad K_2 = 1.5\,\text{J/V} \qquad H_1 = 1\,\text{V/K}$$
$$R_T = 2\,\text{K/J} \qquad C_T = 25\,\text{Js/W}$$

*Find*
(a) The open-loop time constant
(b) The closed-loop time constant
(c) The percentage steady-state error in the output when the desired value is constant.

*Solution*
(a) 50 seconds
(b) 3.125 seconds
(c) 6.25%

*Example 4.13*
Figure 4.45 shows the block diagram representation of a process plant being controlled by a PID controller.

(a) Find an expression for the complete response $C(s)$ when $R_1(s)$ and $R_2(s)$ act simultaneously.
(b) Using the Ziegler–Nichols Process Reaction Method, determine values for $K_1$, $T_i$ and $T_d$ when $T_1 = 10$ seconds and $T_2 = 20$ seconds.

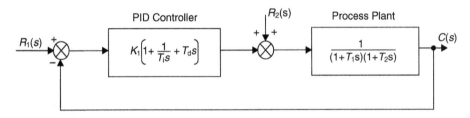

**Fig. 4.45** Process plant under PID control.

(c) Insert the values into the expression found in (a). Using MATLAB, or otherwise, determine the response $c(t)$ when $r_1(t)$ is a unit step and $r_2(t)$ is zero. What is the ratio of successive peaks?

*Solution*

(a) $C(s) = \dfrac{K_1(T_iT_ds^2 + T_is + 1)R_1(s) + T_isR_2(s)}{T_iT_1T_2s^3 + T_i(T_1 + T_2 + K_1T_d)s^2 + T_i(1 + K_1)s + K_1}$

(b) $K_1 = 17.2 \quad T_i = 6$ seconds $\quad T_d = 1.5$ seconds

(c) $C(s) = \dfrac{17.2(9s^2 + 6s + 1)}{s(1200s^3 + 334.8s^2 + 109.2s + 17.2)}$

$\qquad 2.75$

# 5

# Classical design in the s-plane

## 5.1 Stability of dynamic systems

The response of a linear system to a stimulus has two components:

(a) steady-state terms which are directly related to the input
(b) transient terms which are either exponential, or oscillatory with an envelope of exponential form.

If the exponential terms decay as time increases, then the system is said to be *stable*. If the exponential terms increase with increasing time, the system is considered *unstable*. Examples of stable and unstable systems are shown in Figure 5.1. The motions shown in Figure 5.1 are given graphically in Figure 5.2. (Note that (b) in Figure 5.2 does not represent (b) in Figure 5.1.) The time responses shown in Figure 5.2 can be expressed mathematically as:
For (a) (Stable)

$$x_o(t) = Ae^{-\sigma t} \sin(\omega t + \phi) \tag{5.1}$$

For (b) (Unstable)

$$x_o(t) = Ae^{\sigma t} \sin(\omega t + \phi) \tag{5.2}$$

For (c) (Stable)

$$x_o(t) = Ae^{-\sigma t} \tag{5.3}$$

For (d) (Unstable)

$$x_o(t) = Ae^{\sigma t} \tag{5.4}$$

From equations (5.1)–(5.4), it can be seen that the stability of a dynamic system depends upon the sign of the exponential index in the time response function, which is in fact a real root of the characteristic equation as explained in section 5.1.1.

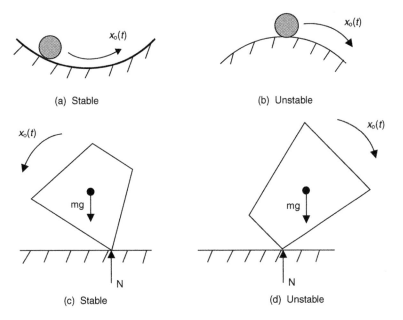

**Fig. 5.1** Stable and unstable systems.

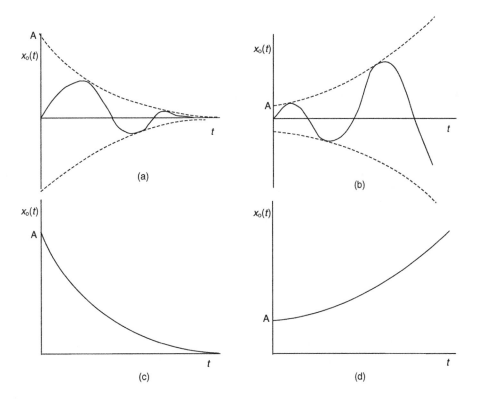

**Fig. 5.2** Graphical representation of stable and unstable time responses.

### 5.1.1   Stability and roots of the characteristic equation

The characteristic equation was defined in section 3.6.2 for a second-order system as

$$as^2 + bs + c = 0 \tag{5.5}$$

The roots of the characteristic equation given in equation (5.5) were shown in section 3.6.2. to be

$$s_1, s_2 = \frac{-b \pm \sqrt{b^2 - 4ac}}{2a} \tag{5.6}$$

These roots determine the transient response of the system and for a second-order system can be written as

(a) Overdamping

$$
\begin{aligned}
s_1 &= -\sigma_1 \\
s_2 &= -\sigma_2
\end{aligned}
\tag{5.7}
$$

(b) Critical damping

$$s_1 = s_2 = -\sigma \tag{5.8}$$

(c) Underdamping

$$s_1, s_2 = -\sigma \pm j\omega \tag{5.9}$$

If the coefficient $b$ in equation (5.5) were to be negative, then the roots would be

$$s_1, s_2 = +\sigma \pm j\omega \tag{5.10}$$

The roots given in equation (5.9) provide a stable response of the form given in Figure 5.2(a) and equation (5.1), whereas the roots in equation (5.10) give an unstable response as represented by Figure 5.2(b) and equation (5.2).

The only difference between the roots given in equation (5.9) and those in equation (5.10) is the sign of the real part. If the real part $\sigma$ is negative then the system is stable, but if it is positive, the system will be unstable. This holds true for systems of any order, so in general it can be stated: 'If any of the roots of the characteristic equation have positive real parts, then the system will be unstable'.

## 5.2   The Routh–Hurwitz stability criterion

The work of Routh (1905) and Hurwitz (1875) gives a method of indicating the presence and number of unstable roots, but not their value. Consider the characteristic equation

$$a_n s^n + a_{n-1} s^{n-1} + \cdots + a_1 s + a_0 = 0 \tag{5.11}$$

The Routh–Hurwitz stability criterion states:

(a) For there to be no roots with positive real parts then there is a necessary, but not sufficient, condition that all coefficients in the characteristic equation have the same sign and that none are zero.

If (a) above is satisfied, then the necessary and sufficient condition for stability is either

(b) all the Hurwitz determinants of the polynomial are positive, or alternatively
(c) all coefficients of the first column of Routh's array have the same sign. The number of sign changes indicate the number of unstable roots.

The Hurwitz determinants are

$$D_1 = a_1 \quad D_2 = \begin{vmatrix} a_1 & a_3 \\ a_0 & a_2 \end{vmatrix} \tag{5.12}$$

$$D_3 = \begin{vmatrix} a_1 & a_3 & a_5 \\ a_0 & a_2 & a_4 \\ & a_1 & a_3 \end{vmatrix} \quad D_4 = \begin{vmatrix} a_1 & a_3 & a_5 & a_7 \\ a_0 & a_2 & a_4 & a_6 \\ & a_1 & a_3 & a_5 \\ & & a_2 & a_4 \end{vmatrix} \quad \text{etc.}$$

Routh's array can be written in the form shown in Figure 5.3.
    In Routh's array Figure 5.3

$$b_1 = \frac{1}{a_{n-1}} \begin{vmatrix} a_{n-1} & a_{n-3} \\ a_n & a_{n-2} \end{vmatrix} \quad b_2 = \frac{1}{a_{n-1}} \begin{vmatrix} a_{n-1} & a_{n-5} \\ a_n & a_{n-4} \end{vmatrix} \quad \text{etc.} \tag{5.13}$$

$$c_1 = \frac{1}{b_1} \begin{vmatrix} b_1 & b_2 \\ a_{n-1} & a_{n-3} \end{vmatrix} \quad c_2 = \frac{1}{b_1} \begin{vmatrix} b_1 & b_3 \\ a_{n-1} & a_{n-5} \end{vmatrix} \quad \text{etc.} \tag{5.14}$$

Routh's method is easy to apply and is usually used in preference to the Hurwitz technique. Note that the array can also be expressed in the reverse order, commencing with row $s^n$.

| | | | |
|---|---|---|---|
| $s^0$ | $p_1$ | | |
| $s^1$ | $q_1$ | | |
| $\vdots$ | $\vdots$ | | |
| $s^{n-3}$ | $c_1$ | $c_2$ | $c_3$ |
| $s^{n-2}$ | $b_1$ | $b_2$ | $b_3$ |
| $s^{n-1}$ | $a_{n-1}$ | $a_{n-3}$ | $a_{n-5}$ |
| $s^n$ | $a_n$ | $a_{n-2}$ | $a_{n-4}$ |

**Fig. 5.3** Routh's array.

*Example 5.1* (See also Appendix 1, A1.5)

Check the stability of the system which has the following characteristic equation

$$s^4 + 2s^3 + s^2 + 4s + 2 = 0 \tag{5.15}$$

*Test 1*: All coefficients are present and have the same sign. Proceed to Test 2, i.e. Routh's array

$$
\begin{array}{c|ccc}
s^0 & 2 & & \\
s^1 & 8 & & \\
s^2 & -1 & 2 & \\
s^3 & 2 & 4 & \\
s^4 & 1 & 1 & 2
\end{array}
\tag{5.16}
$$

The bottom two rows of the array in (5.16) are obtained from the characteristic equation. The remaining coefficients are given by

$$b_1 = \frac{1}{2}\begin{vmatrix} 2 & 4 \\ 1 & 1 \end{vmatrix} = \frac{1}{2}(2 - 4) = -1 \tag{5.17}$$

$$b_2 = \frac{1}{2}\begin{vmatrix} 2 & 0 \\ 1 & 2 \end{vmatrix} = \frac{1}{2}(4 - 0) = 2 \tag{5.18}$$

$$b_3 = 0 \tag{5.19}$$

$$c_1 = -1\begin{vmatrix} -1 & 2 \\ 2 & 4 \end{vmatrix} = -1(-4 - 4) = 8 \tag{5.20}$$

$$c_2 = 0 \tag{5.21}$$

$$d_1 = \frac{1}{8}\begin{vmatrix} 8 & 0 \\ -1 & 2 \end{vmatrix} = \frac{1}{8}(16 - 0) = 2 \tag{5.22}$$

In the array given in (5.16) there are two sign changes in the column therefore there are two roots with positive real parts. Hence the system is unstable.

### 5.2.1 Maximum value of the open-loop gain constant for the stability of a closed-loop system

The closed-loop transfer function for a control system is given by equation (4.4)

$$\frac{C}{R}(s) = \frac{G(s)}{1 + G(s)H(s)} \tag{5.23}$$

In general, the characteristic equation is most easily formed by equating the denominator of the transfer function to zero. Hence, from equation (5.23), the characteristic equation for a closed-loop control system is

$$1 + G(s)H(s) = 0 \tag{5.24}$$

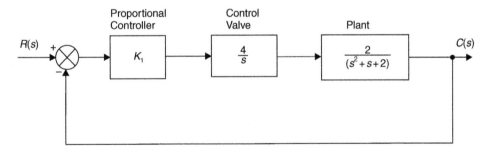

**Fig. 5.4** Closed-loop control system.

*Example 5.2* (See also Appendix 1, *examp52.m*)
Find the value of the proportional controller gain $K_1$ to make the control system shown in Figure 5.4 just unstable.

*Solution*
The open-loop transfer function is

$$G(s)H(s) = \frac{8K_1}{s(s^2 + s + 2)} \tag{5.25}$$

The open-loop gain constant is

$$K = 8K_1 \tag{5.26}$$

giving

$$G(s)H(s) = \frac{K}{s(s^2 + s + 2)} \tag{5.27}$$

From equation (5.24) the characteristic equation is

$$1 + \frac{K}{s(s^2 + s + 2)} = 0 \tag{5.28}$$

or

$$s(s^2 + s + 2) + K = 0 \tag{5.29}$$

which can be expressed as

$$s^3 + s^2 + 2s + K = 0 \tag{5.30}$$

The characteristic equation can also be found from the closed-loop transfer function. Using equation (4.4)

$$\frac{C}{R}(s) = \frac{G(s)}{1 + G(s)H(s)}$$

Given the open-loop transfer function in equation (5.27), where $H(s)$ is unity, then

$$\frac{C}{R}(s) = \frac{\frac{K}{s(s^2+s+2)}}{1 + \frac{K}{s(s^2+s+2)}} \tag{5.31}$$

Multiplying numerator and denominator by $s(s^2 + s + 2)$

$$\frac{C}{R}(s) = \frac{K}{s(s^2 + s + 2) + K} \tag{5.32}$$

$$\frac{C}{R}(s) = \frac{K}{s^3 + s^2 + 2s + K} \tag{5.33}$$

Equating the denominator of the closed-loop transfer function to zero

$$s^3 + s^2 + 2s + K = 0 \tag{5.34}$$

Equations (5.30) and (5.34) are identical, and both are the characteristic equation. It will be noted that all terms are present and have the same sign (Routh's first condition). Proceeding straight to Routh's array

$$\begin{array}{c|cc}
s^0 & K & \\
s^1 & (2 - K) & \\
s^2 & 1 & K \\
s^3 & 1 & 2
\end{array} \tag{5.35}$$

where

$$b_1 = 1 \begin{vmatrix} 1 & K \\ 1 & 2 \end{vmatrix} = (2 - K)$$

$$b_2 = 0$$

$$c_1 = K$$

To produce a sign change in the first column,

$$K \geq 2 \tag{5.36}$$

Hence, from equation (5.26), to make the system just unstable

$$K_1 = 0.25$$

Inserting (5.36) into (5.30) gives

$$s^3 + s^2 + 2s + 2 = 0$$

factorizing gives

$$(s^2 + 2)(s + 1) = 0$$

hence the roots of the characteristic equation are

$$s = -1$$

$$s = 0 \pm j\sqrt{2}$$

and the transient response is

$$c(t) = Ae^{-t} + B\sin(\sqrt{2}t + \phi) \tag{5.37}$$

From equation (5.37) it can be seen that when the proportional controller gain $K_1$ is set to 0.25, the system will oscillate continuously at a frequency of $\sqrt{2}$ rad/s.

## 5.2.2   Special cases of the Routh array

### Case 1: A zero in the first column

If there is a zero in the first column, then further calculation cannot normally proceed since it will involve dividing by zero. The problem is solved by replacing the zero with a small number $\varepsilon$ which can be assumed to be either positive or negative. When the array is complete, the signs of the elements in the first column are evaluated by allowing $\varepsilon$ to approach zero.

*Example 5.3*
$s^4 + 2s^3 + 2s^2 + 4s + 3 = 0$

$$
\begin{array}{c|ccc}
s^0 & 3 & & \\
s^1 & 4 - 6/\varepsilon & & \\
s^2 & \varepsilon & 3 & \\
s^3 & 2 & 4 & \\
s^4 & 1 & 2 & 3
\end{array}
\qquad (5.38)
$$

Irrespective of whether $\varepsilon$ is a small positive or negative number in array (5.38), there will be two sign changes in the first column.

### Case 2: All elements in a row are zero

If all the elements of a particular row are zero, then they are replaced by the derivatives of an *auxiliary polynomial*, formed from the elements of the previous row.

*Example 5.4*
$s^5 + 2s^4 + 6s^3 + 12s^2 + 8s + 16 = 0$

$$
\begin{array}{c|ccc}
s^0 & 16 & & \\
s^1 & 8/3 & & \\
s^2 & 6 & 16 & \\
s^3 & 8 & 24 & \\
s^4 & 2 & 12 & 16 \\
s^5 & 1 & 6 & 8
\end{array}
\qquad (5.39)
$$

The elements of the $s^3$ row are zero in array (5.39). An auxiliary polynomial $P(s)$ is therefore formed from the elements of the previous row $(s^4)$.
i.e.

$$P(s) = 2s^4 + 12s^2 + 16$$

$$\frac{dP(s)}{ds} = 8s^3 + 24s \qquad (5.40)$$

The coefficients of equation (5.40) become the elements of the $s^3$ row, allowing the array to be completed.

## 5.3  Root-locus analysis

### 5.3.1  System poles and zeros

The closed-loop transfer function for any feedback control system may be written in the factored form given in equation (5.41)

$$\frac{C}{R}(s) = \frac{G(s)}{1 + G(s)H(s)} = \frac{K_c(s - z_{c1})(s - z_{c2})\ldots(s - z_{cn})}{(s - p_{c1})(s - p_{c2})\ldots(s - p_{cn})} \tag{5.41}$$

where $s = p_{c1}, p_{c2}, \ldots, p_{cn}$ are closed-loop poles, so called since their values make equation (5.41) infinite (Note that they are also the roots of the characteristic equation) and $s = z_{c1}, z_{c2}, \ldots, z_{cn}$ are closed-loop zeros, since their values make equation (5.41) zero.

The position of the closed-loop poles in the $s$-plane determine the nature of the transient behaviour of the system as can be seen in Figure 5.5. Also, the open-loop transfer function may be expressed as

$$G(s)H(s) = \frac{K(s - z_{01})(s - z_{02})\ldots(s - z_{0n})}{(s - p_{01})(s - p_{02})\ldots(s - p_{0n})} \tag{5.42}$$

where $z_{01}, z_{02}, \ldots, z_{0n}$ are open-loop zeros and $p_{01}, p_{02}, \ldots, p_{0n}$ are open-loop poles.

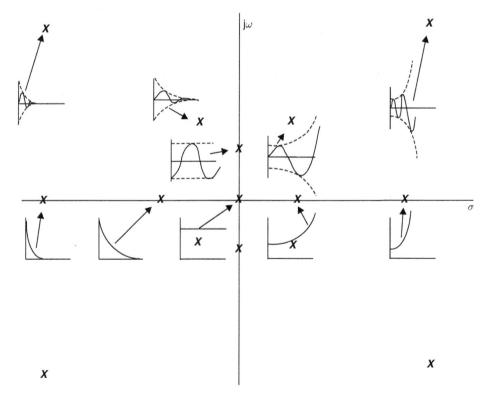

**Fig. 5.5** Effect of closed-loop pole position in the $s$-plane on system transient response.

## 5.3.2   The root locus method

This is a control system design technique developed by W.R. Evans (1948) that determines the roots of the characteristic equation (closed-loop poles) when the open-loop gain-constant $K$ is increased from zero to infinity.

The locus of the roots, or closed-loop poles are plotted in the *s*-plane. This is a complex plane, since $s = \sigma \pm j\omega$. It is important to remember that the real part $\sigma$ is the index in the exponential term of the time response, and if positive will make the system unstable. Hence, any locus in the right-hand side of the plane represents an unstable system. The imaginary part $\omega$ is the frequency of transient oscillation.

When a locus crosses the imaginary axis, $\sigma = 0$. This is the condition of marginal stability, i.e. the control system is on the verge of instability, where transient oscillations neither increase, nor decay, but remain at a constant value.

The design method requires the closed-loop poles to be plotted in the *s*-plane as $K$ is varied from zero to infinity, and then a value of $K$ selected to provide the necessary transient response as required by the performance specification. The loci always commence at open-loop poles (denoted by $x$) and terminate at open-loop zeros (denoted by o) when they exist.

*Example 5.5*
Construct the root-locus diagram for the first-order control system shown in Figure 5.6.

*Solution*
Open-loop transfer function

$$G(s)H(s) = \frac{K}{Ts} \tag{5.43}$$

Open-loop poles

$$s = 0$$

Open-loop zeros:   none
Characteristic equation

$$1 + G(s)H(s) = 0$$

Substituting equation (5.3) gives

$$1 + \frac{K}{Ts} = 0$$
$$\text{i.e.}\quad Ts + K = 0 \tag{5.44}$$

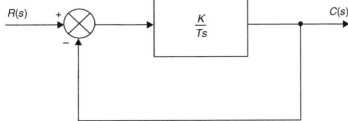

**Fig. 5.6** First-order control system.

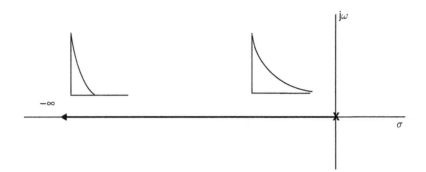

**Fig. 5.7** Root-locus diagram for a first-order system.

Roots of characteristic equation

$$s = -\frac{K}{T} \tag{5.45}$$

When $K$ is varied from zero to infinity the locus commences at the open-loop pole $s = 0$ and terminates at minus infinity on the real axis as shown in Figure 5.7.

From Figure 5.7 it can be seen that the system becomes more responsive as $K$ is increased. In practice, there is an upper limit for $K$ as signals and control elements saturate.

*Example 5.6*
Construct the root-locus diagram for the second-order control system shown in Figure 5.8.
Open-loop transfer function

$$G(s)H(s) = \frac{K}{s(s+4)} \tag{5.46}$$

Open-loop poles

$$s = 0, -4$$

Open-loop zeros:  none
Characteristic equation

$$1 + G(s)H(s) = 0$$

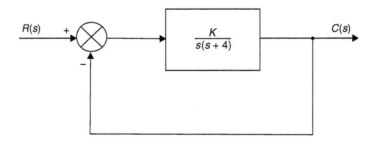

**Fig. 5.8** Second-order control system.

**Table 5.1** Roots of second-order characteristic
equation for different values of $K$

| $K$ | Characteristic equation | Roots |
|---|---|---|
| 0 | $s^2 + 4s = 0$ | $s = 0, -4$ |
| 4 | $s^2 + 4s + 4 = 0$ | $s = -2 \pm j0$ |
| 8 | $s^2 + 4s + 8 = 0$ | $s = -2 \pm j2$ |
| 16 | $s^2 + 4s + 16 = 0$ | $s = -2 \pm j3.46$ |

Substituting equation (5.4) gives

$$1 + \frac{K}{s(s+4)} = 0$$
$$\text{i.e.} \quad s^2 + 4s + K = 0 \tag{5.47}$$

Table 5.1 shows how equation (5.7) can be used to calculate the roots of the
characteristic equation for different values of $K$. Figure 5.9 shows the corresponding
root-locus diagram.

In Figure 5.9, note that the loci commences at the open-loop poles ($s = 0, -4$)
when $K = 0$. At $K = 4$ they branch into the complex space. This is called a breakaway point and corresponds to critical damping.

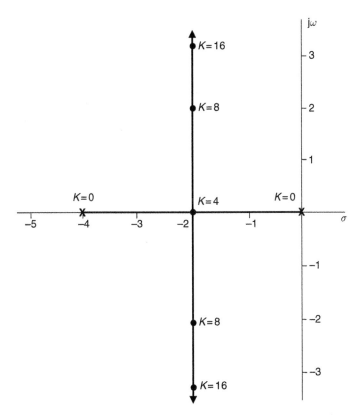

**Fig. 5.9** Root locus diagram for a second-order system.

### 5.3.3  General case for an underdamped second-order system

For the generalized second-order transfer function given in equation (3.43), equating the denominator to zero gives the characteristic equation

$$s^2 + 2\zeta\omega_n s + \omega_n^2 = 0 \tag{5.48}$$

If $\zeta < 1$ in equation (5.48), then the roots of the characteristic equation are

$$s_1, s_2 = -\zeta\omega_n \pm j\omega_n\sqrt{1 - \zeta^2} \tag{5.49}$$

Hence a point $P$ in the $s$-plane can be represented by Figure 5.10.
From Figure 5.10, Radius

$$OP = \sqrt{(-\zeta\omega_n)^2 + \left(\omega_n\sqrt{1 - \zeta^2}\right)^2} \tag{5.50}$$

Simplifying (5.50) gives

$$OP = \omega_n \tag{5.51}$$

Also from Figure 5.10

$$\cos\beta = \frac{|-\zeta\omega_n|}{\omega_n} = \zeta \tag{5.52}$$

Thus, as $\zeta$ is varied from zero to one, point P describes an arc of a circle of radius $\omega_n$, commencing on the imaginary axis ($\beta = 90°$) and finishing on the real axis ($\beta = 0°$).

#### Limits for acceptable transient response in the s-plane
If a system is

(1)  to be stable
(2)  to have acceptable transient response ($\zeta \geq 0.5$)

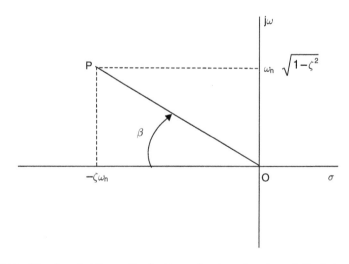

**Fig. 5.10** Roots of the characteristic equation for a second-order system shown in the $s$-plane.

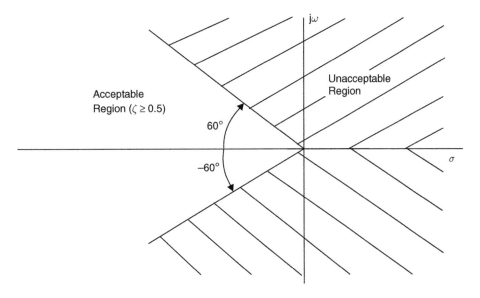

**Fig. 5.11**  Region of acceptable transient response in the *s*-plane for $\zeta \geq 0.5$.

then the closed-loop poles must lie in an area defined by

$$\beta = \pm \cos^{-1} 0.5 = \pm 60° \tag{5.53}$$

This is illustrated in Figure 5.11.

### 5.3.4   Rules for root locus construction

#### Angle and magnitude criteria
The characteristic equation for a closed-loop system (5.24) may also be written as

$$G(s)H(s) = -1 \tag{5.54}$$

Since equation (5.54) is a vector quantity, it can be represented in terms of angle and magnitude as

$$\underline{/G(s)H(s)} = 180° \tag{5.55}$$

$$|G(s)H(s)| = 1 \tag{5.56}$$

#### The angle criterion
Equation (5.55) may be interpreted as 'For a point $s_1$ to lie on the locus, the sum of all angles for vectors between open-loop poles (positive angles) and zeros (negative angles) to point $s_1$ must equal 180°.'

In general, this statement can be expressed as

$$\Sigma \text{ Pole Angles} - \Sigma \text{ Zero Angles} = 180° \tag{5.57}$$

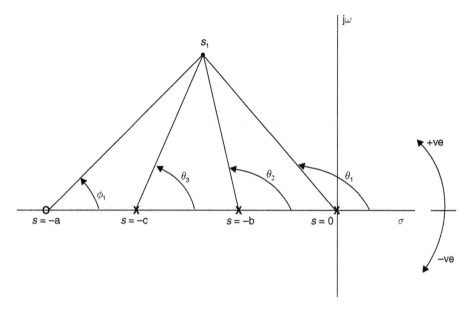

**Fig. 5.12** Application of the angle criterion.

*Example 5.7*
Consider an open-loop transfer function

$$G(s)H(s) = \frac{K(s+a)}{s(s+b)(s+c)}$$

Figure 5.12 shows vectors from open-loop poles and zeros to a trial point $s_1$. From Figure 5.12 and equation (5.57), for $s_1$ to lie on a locus, then

$$(\theta_1 + \theta_2 + \theta_3) - (\phi_1) = 180° \tag{5.58}$$

## The magnitude criterion

If a point $s_1$ lies on a locus, then the value of the open-loop gain constant $K$ at that point may be evaluated by using the magnitude criterion.

Equation (5.56) can be expressed as

$$|K|\left\{\frac{|N(s)|}{|D(s)|}\right\} = 1 \tag{5.59}$$

or

$$|K| = \frac{|D(s)|}{|N(s)|} \tag{5.60}$$

Equation (5.60) may be written as

$$K = \frac{\text{Product of pole vector magnitudes}}{\text{Product of zero vector magnitudes}} \tag{5.61}$$

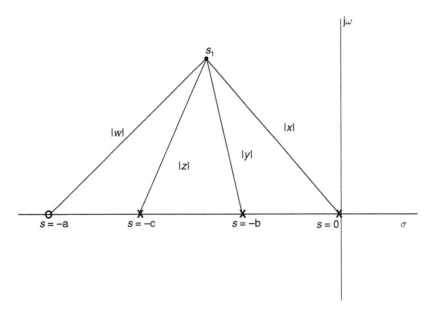

**Fig. 5.13** Application of the magnitude criterion.

For Example 5.7, if $s_1$ lies on a locus, then the pole and zero magnitudes are shown in Figure 5.13. From Figure 5.13 and equation (5.61), the value of the open-loop gain constant $K$ at position $s_1$ is

$$K = \frac{|x||y||z|}{|w|} \tag{5.62}$$

If there are no open-loop zeros in the transfer function, then the denominator of equation (5.62) is unity.

### 5.3.5   Root locus construction rules

1. *Starting points* $(K = 0)$: The root loci start at the open-loop poles.
2. *Termination points* $(K = \infty)$: The root loci terminate at the open-loop zeros when they exist, otherwise at infinity.
3. *Number of distinct root loci*: This is equal to the order of the characteristic equation.
4. *Symmetry of root loci*: The root loci are symmetrical about the real axis.
5. *Root locus asymptotes*: For large values of $k$ the root loci are asymptotic to straight lines, with angles given by

$$\theta = \frac{(1 + 2k)}{(n - m)}$$

   where
   $k = 0, 1, \ldots (n - m - 1)$
   $n$ = no. of finite open-loop poles
   $m$ = no. of finite open-loop zeros

6. *Asymptote intersection*: The asymptotes intersect the real axis at a point given by

$$\sigma_a = \frac{\Sigma \text{ open-loop poles} - \Sigma \text{ open-loop zeros}}{(n-m)}$$

7. *Root locus locations on real axis*: A point on the real axis is part of the loci if the sum of the number of open-loop poles and zeros to the right of the point concerned is odd.
8. *Breakaway points*: The points at which a locus breaks away from the real axis can be calculated using one of two methods:

    (a) Find the roots of the equation

    $$\frac{\mathrm{d}K}{\mathrm{d}s}\bigg|_{s=\sigma_b} = 0$$

    where $K$ has been made the subject of the characteristic equation i.e. $K = \ldots$
    (b) Solving the relationship

    $$\sum_{1}^{n}\frac{1}{(\sigma_b + |p_i|)} = \sum_{1}^{m}\frac{1}{(\sigma_b + |z_i|)}$$

    where $|p_i|$ and $|z_i|$ are the absolute values of open-loop poles and zeros and $\sigma_b$ is the breakaway point.

9. *Imaginary axis crossover*: The location on the imaginary axis of the loci (marginal stability) can be calculated using either:

    (a) The Routh–Hurwitz stability criterion.
    (b) Replacing $s$ by $j\omega$ in the characteristic equation (since $\sigma = 0$ on the imaginary axis).

10. *Angles of departure and arrival*: Computed using the angle criterion, by positioning a trial point at a complex open-loop pole (departure) or zero (arrival).
11. *Determination of points on root loci*: Exact points on root loci are found using the angle criterion.
12. *Determination of K on root loci*: The value of $K$ on root loci is found using the magnitude criterion.

*Example 5.8* (See also Appendix 1, *examp58.m* and *examp58a.m*)
A control system has the following open-loop transfer function

$$G(s)H(s) = \frac{K}{s(s+2)(s+5)}$$

(a) Sketch the root locus diagram by obtaining asymptotes, breakaway point and imaginary axis crossover point. What is the value of $K$ for marginal stability?
(b) Locate a point on the locus that corresponds to a closed-loop damping ratio of 0.5. What is the value of $K$ for this condition? What are the roots of the characteristic equation (closed-loop poles) for this value of $K$?

*Solution*

Part (a)

Open loop poles:   $s = 0, -2, -5$   $n = 3$

Open-loop zeros:   none   $m = 0$

*Asymptote angles ( Rule 5)*

$$\theta_1 = \frac{(1+0)\pi}{3-0} = \frac{\pi}{3} = 60°, \quad k = 0 \tag{5.63}$$

$$\theta_2 = \frac{(1+2)\pi}{3-0} = \pi = 180°, \quad k = 1 \tag{5.64}$$

$$\theta_3 = \frac{(1+4)\pi}{3-0} = \frac{5\pi}{3} = 300°(-60°), \quad k = 2, \quad \text{i.e. } n - m - 1 \tag{5.65}$$

*Asymptote intersection ( Rule 6)*

$$\sigma_a = \frac{\{(0) + (-2) + (-5)\} - 0}{3 - 0} \tag{5.66}$$

$$\sigma_a = -2.33 \tag{5.67}$$

*Characteristic equation*: From equation (5.24)

$$1 + \frac{K}{s(s+2)(s+5)} = 0 \tag{5.68}$$

or

$$s(s+2)(s+5) + K = 0$$

giving

$$s^3 + 7s^2 + 10s + K = 0 \tag{5.69}$$

*Breakaway points ( Rule 8)*

Method (a):  Re-arrange the characteristic equation (5.69) to make $K$ the subject

$$K = -s^3 - 7s^2 - 10s \tag{5.70}$$

$$\frac{dK}{ds} = -3s^2 - 14s - 10 = 0 \tag{5.71}$$

Multiplying through by –1

$$3s^2 + 14s + 10 = 0 \tag{5.72}$$

$$s_1, s_2 = \sigma_b = \frac{-14 \pm \sqrt{14^2 - 120}}{6}$$

$$\sigma_b = -3.79, -0.884 \tag{5.73}$$

Method (b)

$$\frac{1}{\sigma_b} + \frac{1}{\sigma_b + 2} + \frac{1}{\sigma_b + 5} = 0 \tag{5.74}$$

Multiplying through by,

$$\sigma_b(\sigma_b + 2)(\sigma_b + 5)$$

$$(\sigma_b + 2)(\sigma_b + 5) + \sigma_b(\sigma_b + 5) + \sigma_b(\sigma_b + 2) = 0$$

$$\sigma_b^2 + 7\sigma_b + 10 + \sigma_b^2 + 5\sigma_b + \sigma_b^2 + 2\sigma_b = 0 \tag{5.75}$$

$$3\sigma_b^2 + 14\sigma_b + 10 = 0$$

$$\sigma_b = -3.79, -0.884 \tag{5.76}$$

Note that equations (5.72) and (5.75) are identical, and therefore give the same roots. The first root, $-3.79$ lies at a point where there are an even number of open-loop poles to the right, and therefore is not valid. The second root, $-0.884$ has odd open-loop poles to the right, and is valid. In general, method (a) requires less computation than method (b).

*Imaginary axis crossover (Rule 9)*
Method (a) (Routh–Hurwitz)

| $s^0$ | $K$ | |
|---|---|---|
| $s^1$ | $(70 - K)/7$ | |
| $s^2$ | $7$ | $K$ |
| $s^3$ | $1$ | $10$ |

From Routh's array, marginal stability occurs at $K = 70$.

Method (b): Substitute $s = j\omega$ into characteristic equation. From characteristic equation (5.69)

$$(j\omega)^3 + 7(j\omega)^2 + 10(j\omega) + K = 0$$

$$-j\omega^3 - 7\omega^2 + 10j\omega + K = 0 \tag{5.77}$$

Equating imaginary parts gives

$$-\omega^3 + 10\omega = 0$$

$$\omega^2 = 10$$

$$\omega = \pm 3.16 \, \text{rad/s} \tag{5.78}$$

Equating real parts gives

$$-7\omega^2 + K = 0$$

$$8; K = 7\omega^2 = 70 \tag{5.79}$$

Note that method (b) provides both the crossover value (i.e. the frequency of oscillation at marginal stability) and the open-loop gain constant.

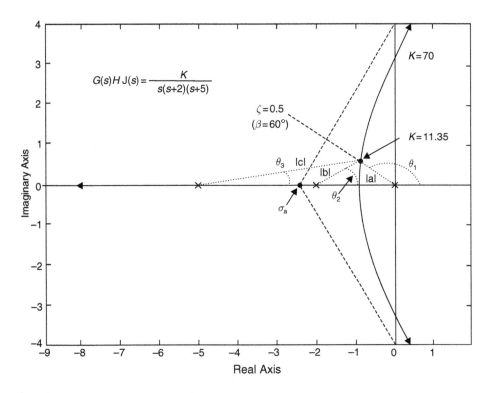

**Fig. 5.14** Sketch of root-locus diagram for Example 5.8.

The root locus diagram is shown in Figure 5.14.

**Part (b)** From equation (5.52), line of constant damping ratio is

$$\beta = \cos^{-1}(\zeta) = \cos^{-1}(0.5) = 60° \qquad (5.80)$$

This line is plotted on Figure 5.14 and trial points along it tested using the angle criterion, i.e.

$$\theta_1 + \theta_2 + \theta_3 = 180°$$

$$\text{At } s = -0.7 + \text{j}1.25 \qquad (5.81)$$

$$120 + 44 + 16 = 180° \qquad (5.82)$$

Hence point lies on the locus.

*Value of open-loop gain constant K:* Applying the magnitude criterion to the above point

$$K = |a||b||c|$$
$$= 1.4 \times 1.8 \times 4.5 = 11.35 \qquad (5.83)$$

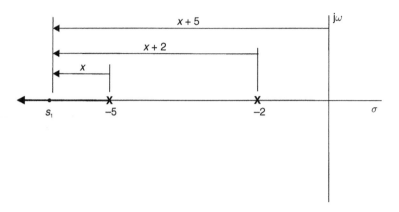

**Fig. 5.15** Determination of real closed-loop pole.

*Closed-loop poles* (For $K = 11.35$): Since the closed-loop system is third-order, there are three closed-loop poles. Two of them are given in equation (5.81). The third lies on the real locus that extends from $-5$ to $-\infty$. Its value is calculated using the magnitude criterion as shown in Figure 5.15.

From Figure 5.15

$$x(x + 2)(x + 5) = 11.35 \tag{5.84}$$

Substituting $x = 0.73$ (i.e. $s_1 = -5.73$) in equation (5.84) provides a solution. Hence the closed-loop poles for $K = 11.35$ are

$$s = -5.73, -0.7 \pm j1.25 \tag{5.85}$$

*Example 5.9* (See also Appendix 1, *examp59.m*)
The open-loop transfer function for a control system is

$$G(s)H(s) = \frac{K}{s(s^2 + 4s + 13)}$$

Find the asymptotes and angles of departure and hence sketch the root locus diagram. Locate a point on the complex locus that corresponds to a damping ratio of 0.25 and hence find

(a) the value of $K$ at this point
(b) the value of $K$ for marginal stability

*Solution*
Open-loop poles:   $s = 0,$

$$\frac{-4 \pm \sqrt{16 - 52}}{2} = -2 \pm j3 \quad n = 3$$

Open-loop zeros:   None   $m = 0$

*Asymptote angles (Rule 5)*

$$\theta_1 = \frac{(1+0)\pi}{3-0} = \frac{\pi}{3} = 60°, \quad k=0 \tag{5.86}$$

$$\theta_2 = \frac{(1+2)\pi}{3-0} = \pi = 180°, \quad k=1 \tag{5.87}$$

$$\theta_3 = \frac{(1+4)\pi}{3-0} = \frac{5\pi}{3} = 300°, \quad k=2, \quad n-m-1 \tag{5.88}$$

*Asymptote intersection (Rule 6)*

$$\sigma_a = \frac{\{(0) + (-2+j3) + (-2-j3)\} - 0}{3} \tag{5.89}$$

$$\sigma_a = -1.333 \tag{5.90}$$

Characteristic equation

$$s^3 + 4s^2 + 13s + K = 0 \tag{5.91}$$

*Breakaway points*: None, due to complex open-loop poles.

*Imaginary axis crossover (Rule 9)*
Method (b)

$$(j\omega)^3 + 4(j\omega)^2 + 13j\omega + K = 0$$

or

$$-j\omega^3 - 4\omega^2 + 13j\omega + K = 0 \tag{5.92}$$

Equating imaginary parts

$$-\omega^3 + 13\omega = 0$$

$$\omega^2 = 13$$

$$\omega = \pm 3.6 \, \text{rad/s} \tag{5.93}$$

Equating real parts

$$-4\omega^2 + K = 0$$

$$K = 52 \tag{5.94}$$

*Angle of departure (Rule 10)*: If angle of departure is $\theta_d$, then from Figure 5.16

$$\theta_a + \theta_b + \theta_d = 180°$$

$$\theta_d = 180 - \theta_a - \theta_b$$

$$\theta_d = 180 - 123 - 90 = -33° \tag{5.95}$$

Locate point that corresponds to $\zeta = 0.25$. From equation (5.52)

$$\beta = \cos^{-1}(0.25) = 75.5° \tag{5.96}$$

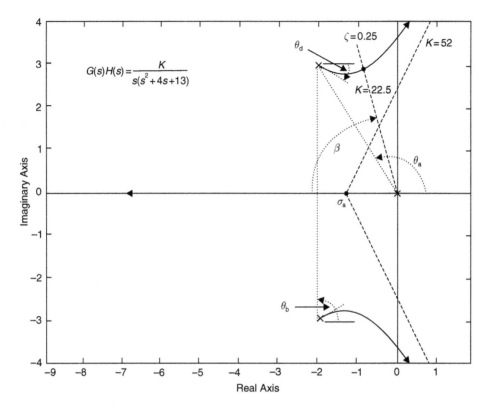

**Fig. 5.16** Root locus diagram for Example 5.9.

Plot line of constant damping ratio on Figure 5.16 and test trial points along it using angle criterion.

At $s = -0.8 + j2.9$

$$104.5 + 79.5 - 4 = 180°$$

Hence point lies on locus.
Applying magnitude criterion

$$K = 3.0 \times 6.0 \times 1.25 = 22.5 \tag{5.97}$$

## 5.4  Design in the *s*-plane

The root locus method provides a very powerful tool for control system design. The objective is to shape the loci so that closed-loop poles can be placed in the *s*-plane at positions that produce a transient response that meets a given performance specification. It should be noted that a root locus diagram does not provide information relating to steady-state response, so that steady-state errors may go undetected, unless checked by other means, i.e. time response.

**Table 5.2** Compensator characteristics

| Compensator | Characteristics |
|---|---|
| PD | One additional zero |
| PI | One additional zero |
| | One additional pole at origin |
| PID | Two additional zeros |
| | One additional pole at origin |

## 5.4.1 Compensator design

A compensator, or controller, placed in the forward path of a control system will modify the shape of the loci if it contains additional poles and zeros. Characteristics of conventional compensators are given in Table 5.2.

In compensator design, hand calculation is cumbersome, and a suitable computer package, such as MATLAB is generally used.

### Case Study

*Example 5.10* (See also Appendix 1, *examp510.m*)
A control system has the open-loop transfer function given in Example 5.8, i.e.

$$G(s)H(s) = \frac{1}{s(s+2)(s+5)}, \quad K = 1$$

A PD compensator of the form

$$G(s) = K_1(s+a) \tag{5.98}$$

is to be introduced in the forward path to achieve a performance specification

Overshoot                       less than 5%
Settling time ($\pm 2\%$)   less than 2 seconds

Determine the values of $K_1$ and $a$ to meet the specification.

### Original controller
The original controller may be considered to be a proportional controller of gain $K$ and the root locus diagram is shown in Figure 5.14. The selected value of $K = 11.35$ is for a damping ratio of 0.5 which has an overshoot of 16.3% in the time domain and is not acceptable. With a damping ratio of 0.7 the overshoot is 4.6% which is within specification. This corresponds to a controller gain of 7.13. The resulting time response for the original system ($K = 11.35$) is shown in Figure 5.20 where the settling time can be seen to be 5.4 seconds, which is outside of the specification. This also applies to the condition $K = 7.13$.

### PD compensator design
With the PD compensator of the form given in equation (5.98), the control problem, with reference, to Figure 5.14, is where to place the zero $a$ on the real axis. Potential locations include:

(i)   Between the poles $s = 0, -2$, i.e. at $s = -1$
(ii)   At $s = -2$ (pole/zero cancellation)
(iii)   Between the poles $s = -2, -5$, i.e at $s = -3$

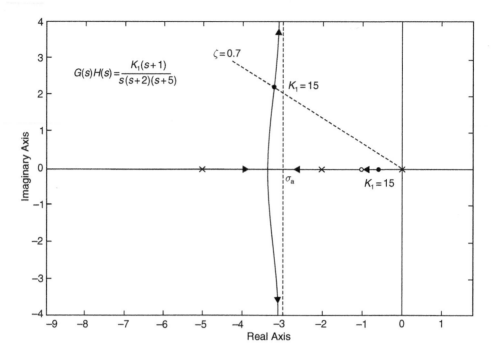

**Fig. 5.17** Root locus diagram for compensator $K_1(s + 1)$.

*Option 1* (zero positioned at $s = -1$): The cascaded compensator and plant transfer function become

$$G(s)H(s) = \frac{K_1(s + 1)}{s(s + 2)(s + 5)} \qquad (5.99)$$

The root locus diagram is shown in Figure 5.17.

It can be seen in Figure 5.17 that the pole at the origin and the zero at $s = -1$ dominate the response. With the complex loci, $\zeta = 0.7$ gives $K_1$ a value of 15. However, this value of $K_1$ occurs at $-0.74$ on the dominant real locus. The time response shown in Figure 5.20 shows the dominant first-order response with the oscillatory second-order response superimposed. The settling time is 3.9 seconds, which is outside of the specification.

*Option 2*: (zero positioned at $s = -2$): The cascaded compensator and plant transfer function is

$$G(s)H(s) = \frac{K_1(s + 2)}{s(s + 2)(s + 5)} \qquad (5.100)$$

The root locus diagram is shown in Figure 5.18. The pole/zero cancellation may be considered as a locus that starts at $s = -2$ and finishes at $s = -2$, i.e. a point on the diagram. The remaining loci breakaway at $s = -2.49$ and look similar to the second-order system shown in Figure 5.9. The compensator gain $K_1$ that corresponds to $\zeta = 0.7$ is 12.8. The resulting time response is shown in Figure 5.20 and has an overshoot of 4.1% and a settling time of 1.7 seconds, which is within specification.

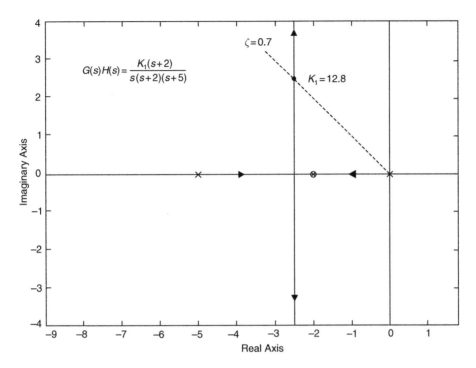

**Fig. 5.18** Root locus diagram for compensator $K_1(s + 2)$.

*Option 3*: (zero positioned at $s = -3$): The cascaded compensator and plant transfer function is

$$G(s)H(s) = \frac{K_1(s + 3)}{s(s + 2)(s + 5)} \tag{5.101}$$

The root locus diagram is shown in Figure 5.19. In this case the real locus occurs between $s = -5$ and $-3$ and the complex dominant loci breakaway at $\sigma_b = -1.15$. Since these loci are further to the right than the previous option, the transient response will be slower. The compensator gain that corresponds to $\zeta = 0.7$ is $K_1 = 5.3$. The resulting time response is shown in Figure 5.20, where the overshoot is 5.3% and the settling time is 3.1 seconds.

*Summary*: Of the three compensators considered, only option 2 met the performance specification. The recommended compensator is therefore

$$G(s) = 12.8(s + 2) \tag{5.102}$$

### Case study

*Example 5.11* (See also Appendix 1, *examp511.m*)
A ship roll stabilization system is shown in Figure 5.21. The system parameters are

Fin time constant     $T_f = 1.0$ seconds
Ship roll natural frequency   $\omega_n = 1.414$ rad/s

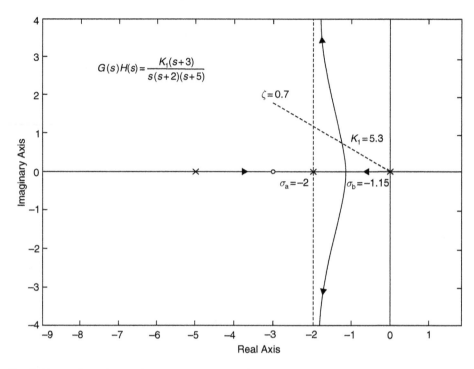

**Fig. 5.19** Root locus diagram for compensator $K_1(s + 3)$.

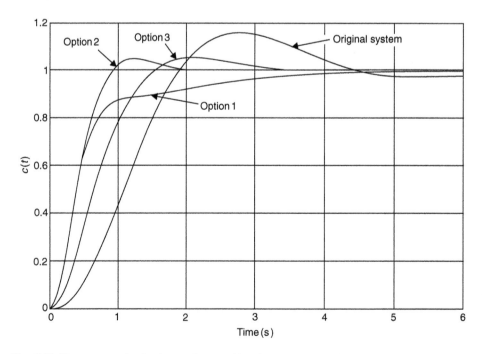

**Fig. 5.20** Time response for the three options considered.

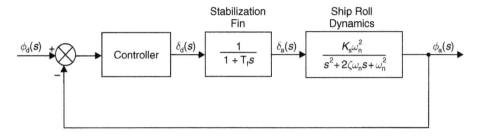

**Fig. 5.21** Ship roll stabilization system.

Ship roll damping ratio  $\zeta = 0.248$
Ship steady-state gain  $K_s = 0.5$

## Performance specification

Without stabilization, the step response of the roll dynamics produces a 45% over-shoot and a settling time of 10 seconds. The stabilization control system is required to provide a step response with an overshoot of less than 25%, a settling time of less than 2 seconds, and zero steady-state error.

(a) *Proportional control*: With a proportional gain $K_1$, the open-loop transfer function is

$$G(s)H(s) = \frac{K_1 K_s \omega_n^2}{(1 + T_f s)(s^2 + 2\zeta\omega_n s + \omega_n^2)} \tag{5.103}$$

Inserting values

$$G(s)H(s) = \frac{K}{(s+1)(s^2 + 0.7s + 2)} \tag{5.104}$$

where

$$K = K_1 K_s \omega_n^2 \tag{5.105}$$

Open-loop poles:  $s = -1, -0.35 \pm \text{j}1.37$   $n = 3$
Open-loop zeros:  None   $m = 0$

*Asymptote angles (Rule 5)*

$$\theta_1 = \frac{(1+0)\pi}{3-0} = \frac{\pi}{3} = 60°, \quad k = 0 \tag{5.106}$$

$$\theta_2 = \frac{(1+2)\pi}{3-0} = \pi = 180°, \quad k = 1 \tag{5.107}$$

$$\theta_3 = \frac{(1+4)}{3-0} = \frac{5\pi}{3} = 300°, \quad k = 2 \tag{5.108}$$

*Asymptote intersection (Rule 6)*

$$\sigma_a = \frac{\{(-1) + (-0.35 + \text{j}1.37) + (-0.35 - \text{j}1.37)\} - 0}{3 - 0} \tag{5.109}$$

$$\sigma_a = -0.57 \tag{5.110}$$

Characteristic equation

$$1 + \frac{K}{(s+1)(s^2 + 0.7s + 2)} = 0 \qquad (5.111)$$

giving

$$s^3 + 1.7s^2 + 2.7s + (2 + K) = 0 \qquad (5.112)$$

Breakaway points:   None

*Imaginary axis crossover (Rule 9)*
From characteristic equation (5.112)

$$(j\omega)^3 + 1.7(j\omega)^2 + 2.7(j\omega) + (2 + K) = 0$$
$$-j\omega^3 - 1.7\omega^2 + 2.7j\omega + (2 + K) = 0$$

Equating imaginary parts

$$-\omega^3 + 2.7\omega = 0$$
$$\omega^2 = 2.7$$
$$\omega = \pm 1.643\,\text{rad/s} \qquad (5.113)$$

Equating real parts

$$-1.7\omega^2 + (2 + K) = 0$$
$$K = 1.7\omega^2 - 2 = 2.59 \qquad (5.114)$$

The root locus diagram is shown in Figure 5.22. It can be seen that proportional control is not appropriate since as the controller gain $K_1$ is increased the complex loci head towards the imaginary axis, making the response even more oscillatory than the open-loop response, until at $K = 2.59$ ($K_1 = 2.59$) the system becomes unstable. Also, since no pure integrator is present in the control loop, there will be significant steady-state errors.

(b) *PID control*: In order to achieve an acceptable response, the complex loci need to be attracted into the left-hand-side of the $s$-plane. This might be achieved by placing a pair of complex zeros to the left of the open-loop poles. In addition, a pure integrator needs to be introduced. This points to a PID controller of the form

$$G(s) = \frac{K_1(s^2 + bs + c)}{s} \qquad (5.115)$$

putting $b = 4$ and $c = 8$ gives a pair of complex zeros

$$s = -2 \pm j2$$

The open-loop transfer function now becomes

$$G(s)H(s) = \frac{K(s^2 + 4s + 8)}{s(s+1)(s^2 + 0.7s + 2)} \qquad (5.116)$$

The root locus diagram is shown in Figure 5.23. The control strategy however, has not worked. The pure integrator and the open-loop pole $s = -1$ produce a

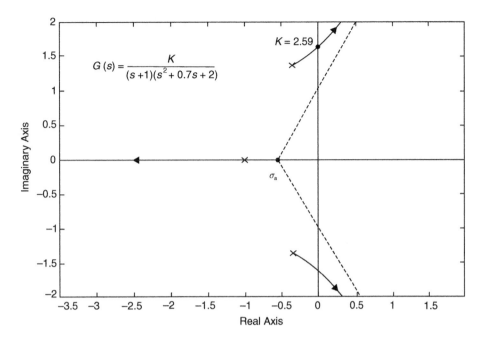

**Fig. 5.22** Proportional control, ship roll stabilization system.

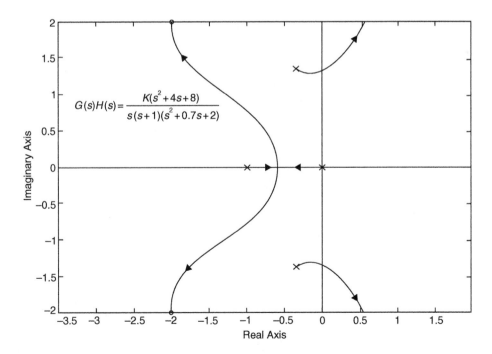

**Fig. 5.23** PID control, ship roll stabilization system.

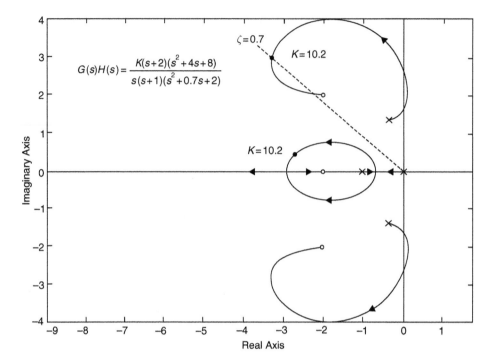

**Fig. 5.24** PIDD control, ship roll stabilization system.

breakaway point at $s = -0.6$. This in turn creates a second pair of complex loci that terminate at the new open-loop zeros, leaving the original complex loci to crossover the imaginary axis as before.

(c) *PIDD control*: If an additional open-loop zero is placed on the real axis, to the left of the open-loop pole $s = -1$, a further breakaway point will occur to the left of the new zero. This should have the effect of bringing one pair of complex loci back to the real axis, whilst allowing the original complex loci to terminate at the complex open-loop zeros. If a new real zero is placed at $s = -2$, the open-loop transfer function becomes

$$G(s)H(s) = \frac{K(s+2)(s^2 + 4s + 8)}{s(s+1)(s^2 + 0.7s + 2)} \tag{5.117}$$

The resulting root-locus diagram is shown in Figure 5.24.

The control strategy for the root-locus diagram shown in Figure 5.24 is called PIDD, because of the additional open-loop zero. The system is unstable between $K = 0.17$ and $K = 1.06$, but exhibits good transient response at $K = 10.2$ on both complex loci.

Figure 5.25 shows the step response for (a) the hull roll action without a stabilizer system, and (b) the hull roll action with a controller/compensator with a control law

$$G(s) = \frac{10.2(s+2)(s^2 + 4s + 8)}{s} \tag{5.118}$$

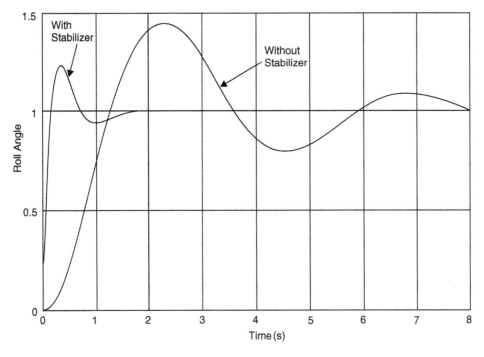

**Fig. 5.25** Ship hull step response with and without stabilizers system.

## System performance

(i) *Without stabilizer system*

| | |
|---|---|
| Rise time (95%) | 1.3 seconds |
| Percentage Overshoot | 45% |
| Settling time (±2%) | 10.0 seconds |

(ii) *With stabilizer system*

| | |
|---|---|
| Rise time (95%) | 0.14 seconds |
| Percentage overshoot | 22.8% |
| Settling time (±2%) | 1.4 seconds |

With the stabilizer system, the step response meets the performance specification.

## 5.5 Further problems

*Example 5.12*
Use the Routh–Hurwitz criterion to determine the number of roots with positive real parts in the following characteristic equations

(a) $s^4 + 3s^3 + 6s^2 + 2s + 5 = 0$   Ans: two
(b) $s^5 + 2s^4 + 3s^3 + 4s^2 + 2s + 1 = 0$   Ans: none

*Example 5.13*
Find the value of the open-loop gain constant $K$ to make the control system, whose open-loop transfer function is given below, just unstable.

$$G(s)H(s) = \frac{K}{s(s+1)(s+8)} \quad \text{Ans: } 72$$

*Example 5.14*
A feedback control system has the following open-loop transfer function

$$G(s)H(s) = \frac{K}{s(s+1)(s+5)}$$

(a) Sketch the root locus by obtaining asymptotes, breakaway point and imaginary axis cross-over point.
(b) A compensating element having a transfer function $G(s) = (s+2)$ is now included in the open-loop transfer function. If the breakaway point is $-0.56$, sketch the new root locus. Comment on stability of the system with, and without the compensator.
(c) Demonstrate that for the compensated system the co-ordinates $-2.375$, $-1.8 \pm j4.0$ lie on the curve. What is the value of $K$ for these points?

*Solution*

(a) $\sigma_a = -2$, $\sigma_b = -0.47$, $\omega = \pm j2.24 \, \text{rad/s}$
(b) With compensator, system stable for all $K$
(c) $K = 23$

*Example 5.15*
A feedback control system employing proportional control has the following open-loop transfer function

$$G(s)H(s) = \frac{K}{(s+1)(s^2+s+1)}$$

(a) Using asymptotes, sketch the root locus diagram for the closed-loop system and find
   (i) the angles of departure from any complex open-loop poles,
   (ii) the frequency of transient oscillation at the onset of instability,
   (iii) the value of $K$ to give the dominant closed-loop poles a damping ratio $\zeta$ of 0.3
(b) To improve the steady-state performance the proportional controller is replaced by a proportional plus integral controller. The forward-path transfer function now becomes

$$G(s) = \frac{K(s+2)}{s(s+1)(s^2+s+1)}$$

Demonstrate that

(i) the two breakaway points occur at

$$\sigma_{b_1} = -0.623$$
$$\sigma_{b_2} = -2.53$$

(ii) the imaginary axis crossover occurs when $K = 0.464$

*Solution*
(a) (i) $\pm 30°$, (ii) $1.414\,\text{rad/s}$, (iii) $K = 0.55$

*Example 5.16*

(a) The laser guided missile shown in Figure 5.26(a) has a pitch moment of inertia of $90\,\text{kg}\,\text{m}^2$. The control fins produce a moment about the pitch mass centre of $360\,\text{Nm}$ per radian of fin angle $\beta$. The fin positional control system has unity gain and possesses a time constant of 0.2 seconds. If all other aerodynamic effects are ignored, find the transfer functions of the control fins and missile (in pole-zero format) in the block diagram given in Figure 5.26(b).

(b) You are to conduct a feasibility study to evaluate various forms of control action. Initially proportional control is to be considered. Using asymptotes only, construct the root locus diagram and give reasons why it would be unsuitable.

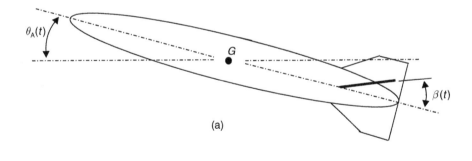

(a)

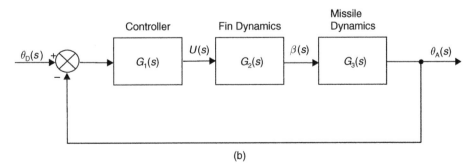

(b)

**Fig. 5.26** Laser guided missile.

(c) An open-loop zero is now introduced at $s = -2$. Again construct the root-locus diagram using asymptotes and comment on the suitability of the system.

(d) Open-loop zeros at $s = -2$ and $s = -3$ are now introduced. Demonstrate that at $s = -2.45$ the complex loci join the real axis prior to terminating at the open-loop zeros. Show also that a trial point on the locus exists when $j\omega = 1.45$ and the damping ratio $\zeta = 0.7$. Sketch the root locus diagram and evaluate the controller gain that corresponds to $\zeta = 0.7$.

*Solution*

(a) $G_2(s) = 5/(s + 5)$   $G_3(s) = 4/s^2$
(b) System unstable for all $K$
(c) System stable for all $K$
(d) Controller gain $= 0.24$

# 6

# Classical design in the frequency domain

## 6.1 Frequency domain analysis

Control system design in the frequency domain can be undertaken using a purely theoretical approach, or alternatively, using measurements taken from the components in the control loop. The technique allows transfer functions of both the system elements and the complete system to be estimated, and a suitable controller/compensator to be designed.

Frequency domain analysis is concerned with the calculation or measurement of the *steady-state* system output when responding to a constant amplitude, variable frequency sinusoidal input. Steady-state errors, in terms of amplitude and phase relate directly to the dynamic characteristics, i.e. the transfer function, of the system.

Consider a harmonic input

$$\theta_i(t) = A_1 \sin \omega t \tag{6.1}$$

This can be expressed in complex exponential form

$$\theta_i(t) = A_1 e^{j\omega t} \tag{6.2}$$

The steady-state response of a linear system will be

$$\theta_o(t) = A_2 \sin(\omega t - \phi) \tag{6.3}$$

or

$$\theta_o(t) = A_2 e^{j(\omega t - \phi)} \tag{6.4}$$

where $\phi$ is the phase relationship between the input and output sinewaves as shown in Figure 6.1. The amplitude ratio $A_2/A_1$ is called the modulus and given the symbol $|G|$. Thus

$$\frac{A_2}{A_1} = |G| \tag{6.5}$$

or

$$A_2 = A_1 |G| \tag{6.6}$$

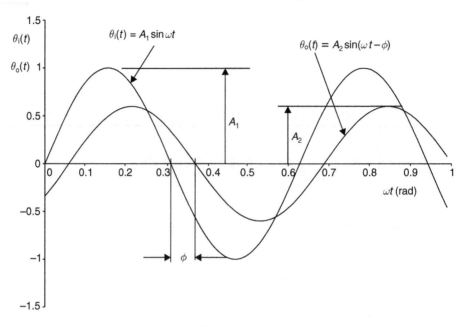

**Fig. 6.1** Steady-state input and output sinusoidal response.

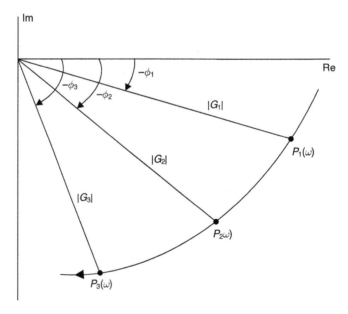

**Fig. 6.2** Harmonic response diagram.

Substituting equation (6.6) into (6.3)

$$\theta_o(t) = A_1|G|e^{j(\omega t - \phi)}$$
$$= A_1|G|e^{j\omega t}e^{-j\phi} \qquad (6.7)$$

$$\theta_o(t) = (A_1 e^{j\omega t})(|G| e^{-j\phi})$$
$$= \theta_i(t) |G| e^{-j\varphi} \tag{6.8}$$

Since $|G|$ and $\phi$ are functions of $\omega$, then equation (6.8) may be written

$$\frac{\theta_o}{\theta_i}(\omega) = |G(\omega)| e^{-j\phi(\omega)} \tag{6.9}$$

For a given value of $\omega$, equation (6.9) represents a point in complex space $P(\omega)$. When $\omega$ is varied from zero to infinity, a locus will be generated in the complex space. This locus, shown in Figure 6.2, is in effect a polar plot, and is sometimes called a harmonic response diagram. An important feature of such a diagram is that its shape is uniquely related to the dynamic characteristics of the system.

## 6.2 The complex frequency approach

Relationship between $s$ and $j\omega$. From equation (6.2)

$$\theta_i(t) = A_1 e^{j\omega t}$$
$$\frac{d\theta}{dt} = j\omega(A_1 e^{j\omega t}) = j\omega\theta_i(t)$$

Taking Laplace transforms

$$s\theta_i(s) = j\omega\theta_i(s) \tag{6.10}$$

or

$$s = j\omega \tag{6.11}$$

Hence, for a sinusoidal input, the steady-state system response may be calculated by substituting $s = j\omega$ into the transfer function and using the laws of complex algebra to calculate the modulus and phase angle.

### 6.2.1 Frequency response characteristics of first-order systems

From equation (3.23)

$$\frac{\theta_o}{\theta_i}(s) = G(s) = \frac{K}{1 + Ts} \tag{6.12}$$

For a sinusoidal input, substitute equation (6.11) into (6.12).

$$\frac{\theta_o}{\theta_i}(j\omega) = G(j\omega) = \frac{K}{1 + j\omega T} \tag{6.13}$$

Rationalize, by multiplying numerator and denominator of equation (6.13) by the conjugate of (6.13), i.e.

$$G(j\omega) = \frac{K(1 - j\omega T)}{(1 + j\omega T)(1 - j\omega T)}$$
$$= \frac{K(1 - j\omega T)}{1 + \omega^2 T^2} \tag{6.14}$$

Equation (6.14) is a complex quantity of the form $a + jb$ where

$$\text{Real part} \quad a = \frac{K}{1 + \omega^2 T^2} \tag{6.15}$$

$$\text{Imaginary part} \quad b = \frac{-K\omega T}{1 + \omega^2 T^2} \tag{6.16}$$

Hence equation (6.14) can be plotted in the complex space (Argand Diagram) to produce a harmonic response diagram as shown in Figure 6.3.

In Figure 6.3 it is convenient to use polar co-ordinates, as they are the modulus and phase angle as depicted in Figure 6.2. From Figure 6.3, the polar co-ordinates are

$$|G(j\omega)| = \sqrt{a^2 + b^2}$$

$$= \sqrt{\left(\frac{K}{1 + \omega^2 T^2}\right)^2 + \left(\frac{-K\omega T}{1 + \omega^2 T^2}\right)^2} \tag{6.17}$$

which simplifies to give

$$|G(j\omega)| = \frac{K}{\sqrt{1 + \omega^2 T^2}} \tag{6.18}$$

Comparing equations (6.14) and (6.18), providing there are no zeros in the transfer function, it is generally true to say

$$|G(j\omega)| = \frac{K}{\sqrt{\text{Denominator of } G(j\omega)}} \tag{6.19}$$

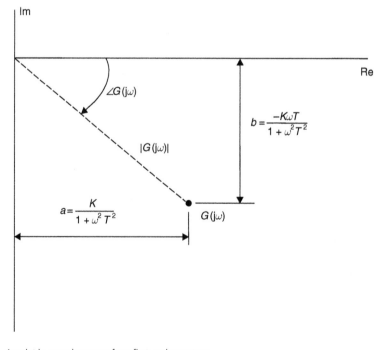

**Fig. 6.3** A point in complex space for a first-order system.

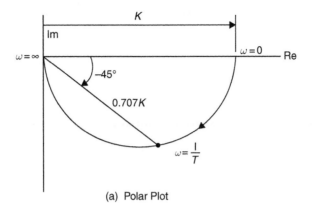

(a) Polar Plot

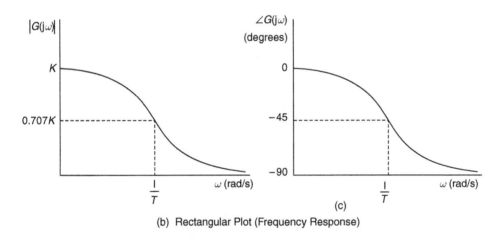

(c)

(b) Rectangular Plot (Frequency Response)

**Fig. 6.4** Graphical display of frequency domain data for a first-order system.

**Table 6.1** Modulus and phase for a first-order system

| $\omega$ (rad/s) | $|G(j\omega)|$ | $\angle G(j\omega)$ (degrees) |
|---|---|---|
| 0 | $K$ | $-0$ |
| $1/T$ | $K/\sqrt{2}$ | $-45$ |
| $\infty$ | 0 | $-90$ |

The argument, or phase angle is

$$\angle G(j\omega) = \tan^{-1}\left(\frac{b}{a}\right) \tag{6.20}$$

$$= \tan^{-1}\left\{\frac{\frac{-K\omega T}{1+\omega^2 T^2}}{\frac{K}{1+\omega^2 T^2}}\right\}$$

which gives

$$\angle G(j\omega) = \tan^{-1}(-\omega T) \tag{6.21}$$

Using equations (6.18) and (6.21), values for the modulus and phase angle may be calculated as shown in Table 6.1. The results in Table 6.1 may be represented as a Polar Plot, Figure 6.4(a) or as a rectangular plot, Figures 6.4(b) and (c). Since the rectangular plots show the system response as a function of frequency, they are usually referred to as frequency response diagrams.

## 6.2.2 Frequency response characteristics of second-order systems

From equation (3.42) the standard form of transfer function for a second-order system is

$$G(s) = \frac{K}{\frac{1}{\omega_n^2} s^2 + \frac{2\zeta}{\omega_n} s + 1} \tag{6.22}$$

Substituting $s = j\omega$

$$G(j\omega) = \frac{K}{\frac{1}{\omega_n^2} (j\omega)^2 + \frac{2\zeta}{\omega_n} (j\omega) + 1} \tag{6.23}$$

or

$$G(j\omega) = \frac{K}{\left\{ 1 - \left( \frac{\omega}{\omega_n} \right)^2 \right\} + j\left\{ 2\zeta \left( \frac{\omega}{\omega_n} \right) \right\}} \tag{6.24}$$

Rationalizing gives

$$G(j\omega) = \frac{K\left[ \left\{ 1 - \left( \frac{\omega}{\omega_n} \right)^2 \right\} - j\left\{ 2\zeta \left( \frac{\omega}{\omega_n} \right) \right\} \right]}{\left\{ 1 - \left( \frac{\omega}{\omega_n} \right)^2 \right\}^2 + \left\{ 2\zeta \left( \frac{\omega}{\omega_n} \right) \right\}^2} \tag{6.25}$$

Using equations (6.17) and (6.19), the modulus is

$$|G(j\omega)| = \frac{K}{\sqrt{\left\{ 1 - \left( \frac{\omega}{\omega_n} \right)^2 \right\}^2 + \left\{ 2\zeta \left( p\frac{\omega}{\omega_n} \right) \right\}^2}} \tag{6.26}$$

And from equation (6.20), the argument is

$$\angle G(j\omega) = \tan^{-1} \left\{ \frac{-2\zeta \left( \frac{\omega}{\omega_n} \right)}{1 - \left( \frac{\omega}{\omega_n} \right)^2} \right\} \tag{6.27}$$

**Table 6.2** Modulus and phase for a second-order system

| $\omega$ (rad/s) | $|G(j\omega)|$ | $\angle G(j\omega)$ (degrees) |
|---|---|---|
| 0 | $K$ | $-0$ |
| $\omega_n$ | $K/2\zeta$ | $-90$ |
| $\infty$ | $0$ | $-180$ |

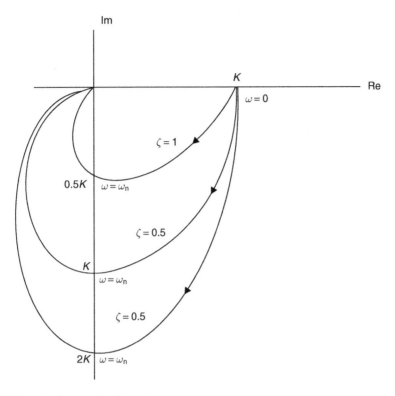

**Fig. 6.5** Polar plot of a second-order system.

From equations (6.26) and (6.27) the modulus and phase may be calculated as shown in Table 6.2. The results in Table 6.2 are a function of $\zeta$ and may be represented as a Polar Plot, Figure 6.5, or by the frequency response diagrams given in Figure 6.6.

## 6.3  The Bode diagram

The Bode diagram is a logarithmic version of the frequency response diagrams illustrated in Figures 6.4(b) and (c), and also Figure 6.6, and consists of

 (i) a log modus–log frequency plot
(ii) a linear phase–log frequency plot.

The technique uses asymptotes to quickly construct frequency response diagrams by hand. The construction of diagrams for high-order systems is achieved by simple graphical addition of the individual diagrams of the separate elements in the system.
    The modulus is plotted on a linear $y$-axis scale in deciBels, where

$$|G(j\omega)| \, dB = 20 \log_{10} |G(j\omega)| \tag{6.28}$$

The frequency is plotted on a logarithmic $x$-axis scale.

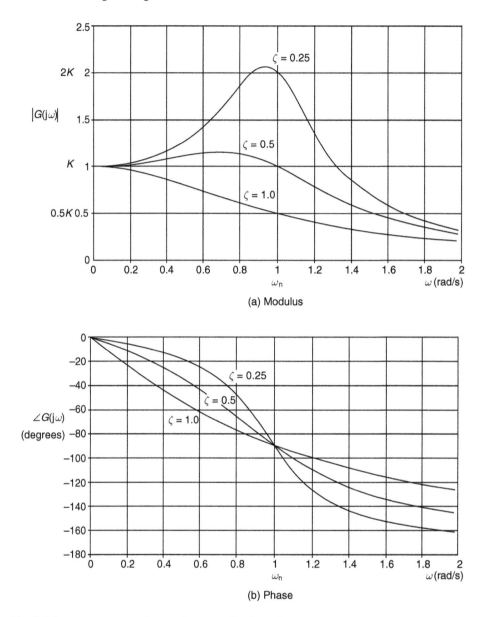

**Fig. 6.6** Frequency response diagrams for a second-order system.

## 6.3.1 Summation of system elements on a Bode diagram

Consider two elements in cascade as shown in Figure 6.7.

$$G_1(j\omega) = |G_1(j\omega)|e^{j\phi_1} \tag{6.29}$$

$$G_2(j\omega) = |G_2(j\omega)|e^{j\phi_2} \tag{6.30}$$

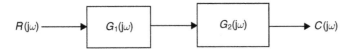

**Fig. 6.7** Summation of two elements in cascade.

$$\frac{C}{R}(j\omega) = G_1(j\omega)G_2(j\omega)$$

$$= |G_1(j\omega)||G_2(j\omega)|e^{j(\phi_1 + \phi_2)} \tag{6.31}$$

Hence

$$\left|\frac{C}{R}(j\omega)\right| = |G_1(j\omega)||G_2(j\omega)|$$

or

$$\left|\frac{C}{R}(j\omega)\right| dB = 20\log_{10}|G_1(j\omega)| + 20\log_{10}|G_2(j\omega)| \tag{6.32}$$

and

$$\angle\frac{C}{R}(j\omega) = \phi_1 + \phi_2 = \angle G_1(j\omega) + \angle G_2(j\omega) \tag{6.33}$$

In general, the complete system frequency response is obtained by summation of the log modulus of the system elements, and also summation of the phase of the system elements.

## 6.3.2 Asymptotic approximation on Bode diagrams

### (a) First-order lag systems
These are conventional first-order systems where the phase of the output lags behind the phase of the input.

(i) *Log modulus plot*: This consists of a low-frequency asymptote and a high-frequency asymptote, which are obtained from equation (6.18).

*Low frequency (LF) asymptote*: When $\omega \to 0$, $|G(j\omega)| \to K$. Hence the LF asymptote is a horizontal line at $K$ dB.

*High frequency (HF) asymptote*: When $\omega \gg 1/T$, equation (6.18) approximates to

$$|G(j\omega)| = \frac{K}{\omega T} \tag{6.34}$$

As can be seen from equation (6.34), each time the frequency doubles (an increase of one octave) the modulus halves, or falls by 6 dB. Or alternatively, each time the frequency increases by a factor of 10 (decade), the modulus falls by 10, or 20 dB. Hence the HF asymptote for a first-order system has a slope which can be expressed as −6 dB per octave, or −20 dB per decade.

From equation (6.34), when $\omega = 1/T$, the HF asymptote has a value of $K$. Hence the asymptotes intersect at $\omega = 1/T$ rad/s. Also at this frequency, from equation (6.18) the exact modulus has a value

$$|G(j\omega)| = \frac{K}{\sqrt{2}}$$

Since $1/\sqrt{2}$ is $-3$ dB, the exact modulus passes 3 dB below the asymptote intersection at $1/T$ rad/s. The asymptotic construction of the log modulus Bode plot for a first-order system is shown in Figure 6.8.

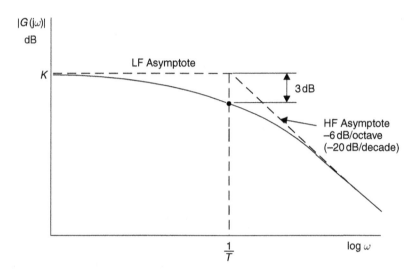

**Fig. 6.8** Bode modulus construction for a first-order system.

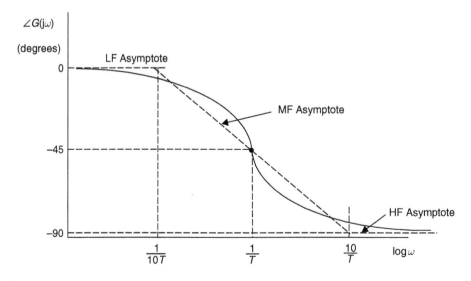

**Fig. 6.9** Bode phase construction for a first-order system.

(ii) *Phase plot*: This has three asymptotes

- A LF horizontal asymptote at $0°$
- A HF horizontal asymptote at $-90°$
- A Mid-Frequency (MF) asymptote that intersects the LF asymptote at $1/10T$ and the HF asymptote at $10/T$ (i.e. a decade either side of $1/T$).

The Bode phase plot for a first-order system is given in Figure 6.9.

## (b) First-order lead systems

These are first-order systems where the phase of the output (in steady-state) leads the phase of the input. The transfer function of a first-order lead system is

$$G(s) = K(1 + Ts) \tag{6.35}$$

and

$$|G(j\omega)| = K\sqrt{1 + \omega^2 T^2} \tag{6.36}$$

$$\angle G(j\omega) = \tan^{-1}(\omega T) \tag{6.37}$$

The Bode diagram, given in Figure 6.10, is the mirror image, about the frequency axis, of the first-order lag system. Note that the transfer function given in equation (6.35) is also that of a PD controller.

## (c) Second-order systems

(i) *Log modulus plot*

*LF asymptote*: A horizontal line at $K$ dB
*HF asymptote*: When $\omega \gg \omega_n$, equation (6.26) approximates to

$$|G(j\omega)| = \frac{K}{\left(\frac{\omega}{\omega_n}\right)^2} \tag{6.38}$$

From equation (6.38), an octave increase in frequency will reduce the modulus by a quarter, or $-12$ dB and a decade frequency increase will reduce the modulus by a factor of 100, or $-40$ dB. Hence the HF asymptote for a second-order system has a slope of $-12$ dB/octave or $-40$ dB/decade. The LF and HF asymptotes intersect at $\omega = \omega_n$. Also at $\omega_n$, the exact value of modulus from equation (6.26) is

$$|G(j\omega)| = \frac{K}{2\zeta}$$

The value of the modulus relative to the LF asymptote is

$$|G(j\omega)|\text{dB} = 20\log_{10}\left(\frac{K/2\zeta}{K}\right) = 20\log_{10}\left(\frac{1}{2\zeta}\right) \tag{6.39}$$

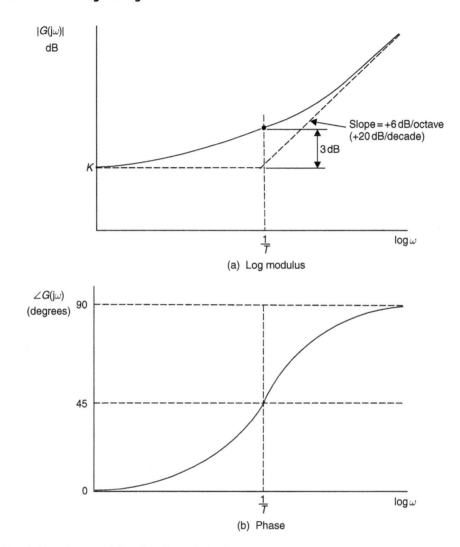

**Fig. 6.10** Bode gain and phase for a first-order lead system.

Hence

$$\zeta = 0.25, \text{ Relative modulus} = +6\,\text{dB}$$
$$\zeta = 0.5, \text{ Relative modulus} = 0\,\text{dB}$$
$$\zeta = 1.0, \text{ Relative modulus} = -6\,\text{dB}$$

(ii) *Phase plot*: This has two asymptotes:

- A LF horizontal asymptote at $0°$
- A HF horizontal asymptote at $-180°$.

The phase curve passes through $-90°$ at $\omega = \omega_n$. Its shape depends upon $\zeta$ and is obtained from the standard curves given in Figure 6.11.

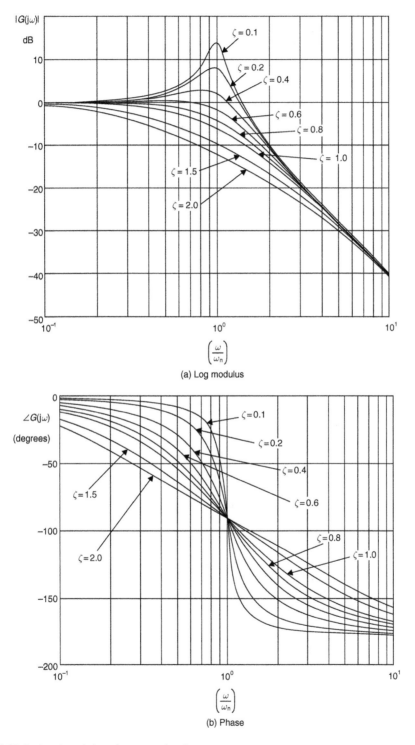

**Fig. 6.11** Bode gain and phase for a second-order system.

### (d) A pure integrator

Consider a pure integrator of the form

$$G(s) = \frac{K}{s} \tag{6.40}$$

now

$$G(j\omega) = \frac{K}{0 + j\omega}$$

Rationalizing

$$G(j\omega) = \frac{K(0 - j\omega)}{\omega^2} \tag{6.41}$$

From equation (6.17)

$$|G(j\omega)| = \sqrt{0 + \frac{K^2\omega^2}{\omega^4}} = \frac{K}{\omega} \tag{6.42}$$

and from equation (6.20)

$$\angle G(j\omega) = \tan^{-1}\left(\frac{-K\omega/\omega^2}{0}\right) = \tan^{-1}(-\infty) = -90° \tag{6.43}$$

It can be seen from equation (6.42) that the modulus will halve in value each time the frequency is doubled, i.e. it has a slope of $-6\,\text{dB/octave}$ ($-20\,\text{dB/decade}$) over the complete frequency range.

Note that

$$|G(j\omega)| = K\,\text{dB} \quad \text{when } \omega = 1$$
$$|G(j\omega)| = 1 = 0\,\text{dB} \quad \text{when } \omega = K$$

The Bode diagram for a pure integrator is shown in Figure 6.12.

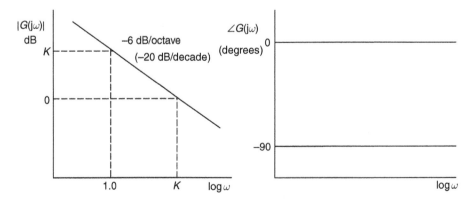

**Fig. 6.12** Bode diagram for a pure integrator.

*Example 6.1* (See also Appendix 1, *examp61.m*)
Construct, using asymptotes, the Bode diagram for

$$G(s) = \frac{2}{1 + 0.5s} \tag{6.44}$$

Low Frequency asymptote is a horizontal line at $K$ dB,

i.e. $+6$ dB

Asymptote intersection (break frequency) occurs at $1/T$, i.e. $2$ rad/s. The Bode diagram is shown in Figure 6.13.

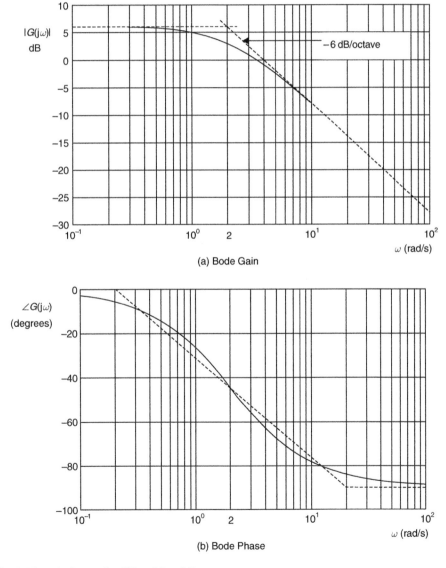

(a) Bode Gain

(b) Bode Phase

**Fig. 6.13** Bode diagram for $G(s) = 2/1 + 0.5s$.

*Example 6.2* (See also Appendix 1, *examp62.m*)
Draw the Bode diagram for

$$G(s) = \frac{4}{0.25s^2 + 0.2s + 1} \qquad (6.45)$$

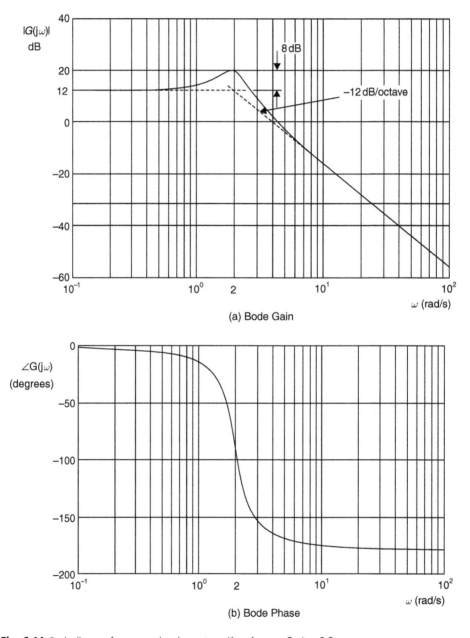

(a) Bode Gain

(b) Bode Phase

**Fig. 6.14** Bode diagram for a second-order system, $K = 4$, $\omega_n = 2$, $\zeta = 0.2$.

Comparing equation (6.45) with the standard form given in (6.22)

$$\frac{1}{\omega_n^2} = 0.25$$

i.e. $\omega_n = 2 \, \text{rad/s}$

$$\frac{2\zeta}{\omega_n} = 0.2$$

i.e. $\zeta = 0.2$

Low Frequency asymptote is a horizontal line at $K \, \text{dB}$

i.e. $20 \log_{10}(4) = +12 \, \text{dB}$

The log modulus relative to the LF asymptote at $\omega = \omega_n$ is given by equation (6.39)

$$|G(j\omega)|_{\omega_n} = 20 \log_{10}\left(\frac{1}{0.4}\right) = 8 \, \text{dB}$$

(Hence the absolute log modulus at $\omega = \omega_n$ is 20 dB). The Bode diagram is given by Figure 6.14. Note in Figure 6.14 that the phase curve was constructed by reading the phase from Figure 6.11(b), an octave either side of $\omega_n$.

*Example 6.3*
Construct, on log-linear graph paper, using asymptotes, and validate using MATLAB or a similar tool, the Bode diagrams for

(a) $G(s) = \dfrac{4}{s(1 + 2s)}$

(b) $G(s) = \dfrac{1}{(1 + 0.5s)(1 + 4s)}$

(c) $G(s) = \dfrac{10(1 + s)}{(1 + 0.2s)(1 + 5s)}$

(d) $G(s) = \dfrac{100}{s(0.25s^2 + 0.1s + 1)}$

## 6.4 Stability in the frequency domain

### 6.4.1 Conformal mapping and Cauchy's theorem

In Chapter 5 the stability of linear control systems were considered in the $s$-plane. Using a process of conformal transformation or mapping, it is possible to map a contour from one complex plane to another. It is therefore possible to transfer stability information from the $s$-plane to another complex plane, the $F(s)$-plane. The relationship between the contours in the two complex planes is given by Cauchy's theorem, which states: 'For a given contour in the $s$-plane that encircles $P$ poles and $Z$ zeros of the function $F(s)$ in a clockwise direction, the resulting

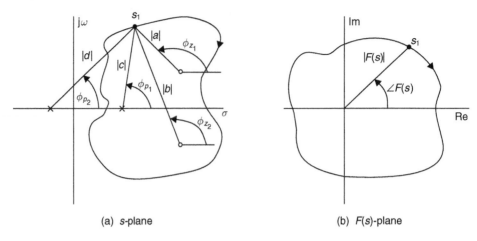

(a) *s*-plane                    (b) *F*(*s*)-plane

**Fig. 6.15** Mapping of a contour from the *s*-plane to the *F*(*s*)-plane.

contour in the $F(s)$-plane encircles the origin a total of $N$ times in a clockwise direction'.
   Where

$$N = Z - P \tag{6.46}$$

Consider a function

$$F(s) = \frac{(s + z_1)(s + z_2)}{(s + p_1)(s + p_2)} \tag{6.47}$$

where $z_1$ and $z_2$ are zeros of $F(s)$ and $p_1$ and $p_2$ are poles. Equation (6.47) can be written as

$$F(s) = |F(s)| \angle F(s)$$

The mapping of a contour from the $s$-plane to the $F(s)$-plane is shown in Figure 6.15. From Figure 6.15

$$|F(s)| = \frac{|a||b|}{|c||d|} \tag{6.48}$$

and

$$\angle F(s) = \phi_{z1} + \phi_{z2} - \phi_{p1} - \phi_{p2} \tag{6.49}$$

As $s_1$ in Figure 6.15(a) is swept clockwise around the contour, it encircles two zeros and one pole. From Cauchy's theorem given in equation (6.46), the number of clockwise encirclements of the origin in Figure 6.15(b) is

$$N = 2 - 1 = 1 \tag{6.50}$$

### 6.4.2  The Nyquist stability criterion

A frequency domain stability criterion developed by Nyquist (1932) is based upon Cauchy's theorem. If the function $F(s)$ is in fact the characteristic equation of a closed-loop control system, then

$$F(s) = 1 + G(s)H(s) \qquad (6.51)$$

Note that the roots of the characteristic equation are the closed-loop poles, which are the zeros of $F(s)$.

In order to encircle any poles or zeros of $F(s)$ that lie in the right-hand side of the $s$-plane, a Nyquist contour is constructed as shown in Figure 6.16. To avoid poles at the origin, a small semicircle of radius $\varepsilon$, where $\varepsilon \to 0$, is included.

Figure 6.17(a) shows the $1 + G(s)H(s)$ plane when $Z - P = 2$, i.e. two clockwise encirclements. However, if the contour is plotted in $G(s)H(s)$ plane as shown in Figure 6.17(b), then it moves one unit to the left, i.e. encircles the $-1$ point.

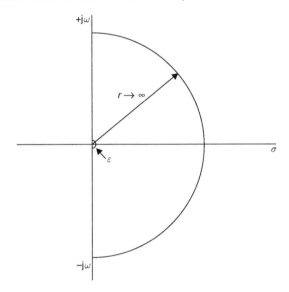

**Fig. 6.16** $s$-plane Nyquist contour.

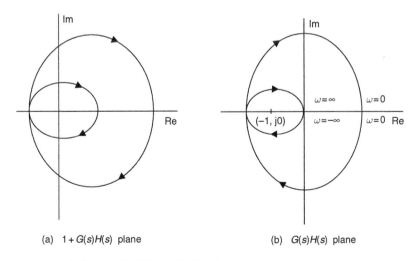

(a)  $1 + G(s)H(s)$  plane          (b)  $G(s)H(s)$  plane

**Fig. 6.17** Contours in the $1 + G(s)H(s)$ and $G(s)H(s)$ planes.

The Nyquist stability criterion can be stated as: 'A closed-loop control system is stable if, and only if, a contour in the $G(s)H(s)$ plane describes a number of counter-clockwise encirclements of the $(-1, j0)$ point, the number of encirclements being equal to the number of poles of $G(s)H(s)$ with positive real parts'.

Hence, because there is a net clockwise encirclement of the $(-1, j0)$ point in Figure 6.17(b) the system is unstable. If, however, there had been a net counter-clockwise encirclement, the system would have been stable, and the number of encirclements would have been equal to the number of poles of $G(s)H(s)$ with positive real parts.

For the condition $P = 0$, the Nyquist criterion is: 'A closed-loop control system is stable if, and only if, a contour in the $G(s)H(s)$ plane does not encircle the $(-1, j0)$ point when the number of poles of $G(s)H(s)$ in the right-hand $s$-plane is zero'.

In practice, only the frequencies $\omega = 0$ to $+\infty$ are of interest and since in the frequency domain $s = j\omega$, a simplified Nyquist stability criterion, as shown in Figure 6.18 is: 'A closed-loop system is stable if, and only if, the locus of the $G(j\omega)H(j\omega)$ function does not enclose the $(-1, j0)$ point as $\omega$ is varied from zero to infinity. Enclosing the $(-1, j0)$ point may be interpreted as passing to the left of the point'. The $G(j\omega)H(j\omega)$ locus is referred to as the Nyquist Diagram.

An important difference between analysis of stability in the $s$-plane and stability in the frequency domain is that, in the former, system models in the form of transfer functions need to be known. In the latter, however, either models or a set of input–output measured open-loop frequency response data from an unknown system may be employed.

### Margins of stability

The closer the open-loop frequency response locus $G(j\omega)H(j\omega)$ is to the $(-1, j0)$ point, the nearer the closed-loop system is to instability. In practice, all control

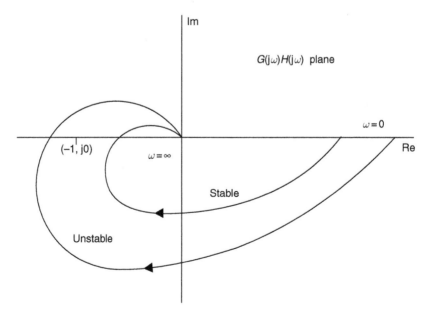

**Fig. 6.18** Nyquist diagram showing stable and unstable contours.

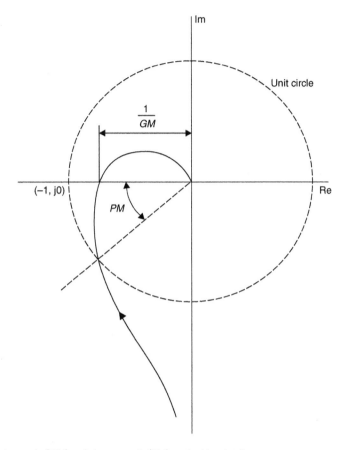

**Fig. 6.19** Gain margin (GM) and phase margin (PM) on the Nyquist diagram.

systems possess a *Margin of Stability*, generally referred to as gain and phase margins. These are shown in Figure 6.19.

*Gain Margin (GM):* The gain margin is the increase in open-loop gain required when the open-loop phase is $-180°$ to make the closed-loop system just unstable.

Nyquist diagram

$$GM = \frac{1}{|G(j\omega)H(j\omega)|_{180}} \tag{6.52}$$

Bode diagram

$$GM = 20\log_{10}\left\{\frac{1}{|G(j\omega)H(j\omega)|_{180}}\right\} \tag{6.53}$$

*Phase Margin (PM):* The phase margin is the change in open-loop phase, required when the open-loop modulus is unity, (or 0 dB on the Bode diagram) to make the closed-loop system just unstable.

$$\text{Phase Margin} = 180 - \angle G(j\omega)H(j\omega)(\text{mod} = 1) \tag{6.54}$$

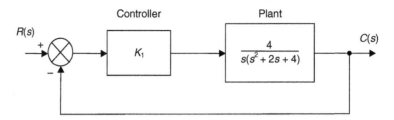

**Fig. 6.20** Closed-loop control system.

*Example 6.4* (See also Appendix 1, *examp64.m*)
Construct the Nyquist diagram for the control system shown in Figure 6.20 and find
the controller gain $K_1$ that

(a) makes the system just unstable. (Confirm using the Routh stability criterion)
(b) gives the system a gain margin of 2 (6 dB). What is now the phase margin?

*Solution*
Open-loop transfer function

$$G(s)H(s) = \frac{K}{s(s^2 + 2s + 4)} \tag{6.55}$$

where

$$K = 4K_1 \tag{6.56}$$

$$G(s)H(s) = \frac{K}{s^3 + 2s^2 + 4s}$$

$$G(j\omega)H(j\omega) = \frac{K}{(j\omega)^3 + 2(j\omega)^2 + 4j\omega}$$

$$= \frac{K}{-2\omega^2 + j(4\omega - \omega^3)}$$

Rationalizing

$$G(j\omega)H(j\omega) = \frac{K\{-2\omega^2 - j(4\omega - \omega^3)\}}{4\omega^4 + (4\omega - \omega^3)^2} \tag{6.57}$$

From equation (6.19)

$$|G(j\omega)H(j\omega)| = \frac{K}{\sqrt{4\omega^4 + (4\omega - \omega^3)^2}} \tag{6.58}$$

From equation (6.20)

$$\angle G(j\omega)H(j\omega) = \tan^{-1}\left\{\frac{-(4\omega - \omega^3)}{-2\omega^2}\right\}$$

$$= \tan^{-1}\left(\frac{4 - \omega^2}{2\omega}\right) \tag{6.59}$$

**Table 6.3** Data for Nyquist diagram for system in Figure 6.20

| $\omega$ (rad/s) | 0.5 | 1.0 | 1.5 | 2.0 | 2.5 | 3.0 | 4.0 | 5.0 |
|---|---|---|---|---|---|---|---|---|
| $|G(j\omega)H(j\omega)|(K=1)$ | 0.515 | 0.278 | 0.191 | 0.125 | 0.073 | 0.043 | 0.018 | 0.0085 |
| $\angle G(j\omega)H(j\omega)$ (deg) | $-105$ | $-124$ | $-150$ | $-180$ | $-204$ | $-220$ | $-236$ | $-245$ |

The Nyquist diagram is constructed, for $K = 1$, at frequencies either side of the 180° point, which, from equation (6.59) can be seen to be $\omega = 2\,\text{rad/s}$. Using equations (6.58) and (6.59), Table 6.3 may be evaluated.

(a) From equation (6.52)

$$GM = 1/0.125 = 8$$

Value of $K$ to make system just unstable

$$K_{\text{unstab}} = K \cdot GM$$
$$= 1 \times 8 = 8$$

From equation (6.56)

$$K_1 \text{ for instability} = 2$$

(b) For a $GM$ of 2, the locus must pass through the $(-0.5, j0)$ point in Figure 6.21. This can be done by multiplying all the modulus values in Table 6.3 by four and re-drawing Figure 6.21. Alternatively, the scales of Figure 6.21 can be multiplied by a factor of four (without re-drawing). Hence the unit circle is $0.25 \times 4$ and the $PM$ can be seen to have a value of 50° when the $GM$ is 2.0.

Value of $K$ to give a $GM$ of 2 is the original $K$ times 4, i.e. $1 \times 4 = 4$. From (6.56)

$$K_1 = 1.0$$

Hence, to give a $GM$ of 2 and a $PM$ of 50°, the controller gain must be set at 1.0. If it is doubled, i.e. multiplied by the $GM$, then the system just becomes unstable. Check using the Routh stability criterion:

Characteristic equation

$$1 + \frac{K}{s(s^2 + 2s + 4)} = 0 \tag{6.60}$$

$$s^3 + 2s^2 + 4s + K = 0 \tag{6.61}$$

Routh's array

$$
\begin{array}{c|cc}
s^0 & K & \\
s^1 & \frac{1}{2}(8 - K) & \\
s^2 & 2 & K \\
s^3 & 1 & 4
\end{array}
$$

Thus when $K \geq 8$ then the system is unstable.

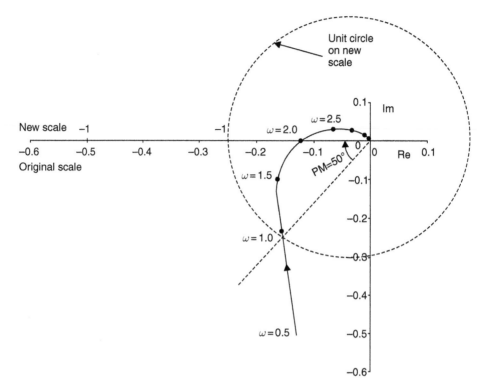

**Fig. 6.21** Nyquist diagram for system in Figure 6.20.

### System type classification

Closed-loop control systems are classified according to the number of pure integrations in the open-loop transfer function. If

$$G(s)H(s) = \frac{K \prod_{i=1}^{m} (s + z_i)}{s^n \prod_{k=1}^{q} (s + p_k)}$$ (6.62)

Then $n$ in equation (6.62) is the 'type number' of the system and $\prod$ denotes the product of the factors. The system 'type' can be observed from the starting point ($\omega \to 0$) of the Nyquist diagram, and the system order from the finishing point ($\omega \to \infty$), see Figure 6.22.

### System 'type' and steady-state errors

From the final value theorem given in equation (3.10) it is possible to define a set of steady-state error coefficients.

1. Position error coefficient

$$K_p = \lim_{s \to 0} G(s)$$

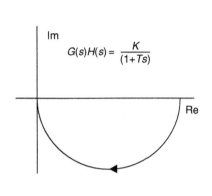

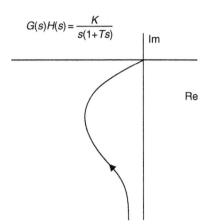

(a) First-order type zero system

(b) Second-order type one system

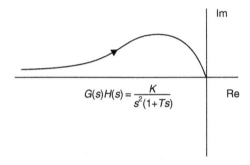

(c) Third-order type two system

**Fig. 6.22** Relationship between system type classification and the Nyquist diagram.

For a step input,

$$e_{ss} = \frac{1}{1 + K_p} \tag{6.63}$$

2. Velocity error coefficient

$$K_v = \lim_{s \to 0} s \, G(s)$$

For a ramp input,

$$e_{ss} = \frac{1}{K_v} \tag{6.64}$$

3. Acceleration error coefficient

$$K_a = \lim_{s \to 0} s^2 \, G(s)$$

For a parabolic input,

$$e_{ss} = \frac{1}{K_a} \tag{6.65}$$

**Table 6.4** Relationship between input function, system type and steady-state error

| Input function | Steady-state error | | |
|---|---|---|---|
| | Type zero | Type one | Type two |
| Step | constant | zero | zero |
| Ramp | increasing | constant | zero |
| Parabolic | increasing | increasing | constant |

Table 6.4 shows the relationship between input function, system type and steady-state error. From Table 6.4 it might appear that it would be desirable to make most systems type two. It should be noted from Figure 6.22(c) that type two systems are unstable for all values of $K$, and will require some form of compensation (see Example 6.6).

In general, type zero are unsatisfactory unless the open-loop gain $K$ can be raised, without instability, to a sufficiently high value to make $1/(1 + K_p)$ acceptably small. Most control systems are type one, the integrator either occurring naturally, or deliberately included in the form of integral control action, i.e. PI or PID.

### Stability on the Bode diagram

In general, it is more convenient to use the Bode diagram in control system design rather than the Nyquist diagram. This is because changes in open-loop gain do not affect the Bode phase diagram at all, and result in the Bode gain diagram retaining its shape, but just being shifted in the $y$-direction.

With Example 6.4 (see also Appendix 1, *examp64a.m* and *examp64b.m*), when the controller gain set to $K_1 = 1.0$, the open-loop transfer function is

$$G(s)H(s) = \frac{4}{s(s^2 + 2s + 4)} \tag{6.66}$$

Equation (6.66) represents a pure integrator and a second-order system of the form

$$G(s)H(s) = \frac{1}{s} \frac{1}{(0.25s^2 + 0.5s + 1)} \tag{6.67}$$

As explained in Figure 6.12 the pure integrator asymptote will pass through 0 dB at 1.0 rad/s (for $K = 1$ in equation (6.67)) and the second-order element has an undamped natural frequency of 2.0 rad/s and a damping ratio of 0.5.

Figure 6.23(a), curve (i), shows the Bode gain diagram for the transfer function given in equation (6.66), which has a gain margin of 6 dB (the amount the open-loop gain has to be increased to make the system just unstable. Figure 6.23(a), curve (ii), shows the effect of increasing $K$ by a factor of two (6 dB) to make the system just unstable. For curve (ii) the open-loop transfer function is

$$G(s)H(s) = \frac{8}{s(s^2 + 2s + 4)} \tag{6.68}$$

Figure 6.23(b) shows the Bode phase diagram which is asymptotic to $-90°$ at low frequencies and $-270°$ at high frequencies, passing through $-180°$ at 2 rad/s. To determine the phase margin, obtain the open-loop phase when the modulus is 0 dB (on curve (i), Figure 6.23(a), this is approximately 1.1 rad/s), and subtract the phase from 180° to give a *PM* of 50° as shown.

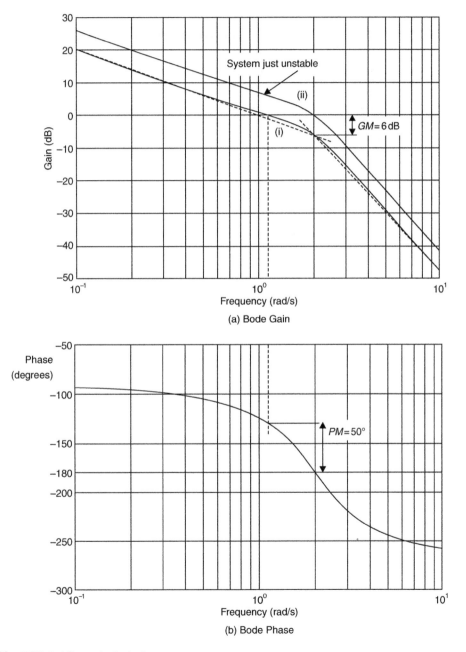

(a) Bode Gain

(b) Bode Phase

**Fig. 6.23** Stability on the Bode diagram.

## 6.5 Relationship between open-loop and closed-loop frequency response

### 6.5.1 Closed-loop frequency response

When designing a control system it is essential to

(a) ensure that the system is stable for all operating regimes
(b) ensure that the closed-loop performance meets the required specification.

Time domain performance specifications are described in section 3.7 and Figure 3.21. Frequency domain performance specifications are given in terms of gain and phase margins to provide adequate stability together with information relating to the closed-loop frequency response. Figure 6.24 shows the closed-loop frequency response of a control system. The closed-loop modulus is usually defined as

$$\left|\frac{C}{R}(j\omega)\right| = M \tag{6.69}$$

*Bandwidth* $(\omega_B)$: This is the frequency at which the closed-loop modulus $M$ has fallen by 3 dB below its zero frequency value. (This is not true for systems that do not operate down to dc levels.)

*Peak modulus* $(M_p)$: This is the maximum value that the closed-loop modulus $M$ rises above its zero frequency value.

*Peak frequency* $(\omega_p)$: This is the frequency that the peak modulus occurs. Note that $\omega_p < \omega_B$.

#### Second-order system closed-loop frequency response
Many closed-loop systems approximate to second-order systems. Equation (6.26) gives the general equation for modulus. If, when $K$ is set to unity, this equation is

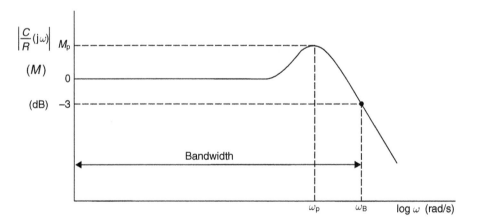

**Fig. 6.24** Closed-loop frequency response.

squared, differentiated with respect to $\omega^2$ and equated to zero (to find a maximum), then

$$M_p = \frac{1}{2\zeta\sqrt{1-\zeta^2}} \tag{6.70}$$

$$\omega_p = \omega_n\sqrt{1-2\zeta^2} \tag{6.71}$$

If, as a rule-of-thumb, $M_p$ is limited to 3 dB ($\sqrt{2}$), then from equations (6.70), (6.71) and Figure 6.11

$$\begin{aligned}\zeta &= 0.38 \\ \omega_p &= 0.84\omega_n \\ \omega_B &= 1.4\omega_n\end{aligned} \tag{6.72}$$

In general, for a unity feedback control system, the closed-loop frequency response is given by equation (6.73)

$$\frac{C}{R}(j\omega) = \frac{G(j\omega)}{1+G(j\omega)} \tag{6.73}$$

Equation (6.73) can be expressed in rectangular co-ordinates as

$$\frac{C}{R}(j\omega) = \frac{X(j\omega)+jY(j\omega)}{1+X(j\omega)+jY(j\omega)} \tag{6.74}$$

Hence

$$\left|\frac{C}{R}(j\omega)\right| = M = \frac{X(j\omega)+jY(j\omega)|}{|1+X(j\omega)+jY(j\omega)|} \tag{6.75}$$

Equation (6.75) can be expressed as an equation of a circle of the form

$$\left(X(j\omega)+\frac{M^2}{M^2-1}\right)^2 + Y(j\omega)^2 = \left(\frac{M}{M^2-1}\right)^2 \tag{6.76}$$

i.e.

$$\begin{aligned}\text{centre} &\quad \left(\frac{-M^2}{M^2-1}, 0\right) \\ \text{radius} &\quad \pm\left(\frac{M}{M^2-1}\right)\end{aligned} \tag{6.77}$$

Also, from equation (6.73)

$$\angle\frac{C}{R}(j\omega) = \angle(X(j\omega)+Y(j\omega)) - \angle(1+X(j\omega)+Y(j\omega)) \tag{6.78}$$

let

$$N = \tan\left\{\angle\frac{C}{R}(j\omega)\right\} \tag{6.79}$$

Equation (6.78) can also be expressed as an equation of a circle of the form

$$\left(X(j\omega) + \frac{1}{2}\right)^2 + \left(Y(j\omega) - \frac{1}{2N}\right)^2 = \frac{1}{4}\left(\frac{N^2 + 1}{N^2}\right) \tag{6.80}$$

i.e.

$$\begin{array}{ll} \text{centre} & -1/2,\ 1/2N \\[2mm] \text{radius} & \dfrac{\sqrt{N^2 + 1}}{2N} \end{array} \tag{6.81}$$

The $M$ and $N$ circles can be superimposed on a Nyquist diagram (called a Hall chart) to directly obtain closed-loop frequency response information.

Alternatively, the closed-loop frequency response can be obtained from a Nyquist diagram using the direct construction method shown in Figure 6.25. From equation (6.73)

$$\left|\frac{C}{R}(j\omega)\right| = \frac{|G(j\omega)|}{|1 + G(j\omega)|} = \frac{|OB|}{|AB|} \tag{6.82}$$

Also from equation (6.73)

$$\angle\frac{C}{R}(j\omega) = \angle G(j\omega) - \angle(1 + G(j\omega)) \tag{6.83}$$

$$= \angle COB - \angle OAB$$

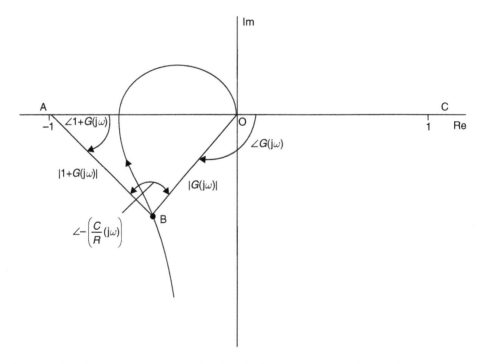

**Fig. 6.25** Closed-loop frequency response from Nyquist diagram using the direct construction method.

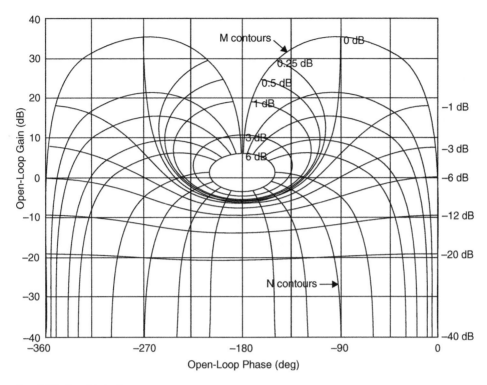

**Fig. 6.26** The Nichols chart.

Hence

$$\angle \frac{C}{R}(j\omega) = -\angle\text{ABO} \tag{6.84}$$

### The Nichols chart
The Nichols chart shown in Figure 6.26 is a rectangular plot of open-loop phase on the $x$-axis against open-loop modulus (dB) on the $y$-axis. $M$ and $N$ contours are superimposed so that open-loop and closed-loop frequency response characteristics can be evaluated simultaneously. Like the Bode diagram, the effect of increasing the open-loop gain constant $K$ is to move the open-loop frequency response locus in the $y$-direction. The Nichols chart is one of the most useful tools in frequency domain analysis.

*Example 6.5*
For the control system given in Example 6.4, determine

(a) The controller gain $K_1$ to give the best flatband response. What is the bandwidth, gain margin and phase margin?
(b) The controller gain $K_1$ to give a peak modulus $M_p$ of 3 dB. What is the bandwidth, gain margin and phase margin?

(c) For the controller gain in (b), what, in the time domain, is the rise-time, settling time and percentage overshoot?

*Solution*

(a) The open-loop transfer function for Example 6.4 is given by equation (6.55)

$$G(s)H(s) = \frac{K}{s(s^2 + 2s + 4)} \tag{6.85}$$

Figure 6.27 (see also Appendix 1, *fig627.m*) shows the Nichols chart for $K = 4$ (controller gain $K_1 = 1$). These are the settings shown in the Bode diagram in Figure 6.23(a), curve (i), and (b), where

$$\text{Gain margin} = 6\,\text{dB}$$

$$\text{Phase margin} = 50°$$

From Figure 6.27 it can be seen that the peak modulus $M_p$ is $4\,\text{dB}$, occurring at $\omega_p = 1.63\,\text{rad/s}$. The bandwidth $\omega_B$ is $2.2\,\text{rad/s}$. For the best flatband response, the open-loop frequency response locus should follow the $0\,\text{dB}$ $M$ contour for as wide

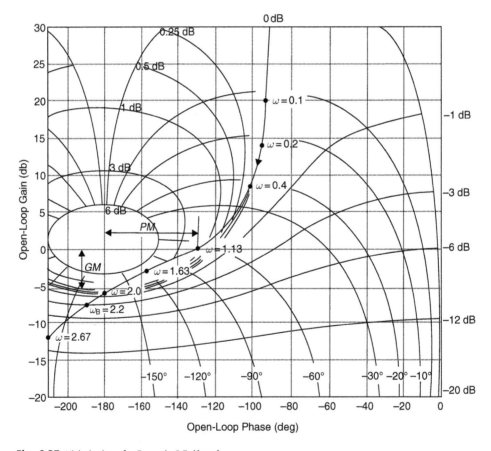

**Fig. 6.27** Nichols chart for Example 6.5, $K = 4$.

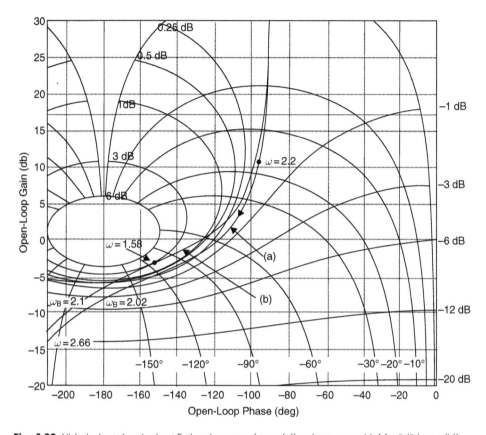

**Fig. 6.28** Nichols chart showing best flatband response (curve (a)) and response with $M_p$=3dB (curve (b)).

a frequency range as possible. This is shown in Figure 6.28, curve (a). To obtain curve (a), the locus has been moved down by 2 dB from that shown in Figure 6.27. This represents a gain reduction of

$$\text{gain reduction factor} = \text{alog}(-2/20) = 0.8 \tag{6.86}$$

Hence, for best flatband response

$$K = 4.0 \times 0.8 = 3.2 \tag{6.87}$$

$$\text{Controller gain } K_1 = K/4 = 3.2/4 = 0.8 \tag{6.88}$$

From Nichols chart

$$\text{Gain margin} = 8.15\,\text{dB}$$
$$\text{Phase margin} = 60° \tag{6.89}$$
$$\text{Bandwidth} = 2.02\,\text{rad/s}$$

(b) To obtain curve (b), the locus has been moved down by 0.5 dB from that shown in Figure 6.27. This represents a gain reduction of

$$\text{gain reduction factor} = \text{alog}(-0.5/20) = 0.944 \tag{6.90}$$

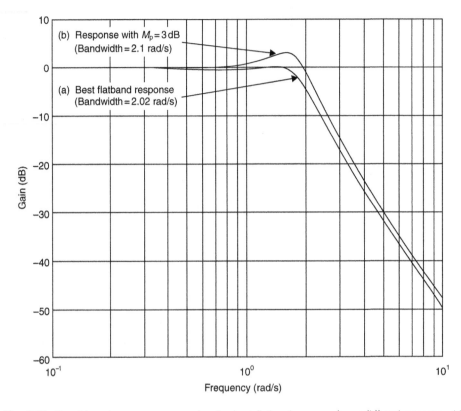

**Fig. 6.29** Closed-loop frequency response showing best flatband response (curve (b)) and response with $M_p$=3dB (curve (a)).

Hence, for a peak modulus of $M_p = 3\,\text{dB}$,

$$K = 4.0 \times 0.944 = 3.8 \tag{6.91}$$

$$\text{Controller gain } K_1 = K/4 = 3.8/4 = 0.95 \tag{6.92}$$

From Nichols chart

$$\text{Gain margin} = 6.36\,\text{dB}$$
$$\text{Phase margin} = 53° \tag{6.93}$$
$$\text{Bandwidth} = 2.1\,\text{rad/s}$$

Figure 6.29 (see also Appendix 1, *fig629.m*) shows the closed-loop modulus frequency response. Curve (a) is the best flatband response, curve (b) is the response with $M_p$ set to 3 dB.

## 6.6 Compensator design in the frequency domain

In section 4.5, controllers, particularly PID controllers for closed-loop systems were discussed. In Chapter 5 it was demonstrated how compensators could be designed in the $s$-plane to improve system performance. In a similar manner, it is possible to design compensators (that are usually introduced in the forward path) using frequency domain techniques.

The general approach is to re-shape the open-loop frequency response locus $G(j\omega)H(j\omega)$ in such a manner that the closed-loop frequency response meets a given frequency domain specification in terms of bandwidth and peak modulus.

*PD cascade compensation*: In Chapter 5, case-study Example 5.10, it was demonstrated how a cascaded PD compensator could improve both system performance and stability. However, in this chapter, Figure 6.10 gives the frequency response characteristics of a PD controller/compensator. The important thing to note about Figure 6.10 is that, in theory, above $\omega = 1/T$, the log modulus increases at $+6\,\text{dB/octave}$ for evermore. In practice this will not happen because eventually system elements will saturate. But what will happen, however, is that any high frequency noise in the system will be greatly amplified. It therefore becomes necessary, in a practical application, both with PD and PID controllers, to introduce, at some suitable high frequency, a low-pass filter.

## 6.6.1   Phase lead compensation

A phase lead compensator is different from the first-order lead system given in equation (6.35) and Figure 6.10 because it contains both numerator and denominator first-order transfer functions.

### (a) Passive lead compensation
A passive lead network (using two resistors and one capacitor) has a transfer function of the form

$$G(s) = \frac{1}{\alpha}\frac{(1 + \alpha Ts)}{(1 + Ts)} \tag{6.94}$$

There are two disadvantages of passive lead compensation:

 (i) the time constants are linked
 (ii) the gain constant $1/\alpha$ is always less than unity (called insertion loss) and additional amplification of value $\alpha$ is required to maintain the value of the open-loop gain $K$.

### (b) Active lead compensation
An active lead compensation network is shown in Figure 6.30. For an inverting operational amplifier

$$\frac{V_o}{V_i}(s) = -\frac{Z_f}{Z_i} \tag{6.95}$$

where $Z_i$ = input impedance and $Z_f$ = feedback impedance.
    Now

$$\frac{1}{Z_i} = \frac{1}{R_1} + C_1 s = \frac{1 + R_1 C_1 s}{R_1}$$

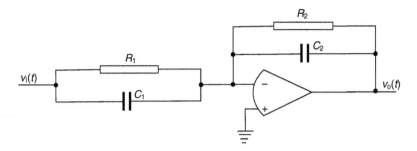

**Fig. 6.30** Active lead compensation network.

Hence

$$Z_i = \frac{R_1}{1 + R_1 C_1 s} \tag{6.96}$$

and

$$Z_f = \frac{R_2}{1 + R_2 C_2 s} \tag{6.97}$$

Inserting equations (6.96) and (6.97) into (6.95)

$$\frac{V_o}{V_i}(s) = \frac{-R_2}{R_1} \left\{ \frac{1 + R_1 C_1 s}{1 + R_2 C_2 s} \right\} \tag{6.98}$$

or, in general

$$G(s) = K_1 \left\{ \frac{1 + T_1 s}{1 + T_2 s} \right\} \tag{6.99}$$

Thus from equation (6.99) it can be seen that the system designer has complete flexibility since, $K_1$, $T_1$ and $T_2$ are not linked. For a lead network, $T_1$ must be greater than $T_2$. The Bode diagram for an active lead network is shown in Figure 6.31.

From equation (6.99)

$$G(j\omega) = K_1 \left\{ \frac{(1 + j\omega T_1)}{(1 + j\omega T_2)} \right\} = \frac{K_1(1 + j\omega T_1)(1 - j\omega T_2)}{(1 + \omega^2 T_2^2)} \tag{6.100}$$

expanding

$$G(j\omega) = \frac{K_1(1 - j\omega T_2 + j\omega T_1 + T_1 T_2 \omega^2)}{(1 + \omega^2 T_2^2)} \tag{6.101}$$

giving

$$G(j\omega) = \frac{K_1\{(1 + T_1 T_2 \omega^2) + j\omega(T_1 - T_2)\}}{(1 + \omega^2 T_2^2)} \tag{6.102}$$

From equation (6.20)

$$\tan \phi = \frac{\omega(T_1 - T_2)}{(1 + T_1 T_2 \omega^2)} \tag{6.103}$$

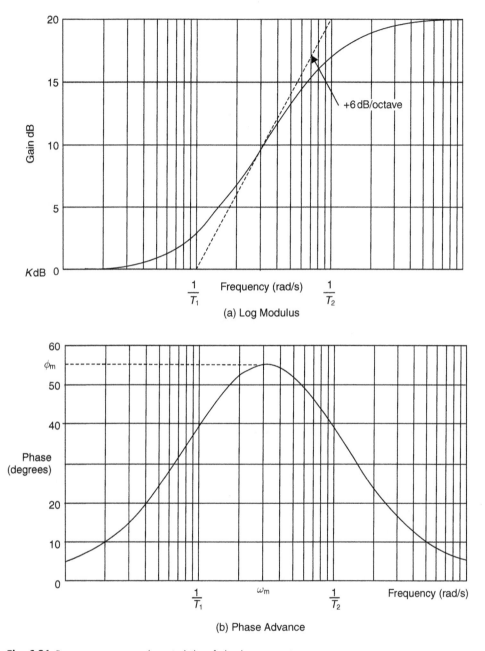

**Fig. 6.31** Frequency response characteristics of a lead compensator.

To find $\omega_{\mathrm{m}}$, differentiate equation (6.103) with respect to $\omega$, and equate to zero. This gives

$$\omega_{\mathrm{m}} = \frac{1}{\sqrt{T_1 T_2}} \tag{6.104}$$

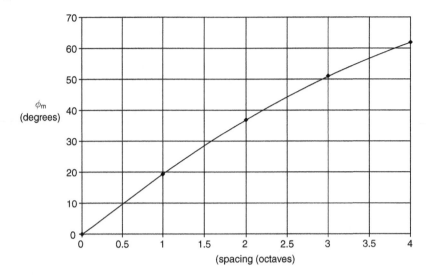

**Fig. 6.32** Relationship between $\phi_m$ and the spacing of $1/T_1$ and $1/T_2$ in octaves.

Substituting equation (6.104) into (6.103) to give

$$\phi_m = \tan^{-1}\left\{\frac{T_1 - T_2}{2\sqrt{T_1 T_2}}\right\} \qquad (6.105)$$

The value of $\phi_m$ depends upon the spacing of $1/T_1$ and $1/T_2$ on the $\log\omega$ axis, see Figure 6.32.

### Design procedure for lead compensation
1. Set $K$ to a suitable value so that any steady-state error criteria are met.
2. Plot the open-loop frequency response and obtain the phase margin and the modulus crossover frequency. (i.e. the frequency at which the modulus passes through 0 dB)
3. Set $\omega_m$ to the modulus crossover frequency and estimate the phase advance $\phi_m$ required to provide a suitable phase margin. From equations (6.104) and (6.105), determine $T_1$ and $T_2$.
4. Plot the compensated system open-loop frequency response. Note that the modulus crossover frequency has now increased. Reduce the compensator gain $K_1$ so that the modulus crossover frequency returns to its original value, and the desired phase margin is met.

### Case study

*Example 6.6*
The laser guided missile shown in Figure 5.26 has an open-loop transfer function (combining the fin dynamics and missile dynamics) of

$$G(s)H(s) = \frac{20}{s^2(s + 5)} \qquad (6.106)$$

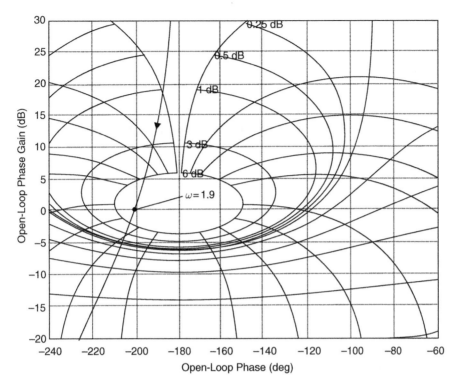

**Fig. 6.33** Nichols chart for uncompensated laser guided missile.

Design a cascade lead compensator that will ensure stability and provide a phase margin of at least 30°, a bandwidth greater than 5 rad/s and a peak closed-loop modulus $M_p$ of less than 6 dB.

*Solution*
The open-loop transfer function is third-order type 2, and is unstable for all values of open-loop gain $K$, as can be seen from the Nichols chart in Figure 6.33. From Figure 6.33 it can be seen that the zero modulus crossover occurs at a frequency of 1.9 rad/s, with a phase margin of $-21°$. A lead compensator should therefore have its maximum phase advance $\phi_m$ at this frequency. However, inserting the lead compensator in the loop will change (increase) the modulus crossover frequency.

### Lead compensator design one
Place $\omega_m$ at the modulus crossover frequency of 2 rad/s and position the compensator corner frequencies an octave below, and an octave above this frequency. Set the compensator gain to unity. Hence

$$\omega_m = 2 \text{ rad/s} \quad 1/T_1 = 1 \text{ rad/s} \quad 1/T_2 = 4 \text{ rad/s}$$
$$K = 1.0 \quad \phi_m = 36.9°$$

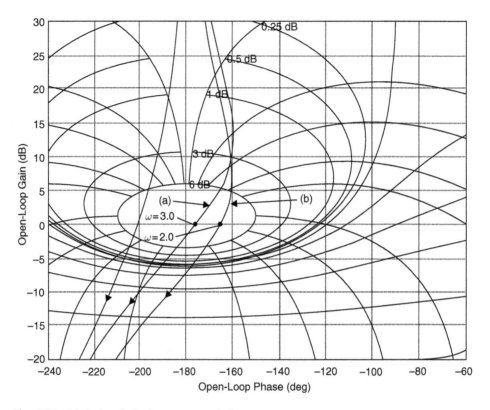

**Fig. 6.34** Nichols chart for lead compensator, design one.

The compensator is therefore

$$G(s) = \frac{(1+s)}{(1+0.25s)} \tag{6.107}$$

The Nichols chart for the uncompensated and compensated system (curve (a)) is shown in Figure 6.34 (see also Appendix 1, *fig634.m*). From Figure 6.34, curve (a)

$$\text{Gain margin} = 2\,\text{dB}$$
$$\text{Phase margin} = 4°$$
$$\text{Modulus crossover frequency} = 3.0\,\text{rad/s}$$

Figure 6.35 shows the Bode gain and phase for both compensated and uncompensated systems. From Figure 6.35, it can be seen that by reducing the open-loop gain by 5.4 dB, the original modulus crossover frequency, where the phase advance is a maximum, can be attained.

$$\text{Gain reduction} = \text{alog}\left(\frac{-5.4}{20}\right) = 0.537 \tag{6.108}$$

Hence the lead compensator transfer function is

$$G(s) = \frac{0.537(1+s)}{(1+0.25s)} \qquad (6.109)$$

The open-loop frequency response contours for the compensator given in equation (6.109) are curves (b) in Figures 6.34 and 6.35 which produce

$$\text{Gain margin} = 7\,dB$$
$$\text{Phase margin} = 15°$$
$$\text{Modulus crossover frequency} = 2\,\text{rad/s}$$

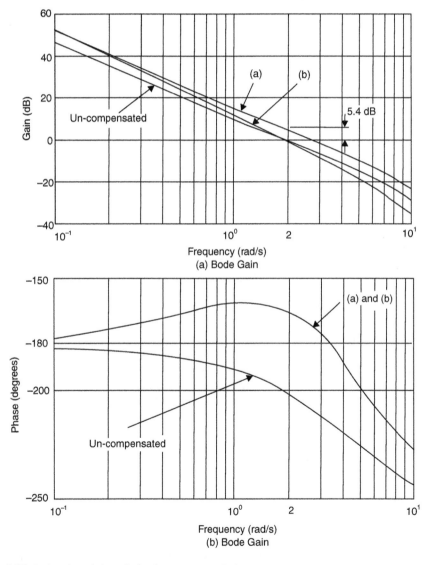

**Fig. 6.35** Bode gain and phase for lead compensator, design one.

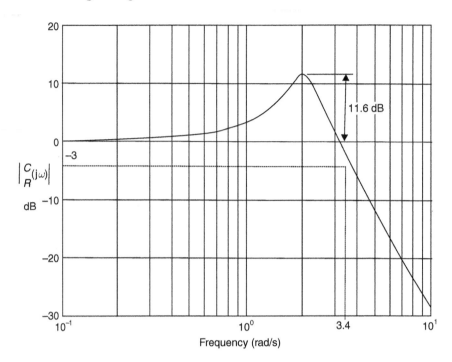

**Fig. 6.36** Closed-loop frequency response for lead compensator one.

Figure 6.36 shows the closed-loop frequency response using lead compensator one and defined by equation (6.109) (modulus only). From Figure 6.36

$$\text{Peak modulus } M_p = 11.6\,\text{dB}$$
$$\text{Bandwidth} = 3.4\,\text{rad/s}$$

This design does not meet the performance specification.

### Lead compensator design two

From Figure 6.34 it can be seen that to achieve the desired phase margin of at least 30° then the compensator must provide in the order of an additional 20° of phase advance, i.e. 57° in total, at the modulus crossover frequency.

From Figure 6.32, this suggests four octaves between the corner frequencies. Let $1/T_1$ remain at 1 rad/s and Position $1/T_2$ at 16 rad/s (4 octaves higher). This provides

$$\phi_m = 61.9°$$
$$\omega_m = 4\,\text{rad/s}$$

The design two compensator is therefore

$$G(s) = \frac{(1+s)}{(1+0.0625s)} \tag{6.110}$$

The open-loop Bode gain and phase with the lead compensator given in equation (6.110) inserted in the control loop is shown in Figure 6.37. From Figure 6.37 curve (i) it can be seen that the modulus crossover frequency is 3.37 rad/s, and the phase margin is (180 − 152.4), or 27.6°. This is close to, but does not quite achieve the specification. However, from Figure 6.37, the maximum phase advance of −145.3° occurs at 1.9 rad/s. At this frequency, the open-loop gain $K$ is 6.8 dB. Therefore, if the open-loop gain is reduced by this amount as shown in Figure 6.37, curve (ii) then the modulus crossover frequency becomes 1.9 rad/s and the phase margin is (180 − 145.3), or 34.7°, which is within specification. The compensator gain $K_1$ therefore becomes

$$K_1 = \text{alog}\left(\frac{-6.8}{20}\right) = 0.457 \quad\quad (6.111)$$

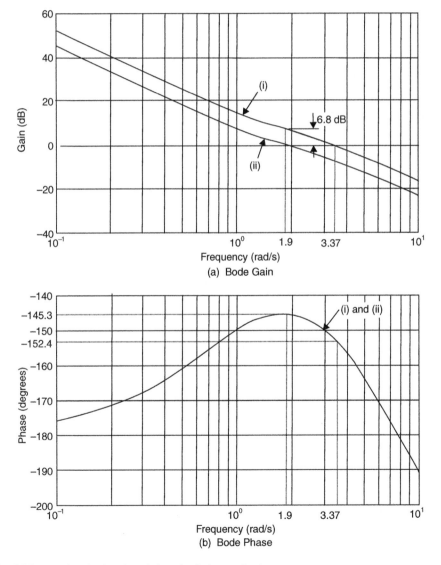

**Fig. 6.37** Open-loop bode gain and phase for design two lead compensator.

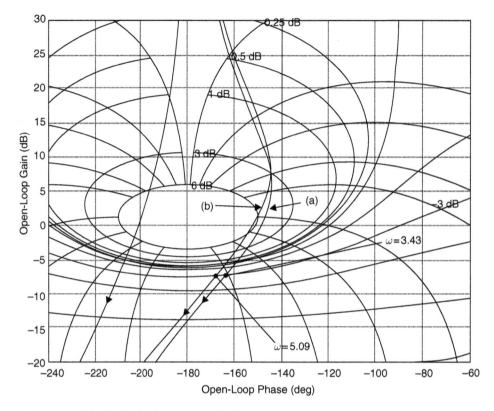

**Fig. 6.38** Nichols chart for lead compensator design two.

Figure 6.38, curve (a) shows the Nichols chart for the value of $K_1$ given in equation (6.111). It can be seen that $M_p = 5\,dB$, but the bandwidth is 3.43 rad/s, which is outside of specification. However, because of the shape of the locus, it is possible to reduce the gain margin (18.6 dB in curve (a)) which will increase the bandwidth, but not significantly change the peak modulus, or the phase margin. Figure 6.38, curve (b) shows the open-loop gain $K$ increased by 4.85 dB. This now has a peak modulus of 5.5 dB, a phase margin of 30.6° and a bandwidth of 5.09 rad/s, all of which are within specification. The new compensator gain $K_1$ is therefore

$$K_1 = 0.457 \times \mathrm{alog}\left(\frac{4.85}{20}\right)$$

$$= 0.457 \times 1.748 = 0.8 \tag{6.112}$$

The final lead compensator is

$$G(s) = \frac{0.8(1+s)}{(1+0.0625s)} \tag{6.113}$$

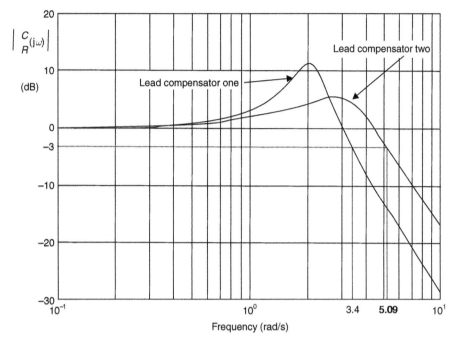

**Fig. 6.39** Closed-loop frequency response for both lead compensator designs.

System frequency domain performance

$$\text{Closed-loop peak } M_{\text{p}} = 5.5\,\text{dB}$$
$$\text{Gain margin} = 13.75\,\text{dB}$$
$$\text{Bandwidth} = 5.09\,\text{rad/s}$$
$$\text{Phase margin} = 30.6°$$

Figure 6.39 shows, for both lead compensator designs, the closed-loop frequency response characteristics for the system.

## 6.6.2  Phase lag compensation

Using passive components, a phase lag compensator may be constructed, whose transfer function is of the form

$$G(s) = \frac{(1 + Ts)}{(1 + \alpha Ts)} \tag{6.114}$$

where $\alpha$ is a number greater than unity. Passive lag networks suffer the same disadvantages of passive lead networks as discussed earlier.

The active network shown in Figure 6.30 has the transfer function given in equation (6.99)

$$G(s) = \frac{K_1(1 + T_1 s)}{(1 + T_2 s)} \tag{6.115}$$

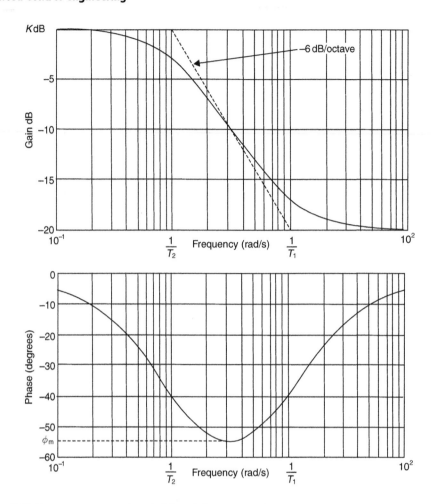

**Fig. 6.40** Frequency response characteristics of a lag compensator.

When $T_2$ is greater than $T_1$ equation (6.115) is an active lag network, whose Bode diagram is shown in Figure 6.40. The relationships between $T_1$, $T_2$, $\omega_m$ and $\phi_m$ are as given in equations (6.104) and (6.105), except, in this case, $\phi_m$ is negative. The same comment applies to Figure 6.32 which shows the relationship between the spacing of reciprocals of $T_2$ and $T_1$ and $\phi_m$.

### Design procedure for lag compensation

1. Set $K$ to a suitable value so that any steady-state error criteria are met.
2. Identify what modulus attenuation is required to provide an acceptable phase margin and hence determine the spacing between $1/T_2$ and $1/T_1$ (i.e. 6 dB attenuation requires a one octave spacing, 12 dB attenuation needs a two octave spacing, etc.).
3. Position $1/T_1$ one decade below the compensated modulus crossover frequency, and hence calculate $\omega_m$ using equation (6.104).
4. Adjust compensator gain $K_1$ if necessary.

## Case study

*Example 6.7*
A process plant has an open-loop transfer function

$$G(s)H(s) = \frac{30}{(1 + 0.5s)(1 + s)(1 + 10s)} \tag{6.116}$$

As it stands, when the loop is closed, the system is on the verge of instability, with a *GM* of 1.4 dB, a *PM* of 4° and a modulus crossover frequency of 1.4 rad/s. Reducing the open-loop gain *K* by 12 dB (i.e. *K* = 7.5) provides an acceptable *GM* of 13.5 dB, *PM* of 52° with a modulus crossover frequency of 0.6 rad/s. However, this gain setting produces an unacceptable step steady-state error of 12%. Design a lag compensator that maintains the open-loop gain *K* at 30, but provides gain and phase margins, similar to setting *K* at 7.5. What is now the steady-state step error?

*Solution*
Required modulus attenuation is 12 dB. This reduces the modulus crossover frequency from 1.4 to 0.6 rad/s.

Position $1/T_1$ one decade below 0.6 rad/s i.e. 0.06 rad/s. For a 12 dB attenuation, two octaves are required in the compensator, thus $1/T_2$ is positioned at 0.015 rad/s. From equation (6.104) $\omega_m$ is 0.03 rad/s, and from equation (6.105) (using a negative value), $\phi_m = -36.9°$.

Hence the required lag compensator is

$$G(s) = \frac{K_1(1 + 16.67s)}{(1 + 66.67s)} \tag{6.117}$$

The compensated and uncompensated open-loop frequency response is shown in Figure 6.41. From this Figure the compensated gain margin is 12.5 dB, and the phase margin is 48°. In equation (6.117), $K_1$ does not need to be adjusted, and can be set to unity. When responding to a step input, the steady-state error is now 4.6%.

## 6.7  Relationship between frequency response and time response for closed-loop systems

There are a few obvious relationships between the frequency response and time response of closed-loop systems:

(i) As bandwidth increases, the time response will be more rapid, i.e. the settling time will decrease.
(ii) The larger the closed-loop peak $M_p$, the more oscillatory will be the time response.

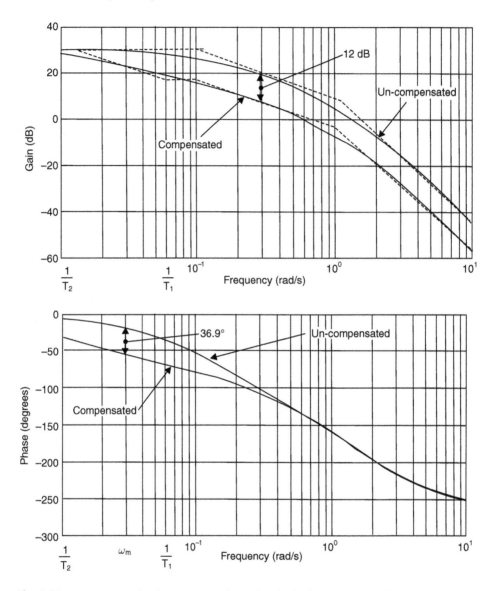

**Fig. 6.41** Lag compensated and uncompensated open-loop bode diagram for Example 6.7.

Since many closed-loop systems approximate to second-order systems, a few interesting observations can be made. For the case when the frequency domain specification has limited the value of $M_p$ to 3 dB for a second-order system, then from equation (6.72)

$$\zeta = 0.38$$
$$\omega_p = 0.84\omega_n \quad \text{(frequency that } M_p \text{ occurs)} \qquad (6.118)$$
$$\omega_B = 1.4\omega_n \quad \text{(bandwidth)}$$

Equation (3.73) gives the time for a second-order system to settle down to within a tolerance band of $\pm 2\%$

$$t_s = \left(\frac{1}{\zeta\omega_n}\right)\ln 50$$

or

$$t_s = \frac{3.912}{\zeta\omega_n} \tag{6.119}$$

Inserting the values in equation (6.118) into equation (6.119) gives

$$t_s = \frac{10.29}{\omega_n} = \frac{14.4}{\omega_B} \tag{6.120}$$

Thus the settling time is inversely proportional to the bandwidth. Comparing equation (6.70) with equation (3.68) gives

$$\% \text{ overshoot} = e^{-2\zeta^2 \pi M_p} \times 100 \tag{6.121}$$

Hence a closed-loop system with an undamped natural frequency of 1.0 rad/s and a damping ratio of 0.38 has the following performance:

- Frequency domain
  $M_p$ (equation (6.70)) $= 1.422 = 3.06$ dB
  $\omega_B$ (equation (6.72)) $= 1.4$ rad/s.

- Time domain
  $t_s$ (equation (6.120)) $= 10.29$ seconds
  $\%$ overshoot (equation (6.121)) $= 27.5\%$

## 6.8 Further problems

*Example 6.8*
A spring–mass–damper system has a mass of 20 kg, a spring of stiffness 8000 N/m and a damper with a damping coefficient of 80 Ns/m. The system is excited by a constant amplitude harmonic forcing function of the form

$$F(t) = 160 \sin \omega t$$

(a) Determine the system transfer function relating $F(t)$ and $x(t)$ and calculate values for $\omega_n$ and $\zeta$.
(b) What are the amplitudes of vibration when $\omega$ has values of 1.0, 20 and 50 rad/s.
(c) Find the value of the damping coefficient to give critical damping and hence, with this value, determine again the amplitudes of vibration for the angular frequencies specified in (b).

*Solution*
(a) $\omega_n = 20$ rad/s, $\zeta = 0.1$
(b) 0.02 m, 0.1 m and 0.0038 m
(c) $C = 800$ Ns/m 0.02 m, 0.01 m and 0.0028 m

*Example 6.9*
Construct, using asymptotes and standard second-order phase diagrams, the Bode diagrams for

(i) $G(s) = \dfrac{12}{(1 + 2s)}$

(ii) $G(s) = \dfrac{2}{(0.0025s^2 + 0.01s + 1)}$

(iii) $G(s) = 4\left(\dfrac{1 + 2s}{1 + 0.5s}\right)$

(iv) $G(s)H(s) = \dfrac{0.5}{s(s^2 + 0.5s + 1)}$

When the loop is closed, will the system in (iv) be stable or unstable?

*Solution*
The system will have marginal stability.

*Example 6.10*
A control system has an open-loop transfer function

$$G(s)H(s) = \frac{K}{s(0.25s^2 + 0.25s + 1)}$$

Set $K = 1$ and plot the Nyquist diagram by calculating values of open-loop modulus and phase for angular frequency values from 0.8 to 3.0 rad/s in increments of 0.2 rad/s. Hence find the value of $K$ to give a gain margin of 2 (6 dB). What is the phase margin at this value of $K$?

*Solution*
$K = 0.5$, Phase margin $= 82°$

*Example 6.11*
An open-loop frequency response test on an unknown system produced the following results:

| $\omega$ (rad/s) | 0.2 | 0.4 | 0.8 | 1.6 | 3.0 | 4.0 | 4.6 | 5 | 6 | 8 | 10 | 20 | 40 |
|---|---|---|---|---|---|---|---|---|---|---|---|---|---|
| $\lvert G(j\omega)H(j\omega)\rvert$ (dB) | 28 | 22 | 16 | 10.7 | 7.5 | 7.3 | 7.0 | 6.0 | 0.9 | −9.3 | −28 | −36 | −54 |
| $\angle G(j\omega)H(j\omega)$ (deg) | −91 | −92 | −95 | −100 | −115 | −138 | −162 | −180 | −217 | −244 | −259 | −262 | −266 |

Plot the Bode diagram on log-linear paper and determine

(a) The open-loop transfer function.
(b) The open-loop gain constant $K$ to give a gain margin of 4.4 dB. What is the phase margin for this value of $K$?
(c) The closed-loop transfer function (unity feedback) for the value of $K$ found in (b).
(d) The closed-loop peak modulus $M_p$ and bandwidth.

**Table 6.5** Open-loop frequency response data

| $\omega$ (rad/s) | 0.1 | 0.3 | 0.7 | 1.0 | 1.5 | 2.0 | 3.0 | 5.0 | 10.0 |
|---|---|---|---|---|---|---|---|---|---|
| $\|G(j\omega)H(j\omega)\|$ (dB) | 17 | 7 | 0.5 | $-2$ | $-5$ | $-9$ | $-18.5$ | $-33$ | $-51$ |
| $\angle G(j\omega)H(j\omega)$ (deg) | $-92$ | $-98$ | $-112$ | $-123$ | $-150$ | $-180$ | $-220$ | $-224$ | $-258$ |

*Solution*

(a) $G(s)H(s) = \dfrac{5.0}{s(0.04s^2 + 0.1s + 1)}$

(b) $K = 1.5$, Phase margin $= 79°$

(c) $\dfrac{C}{R}(s) = \dfrac{37.5}{s^3 + 2.5s^2 + 25s + 37.5}$

(d) $M_p = 4.6\,\text{dB}$, $\omega_B = 5.6\,\text{rad/s}$.

*Example 6.12*
(a) An open-loop frequency response test on a unity feedback control system produced the data given in Table 6.5. Plot the Bode Diagram and determine the system open-loop and closed-loop transfer functions. What are the phase and gain margins?
(b) A phase lead compensation network of the form

$$G(s) = \frac{1}{\alpha}\frac{(1 + \alpha Ts)}{(1 + Ts)}$$

is to be introduced in the forward path. The maximum phase advance $\phi_m$ is to be $37°$ and is to occur at $\omega_m = 2\,\text{rad/s}$. Determine the expression for the phase angle $\phi$ and hence prove that $\phi_m$ and $\omega_m$ are as given below. Find from these expressions the values of $\alpha$ and $T$ and calculate values for $\phi$ when $\omega = 1, 1.5, 2, 3$ and $5\,\text{rad/s}$. Plot the compensator frequency response characteristics.

$$\phi_m = \tan^{-1}\left[\frac{\alpha - 1}{2\sqrt{\alpha}}\right]$$

$$\omega_m = \frac{1}{T\sqrt{\alpha}}$$

(c) Produce a table using the frequencies specified in part (a) for the compete open-loop frequency response including the compensation network and an amplifier to make up the insertion loss of $1/\alpha$.

Plot these results on a Nichols chart and determine

(i) Maximum closed-loop peak modulus, $M_p$
(ii) Bandwidth (to $-3\,\text{dB}$ point)

*Solution*

(a) $G(s)H(s) = \dfrac{0.7}{s(0.25s^2 + 0.5s + 1)}$

$\dfrac{C}{R}(s) = \dfrac{1}{0.356s^3 + 0.714s^2 + 1.429s + 1}$

Phase margin $= 66°$ and Gain margin $= 9\,\text{dB}$

(b) $G(s) = \dfrac{1}{4}\left\{\dfrac{(1+s)}{(1+0.25s)}\right\}$

(c) $M_\text{p} = 1\,\text{dB}$ and Bandwidth $= 2.7\,\text{rad/s}$

*Example 6.13*

(a) A unity feedback control system has an open-loop transfer function

$$G(s)H(s) = \dfrac{1}{s(1+s)(1+0.5s)}$$

Construct, using asymptotes, the Bode diagram and read off values of open-loop modulus and phase for the following frequencies

$\omega$ (rad/s) $= 0.1, 0.5, 1.0, 1.4, 2.0, 4.0, 6.0$ and $10.0$

You may assume that at frequencies of 1.0 and 2.0 rad/s the open-loop phase angles are $-162°$ and $-198°$ respectively.

Plot the results between 0.1 and 2.0 rad/s on a Nichols Chart and determine

  (i) the phase and gain margins
 (ii) the maximum closed-loop modulus $M_\text{p}$
(iii) the bandwidth to the $-3\,\text{dB}$ point

(b) The performance specification calls for a maximum closed-loop modulus of $+1\,\text{dB}$ and a bandwidth of at least 1.8 rad/s. In order to achieve this, the following active lead compensation element is placed in the forward path

$$G(s) = \dfrac{K_1(1 + T_1 s)}{(1 + T_2 s)}$$

Show that the phase advance $\phi$ is given by

$$\phi = \tan^{-1}\left[\dfrac{\omega(T_1 - T_2)}{1 + T_1 T_2 \omega^2}\right]$$

The frequency of maximum phase advance is to occur at the frequency that corresponds to $-180°$ on the Bode diagram constructed in section (a). The lower break frequency $1/T_1$ is to be half this value and the upper break frequency $1/T_2$ is to be twice this value. Evaluate $T_1$ and $T_2$ and calculate values of $\phi$ for the frequencies specified in section (a). Construct the Bode diagram for the compensation element for the condition $K_1 = 1$, and read off values of modulus at the same frequencies as the calculated phase values.

(c) Using the tables of modulus and phase for the plant and compensator found in sections (a) and (b), determine values for the new overall open-loop modulus and phase when the compensator is inserted in the forward path.

Plot these results on a Nichols Chart and adjust the compensator gain $K_1$ so that the system achieves the required performance specification.
   What are now the values of

   (i) the phase and gain margins
  (ii) the maximum closed-loop modulus, $M_p$
 (iii) the bandwidth
 (iv) the compensator gain constant $K_1$

*Solution*
(a)   (i) Phase margin $= 32°$ and Gain margin $= 9.6\,\mathrm{dB}$
      (ii) $M_p = 5\,\mathrm{dB}$
     (iii) Bandwidth $= 1.3\,\mathrm{rad/s}$

(b) $G(s) = \dfrac{K_1(1 + 1.429s)}{(1 + 0.357s)}$

(c)   (i) Phase margin $= 47°$ and Gain margin $= 13.5\,\mathrm{dB}$
      (ii) $M_p = 1\,\mathrm{dB}$
     (iii) Bandwidth $= 1.85\,\mathrm{rad/s}$
     (iv) $K_1 = 0.861$

# 7

# Digital control system design

As a result of developments in microprocessor technology, the implementation of control algorithms is now invariably through the use of embedded microcontrollers rather than employing analogue devices. A typical system using microprocessor control is shown in Figure 7.1.

In Figure 7.1

- RAM is Random Access Memory and is used for general purpose working space during computation and data transfer.
- ROM, PROM, EPROM is Read Only Memory, Programmable Read Only Memory and Erasable Programmable Read Only Memory and are used for rapid sources of information that seldom, or never need to be modified.
- A/D Converter converts analogue signals from sensors into digital form at a given sampling period $T$ seconds and given resolution (8 bits, 16 bits, 24 bits, etc.)
- D/A Converter converts digital signals into analogue signals suitable for driving actuators and other devices.

The elements of a microprocessor controller (microcontroller) are shown in Figure 7.2. Figure 7.2 shows a Central Processing Unit (CPU) which consists of

- the Arithmetic Logic Unit (ALU) which performs arithmetic and logical operations on the data

and a number of registers, typically

- Program Counter – incremented each time an instruction is executed
- Accumulator(s) – can undertake arithmetic operations
- Instruction register – holds current instruction
- Data address register – holds memory address of data

Control algorithms are implemented in either high level or low level language. The lowest level of code is executable machine code, which is a sequence of binary words that is understood by the CPU. A higher level of language is an assembler, which employs meaningful mnemonics and names for data addresses. Programs written in assembler are rapid in execution. At a higher level still are languages

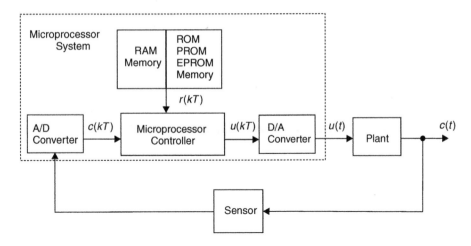

**Fig. 7.1** Microprocessor control of a plant.

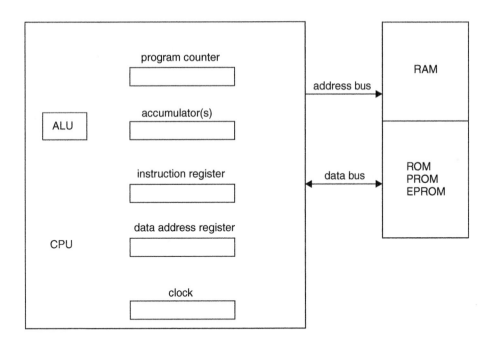

**Fig. 7.2** Elements of a microprocessor controller.

such as C and C++, which are rapidly becoming industry standard for control software.

The advantages of microprocessor control are

- Versatility – programs may easily be changed
- Sophistication – advanced control laws can be implemented.

The disadvantages of microprocessor control are

- Works in discrete time – only snap-shots of the system output through the A/D converter are available. Hence, to ensure that all relevant data is available, the frequency of sampling is very important.

## 7.2  Shannon's sampling theorem

Shannon's sampling theorem states that 'A function $f(t)$ that has a bandwidth $\omega_b$ is uniquely determined by a discrete set of sample values provided that the sampling frequency is greater than $2\omega_b$'. The sampling frequency $2\omega_b$ is called the Nyquist frequency.

It is rare in practise to work near to the limit given by Shannon's theorem. A useful rule of thumb is to sample the signal at about ten times higher than the highest frequency thought to be present.

If a signal is sampled below Shannon's limit, then a lower frequency signal, called an alias may be constructed as shown in Figure 7.3.

To ensure that aliasing does not take place, it is common practice to place an anti-aliasing filter before the A/D converter. This is an analogue low-pass filter with a break-frequency of $0.5\omega_s$ where $\omega_s$ is the sampling frequency ($\omega_s > 10\omega_b$). The higher $\omega_s$ is in comparison to $\omega_b$, the more closely the digital system resembles an analogue one and as a result, the more applicable are the design methods described in Chapters 5 and 6.

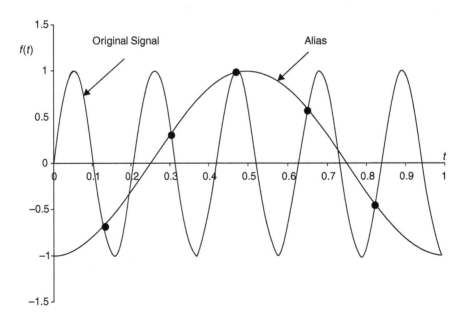

**Fig. 7.3** Construction of an alias due to undersampling.

## 7.3 Ideal sampling

An ideal sample $f^*(t)$ of a continuous signal $f(t)$ is a series of zero width impulses spaced at sampling time $T$ seconds apart as shown in Figure 7.4.

The sampled signal is represented by equation (7.1).

$$f^*(t) = \sum_{k=-\infty}^{\infty} f(kT)\delta(t - kT) \tag{7.1}$$

where $\delta(t - kT)$ is the unit impulse function occurring at $t = kT$.

A sampler (i.e. an A/D converter) is represented by a switch symbol as shown in Figure 7.5. It is possible to reconstruct $f(t)$ approximately from $f^*(t)$ by the use of a hold device, the most common of which is the zero-order hold (D/A converter) as shown in Figure 7.6. From Figure 7.6 it can be seen that a zero-order hold converts a series of impulses into a series of pulses of width $T$. Hence a unit impulse at time $t$ is converted into a pulse of width $T$, which may be created by a positive unit step at time $t$, followed by a negative unit step at time $(t - T)$, i.e. delayed by $T$.

The transfer function for a zero-order hold is

$$\mathscr{L}[f(t)] = \frac{1}{s} - \frac{1}{s}e^{-Ts}$$

$$G_h(s) = \frac{1 - e^{-Ts}}{s} \tag{7.2}$$

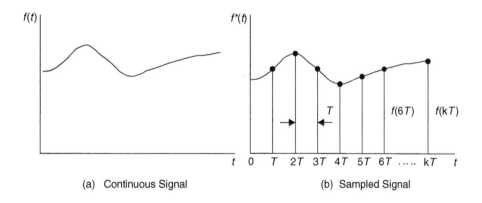

(a) Continuous Signal      (b) Sampled Signal

**Fig. 7.4** The sampling process.

**Fig. 7.5** A sampler.

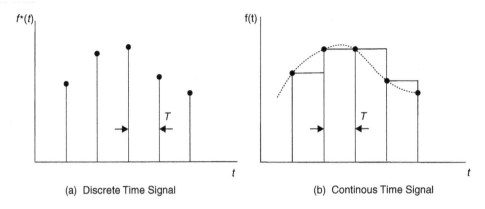

(a) Discrete Time Signal          (b) Continous Time Signal

**Fig. 7.6** Construction of a continuous signal using a zero-order hold.

## 7.4 The z-transform

The $z$-transform is the principal analytical tool for single-input–single-output discrete-time systems, and is analogous to the Laplace transform for continuous systems.

Conceptually, the symbol $z$ can be associated with discrete time shifting in a difference equation in the same way that $s$ can be associated with differentiation in a differential equation.

Taking Laplace transforms of equation (7.1), which is the ideal sampled signal, gives

$$F^*(s) = \mathscr{L}[f^*(t)] = \sum_{k=0}^{\infty} f(kT)e^{-kTs} \tag{7.3}$$

or

$$F^*(s) = \sum_{k=0}^{\infty} f(kT)\left(e^{sT}\right)^{-k} \tag{7.4}$$

Define $z$ as

$$z = e^{sT} \tag{7.5}$$

then

$$F(z) = \sum_{k=0}^{\infty} f(kT)z^{-k} = Z[f(t)] \tag{7.6}$$

In 'long-hand' form equation (7.6) is written as

$$F(z) = f(0) + f(T)z^{-1} + f(2T)z^{-2} + \cdots + f(kT)z^{-k} \tag{7.7}$$

*Example 7.1*
Find the $z$-transform of the unit step function $f(t) = 1$.

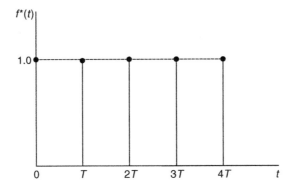

**Fig. 7.7** z-Transform of a sampled unit step function.

*Solution*
From equations (7.6) and (7.7)

$$Z[1(t)] = \sum_{k=0}^{\infty} 1(kT)z^{-k} \tag{7.8}$$

or

$$F(z) = 1 + z^{-1} + z^{-2} + \ldots + z^{-k} \tag{7.9}$$

Figure 7.7 shows a graphical representation of equation (7.9).
Equation (7.9) can be written in 'closed form' as

$$Z[1(t)] = \frac{z}{z-1} = \frac{1}{1-z^{-1}} \tag{7.10}$$

Equations (7.9) and (7.10) can be shown to be the same by long division

$$
\begin{array}{r}
1 + z^{-1} + z^{-2} + \cdots \\
z - 1 \overline{\smash{\big)}\ z \quad\ 0 \quad\ \ 0} \\
\underline{z - 1} \\
0 + 1 \\
\underline{1 - z^{-1}} \\
0 + z^{-1} \\
z^{-1} - z^{-2}
\end{array}
\tag{7.11}
$$

Table 7.1 gives Laplace and z-transforms of common functions.
z-transform Theorems:

(a) Linearity

$$Z[f_1(t) \pm f_2(t)] = F_1(z) \pm F_2(z) \tag{7.12}$$

**Table 7.1** Common Laplace and $z$-transforms

| | $f(t)$ or $f(kT)$ | $F(s)$ | $F(z)$ |
|---|---|---|---|
| 1 | $\delta(t)$ | 1 | 1 |
| 2 | $\delta(t - kT)$ | $e^{-kTs}$ | $z^{-k}$ |
| 3 | $1(t)$ | $\dfrac{1}{s}$ | $\dfrac{z}{z-1}$ |
| 4 | $t$ | $\dfrac{1}{s^2}$ | $\dfrac{Tz}{(z-1)^2}$ |
| 5 | $e^{-at}$ | $\dfrac{1}{(s+a)}$ | $\dfrac{z}{z - e^{-aT}}$ |
| 6 | $1 - e^{-at}$ | $\dfrac{a}{s(s+a)}$ | $\dfrac{z(1 - e^{-aT})}{(z-1)(z - e^{-aT})}$ |
| 7 | $\dfrac{1}{a}(at - 1 + e^{-at})$ | $\dfrac{a}{s^2(s+a)}$ | $\dfrac{z\{(aT - 1 + e^{-aT})z + (1 - e^{-aT} - aTe^{-aT})\}}{a(z-1)^2(z - e^{-aT})}$ |
| 8 | $\sin \omega t$ | $\dfrac{\omega}{s^2 + \omega^2}$ | $\dfrac{z \sin \omega T}{z^2 - 2z \cos \omega T + 1}$ |
| 9 | $\cos \omega t$ | $\dfrac{s}{s^2 + \omega^2}$ | $\dfrac{z(z - \cos \omega T)}{z^2 - 2z \cos \omega T + 1}$ |
| 10 | $e^{-at} \sin \omega t$ | $\dfrac{\omega}{(s+a)^2 + \omega^2}$ | $\dfrac{ze^{-aT} \sin \omega T}{z^2 - 2ze^{-aT} \cos \omega T + e^{-2aT}}$ |
| 11 | $e^{-at} \cos \omega t$ | $\dfrac{(s+a)}{(s+a)^2 + \omega^2}$ | $\dfrac{z^2 - ze^{-aT} \cos \omega T}{z^2 - 2ze^{-aT} \cos \omega T + e^{-2aT}}$ |

(b) Initial Value Theorem

$$f(0) = \lim_{z \to \infty} F(z) \qquad (7.13)$$

(c) Final Value Theorem

$$f(\infty) = \lim_{z \to 1} \left[ \left( \frac{z-1}{z} \right) F(z) \right] \qquad (7.14)$$

## 7.4.1  Inverse transformation

The discrete time response can be found using a number of methods.

### (a) Infinite power series method

*Example 7.2*
A sampled-data system has a transfer function

$$G(s) = \frac{1}{s+1}$$

If the sampling time is one second and the system is subject to a unit step input function, determine the discrete time response. (N.B. normally, a zero-order hold would be included, but, in the interest of simplicity, has been omitted.) Now

$$X_o(z) = G(z)X_i(z) \tag{7.15}$$

from Table 7.1

$$X_o(z) = \left(\frac{z}{z - e^{-T}}\right)\left(\frac{z}{z - 1}\right) \tag{7.16}$$

for $T = 1$ second

$$X_o(z) = \left(\frac{z}{z - 0.368}\right)\left(\frac{z}{z - 1}\right)$$

$$= \frac{z^2}{z^2 - 1.368z + 0.368} \tag{7.17}$$

By long division

$$
\begin{array}{r}
1 + 1.368z^{-1} + 1.503z^{-2} + \cdots \\
z^2 - 1.368z + 0.368 \overline{)z^2 \qquad 0 \qquad\quad 0 \qquad 0} \\
\underline{z^2 - 1.368z + 0.368} \\
0 + 1.368z - 0.368 \\
\underline{1.368z - 1.871 + 0.503z^{-1}} \\
0 + 1.503 - 0.503z^{-1} \\
\underline{1.503 - 2.056z^{-1} + 0.553z^{-2}} \qquad (7.18)
\end{array}
$$

Thus

$$x_o(0) = 1$$

$$x_o(1) = 1.368$$

$$x_o(2) = 1.503$$

### (b) Difference equation method

Consider a system of the form

$$\frac{X_o}{X_i}(z) = \frac{b_0 + b_1 z^{-1} + b_2 z^{-2} + \cdots}{1 + a_1 z^{-1} + a_2 z^{-2} + \cdots} \tag{7.19}$$

Thus

$$(1 + a_1 z^{-1} + a_2 z^{-2} + \cdots)X_o(z) = (b_0 + b_1 z^{-1} + b_2 z^{-2} + \cdots)X_i(z) \tag{7.20}$$

or

$$X_o(z) = (-a_1 z^{-1} - a_2 z^{-2} - \cdots)X_o(z) + (b_0 + b_1 z^{-1} + b_2 z^{-2} + \cdots)X_i(z) \tag{7.21}$$

Equation (7.21) can be expressed as a difference equation of the form

$$x_o(kT) = -a_1 x_o(k-1)T - a_2 x_o(k-2)T - \cdots$$

$$+ b_0 x_i(kT) + b_1 x_i(k-1)T + b_2 x_i(k-2)T + \cdots \tag{7.22}$$

In Example 7.2

$$\frac{X_o}{X_i}(s) = \frac{1}{1+s}$$

$$= \frac{z}{z - e^{-T}} = \frac{z}{z - 0.368} \tag{7.23}$$

Equation (7.23) can be written as

$$\frac{X_o}{X_i}(z) = \frac{1}{1 - 0.368z^{-1}} \tag{7.24}$$

Equation (7.24) is in the same form as equation (7.19). Hence

$$(1 - 0.368z^{-1})X_o(z) = X_i(z)$$

or

$$X_o(z) = 0.368z^{-1}X_o(z) + X_i(z) \tag{7.25}$$

Equation (7.25) can be expressed as a difference equation

$$x_o(kT) = 0.368x_o(k - 1)T + x_i(kT) \tag{7.26}$$

Assume that $x_o(-1) = 0$ and $x_i(kT) = 1$, then from equation (7.26)

$$x_o(0) = 0 + 1 = 1, \quad k = 0$$
$$x_o(1) = (0.368 \times 1) + 1 = 1.368, \quad k = 1$$
$$x_o(2) = (0.368 \times 1.368) + 1 = 1.503, \quad k = 2 \quad \text{etc.}$$

These results are the same as with the power series method, but difference equations are more suited to digital computation.

### 7.4.2  The pulse transfer function

Consider the block diagrams shown in Figure 7.8. In Figure 7.8(a) $U^*(s)$ is a sampled input to $G(s)$ which gives a continuous output $X_o(s)$, which when sampled by a

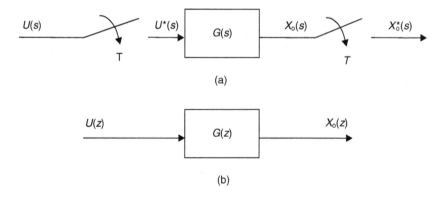

(a)

(b)

**Fig. 7.8** Relationship between $G(s)$ and $G(z)$.

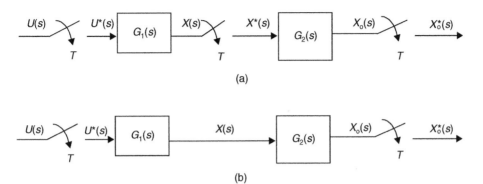

**Fig. 7.9** Blocks in cascade.

synchronized sampler becomes $X_o^*(s)$. Figure 7.8(b) shows the pulse transfer function where $U(z)$ is equivalent to $U^*(s)$ and $X_o(z)$ is equivalent to $X_o^*(s)$.

From Figure 7.8(b) the pulse transfer function is

$$\frac{X_o}{U}(z) = G(z) \tag{7.27}$$

*Blocks in Cascade*: In Figure 7.9(a) there are synchronized samplers either side of blocks $G_1(s)$ and $G_2(s)$. The pulse transfer function is therefore

$$\frac{X_o}{U}(z) = G_1(z)G_2(z) \tag{7.28}$$

In Figure 7.9(b) there is no sampler between $G_1(s)$ and $G_2(s)$ so they can be combined to give $G_1(s)G_2(s)$, or $G_1G_2(s)$. Hence the output $X_o(z)$ is given by

$$X_o(z) = Z\{G_1G_2(s)\}U(z) \tag{7.29}$$

and the pulse transfer function is

$$\frac{X_o}{U}(z) = G_1G_2(z) \tag{7.30}$$

Note that $G_1(z)G_2(z) \neq G_1G_2(z)$.

*Example 7.3* (See also Appendix 1, *examp73.m*)
A first-order sampled-data system is shown in Figure 7.10.
Find the pulse transfer function and hence calculate the response to a unit step and unit ramp. $T = 0.5$ seconds. Compare the results with the continuous system response $x_o(t)$. The system is of the type shown in Figure 7.9(b) and therefore

$$G(s) = G_1G_2(s)$$

Inserting values

$$G(s) = (1 - e^{-Ts})\left\{\frac{1}{s(s+1)}\right\} \tag{7.31}$$

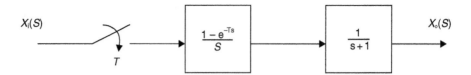

**Fig. 7.10** First-order sampled-data system.

Taking $z$-transforms using Table 7.1.

$$G(z) = (1 - z^{-1})\left\{\frac{z(1 - e^{-T})}{(z-1)(z - e^{-T})}\right\}$$  (7.32)

or

$$G(z) = \left(\frac{z-1}{z}\right)\left\{\frac{z(1 - e^{-T})}{(z-1)(z - e^{-T})}\right\}$$  (7.33)

which gives

$$G(z) = \left(\frac{1 - e^{-T}}{z - e^{-T}}\right)$$  (7.34)

For $T = 0.5$ seconds

$$G(z) = \left(\frac{0.393}{z - 0.607}\right)$$  (7.35)

hence

$$\frac{X_o}{X_i}(z) = \left(\frac{0.393z^{-1}}{1 - 0.607z^{-1}}\right)$$  (7.36)

which is converted into a difference equation

$$x_o(kT) = 0.607x_o(k - 1)T + 0.393x_i(k - 1)T$$  (7.37)

Table 7.2 shows the discrete response $x_o(kT)$ to a unit step function and is compared with the continuous response (equation 3.29) where

$$x_o(t) = (1 - e^{-t})$$  (7.38)

From Table 7.2, it can be seen that the discrete and continuous step response is identical. Table 7.3 shows the discrete response $x(kT)$ and continuous response $x(t)$ to a unit ramp function where $x_o(t)$ is calculated from equation (3.39)

$$x_o(t) = t - 1 + e^{-t}$$  (7.39)

In Table 7.3 the difference between $x_o(kT)$ and $x_o(t)$ is due to the sample and hold. It should also be noted that with the discrete response $x(kT)$, there is only knowledge of the output at the sampling instant.

**Table 7.2** Comparison between discrete and continuous step response

| $k$ | $kT$ (seconds) | $x_i(kT)$ | $x_o(kT)$ | $x_o(t)$ |
|---|---|---|---|---|
| −1 | − 0.5 | 0 | 0 | 0 |
| 0 | 0 | 1 | 0 | 0 |
| 1 | 0.5 | 1 | 0.393 | 0.393 |
| 2 | 1.0 | 1 | 0.632 | 0.632 |
| 3 | 1.5 | 1 | 0.776 | 0.776 |
| 4 | 2.0 | 1 | 0.864 | 0.864 |
| 5 | 2.5 | 1 | 0.918 | 0.918 |
| 6 | 3.0 | 1 | 0.950 | 0.950 |
| 7 | 3.5 | 1 | 0.970 | 0.970 |
| 8 | 4.0 | 1 | 0.982 | 0.982 |

**Table 7.3** Comparison between discrete and continuous ramp response

| $k$ | $kT$ (seconds) | $x_i(kT)$ | $x_o(kT)$ | $x_o(t)$ |
|---|---|---|---|---|
| −1 | − 0.5 | 0 | 0 | 0 |
| 0 | 0 | 0 | 0 | 0 |
| 1 | 0.5 | 0.5 | 0 | 0.107 |
| 2 | 1.0 | 1.0 | 0.304 | 0.368 |
| 3 | 1.5 | 1.5 | 0.577 | 0.723 |
| 4 | 2.0 | 2.0 | 0.940 | 1.135 |
| 5 | 2.5 | 2.5 | 1.357 | 1.582 |
| 6 | 3.0 | 3.0 | 1.805 | 2.050 |
| 7 | 3.5 | 3.5 | 2.275 | 2.530 |
| 8 | 4.0 | 4.0 | 2.757 | 3.018 |

### 7.4.3 The closed-loop pulse transfer function

Consider the error sampled system shown in Figure 7.11. Since there is no sampler between $G(s)$ and $H(s)$ in the closed-loop system shown in Figure 7.11, it is a similar arrangement to that shown in Figure 7.9(b). From equation (4.4), the closed-loop pulse transfer function can be written as

$$\frac{C}{R}(z) = \frac{G(z)}{1 + GH(z)} \tag{7.40}$$

In equation (7.40)

$$GH(z) = Z\{GH(s)\} \tag{7.41}$$

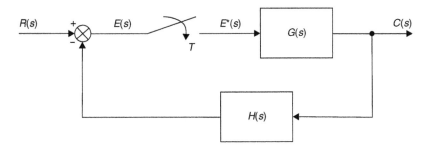

**Fig. 7.11** Closed-loop error sampled system.

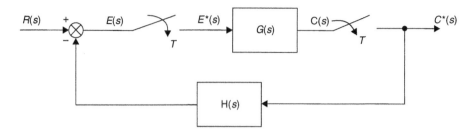

**Fig. 7.12** Closed-loop error and output sampled system.

Consider the error and output sampled system shown in Figure 7.12. In Figure 7.12, there is now a sampler between $G(s)$ and $H(s)$, which is similar to Figure 7.9(a). From equation (4.4), the closed-loop pulse transfer function is now written as

$$\frac{C}{R}(z) = \frac{G(z)}{1 + G(z)H(z)} \tag{7.42}$$

## 7.5 Digital control systems

From Figure 7.1, a digital control system may be represented by the block diagram shown in Figure 7.13.

*Example 7.4* (See also Appendix 1, *examp74.m*)
Figure 7.14 shows a digital control system. When the controller gain $K$ is unity and the sampling time is 0.5 seconds, determine

(a) the open-loop pulse transfer function
(b) the closed-loop pulse transfer function
(c) the difference equation for the discrete time response
(d) a sketch of the unit step response assuming zero initial conditions
(e) the steady-state value of the system output

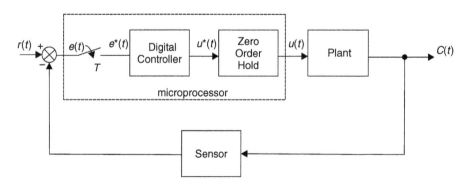

**Fig. 7.13** Digital control system.

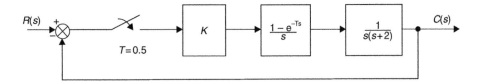

**Fig. 7.14** Digital control system for Example 7.4.

*Solution*

(a)   $G(s) = K\left(\dfrac{1 - e^{-Ts}}{s}\right)\left\{\dfrac{1}{s(s+2)}\right\}$                (7.43)

Given $K = 1$

$$G(s) = \left(1 - e^{-Ts}\right)\left\{\dfrac{1}{s^2(s+2)}\right\}$$                (7.44)

Partial fraction expansion

$$\dfrac{1}{s^2(s+2)} = \left\{\dfrac{A}{s} + \dfrac{B}{s^2} + \dfrac{C}{(s+2)}\right\}$$                (7.45)

or

$$1 = s(s+2)A + (s+2)B + s^2C$$                (7.46)

Equating coefficients gives

$$A = -0.25$$
$$B = 0.5$$
$$C = 0.25$$

Substituting these values into equation (7.44) and (7.45)

$$G(s) = \left(1 - e^{-Ts}\right)\left\{\dfrac{-0.25}{s} + \dfrac{0.5}{s^2} + \dfrac{0.25}{(s+2)}\right\}$$                (7.47)

or

$$G(s) = 0.25\left(1 - e^{-Ts}\right)\left\{-\dfrac{1}{s} + \dfrac{2}{s^2} + \dfrac{1}{(s+2)}\right\}$$                (7.48)

Taking $z$-transforms

$$G(z) = 0.25\left(1 - z^{-1}\right)\left\{\dfrac{-z}{(z-1)} + \dfrac{2Tz}{(z-1)^2} + \dfrac{z}{(z - e^{-2T})}\right\}$$                (7.49)

Given $T = 0.5$ seconds

$$G(z) = 0.25\left(\dfrac{z-1}{z}\right)z\left\{\dfrac{-1}{(z-1)} + \dfrac{2 \times 0.5}{(z-1)^2} + \dfrac{1}{(z - 0.368)}\right\}$$                (7.50)

Hence

$$G(z) = 0.25(z-1) \left\{ \frac{-1(z-1)(z-0.368) + (z-0.368) + (z-1)^2}{(z-1)^2(z-0.368)} \right\} \quad (7.51)$$

$$G(z) = 0.25 \left\{ \frac{-z^2 + 1.368z - 0.368 + z - 0.368 + z^2 - 2z + 1}{(z-1)(z-0.368)} \right\} \quad (7.52)$$

which simplifies to give the open-loop pulse transfer function

$$G(z) = \left( \frac{0.092z + 0.066}{z^2 - 1.368z + 0.368} \right) \quad (7.53)$$

*Note*: This result could also have been obtained at equation (7.44) by using $z$-transform number 7 in Table 7.1, but the solution demonstrates the use of partial fractions.

(b) The closed-loop pulse transfer function, from equation (7.40) is

$$\frac{C}{R}(z) = \frac{\left( \frac{0.092z + 0.066}{z^2 - 1.368z + 0.368} \right)}{\left( 1 + \frac{0.092z + 0.066}{z^2 - 1.368z + 0.368} \right)} \quad (7.54)$$

which simplifies to give the closed-loop pulse transfer function

$$\frac{C}{R}(z) = \frac{0.092z + 0.066}{z^2 - 1.276z + 0.434} \quad (7.55)$$

or

$$\frac{C}{R}(z) = \frac{0.092z^{-1} + 0.066z^{-2}}{1 - 1.276z^{-1} + 0.434z^{-2}} \quad (7.56)$$

(c) Equation (7.56) can be expressed as a difference equation

$$c(kT) = 1.276c(k-1)T - 0.434c(k-2)T + 0.092r(k-1)T + 0.066r(k-2)T \quad (7.57)$$

(d) Using the difference equation (7.57), and assuming zero initial conditions, the unit step response is shown in Figure 7.15.

Note that the response in Figure 7.15 is constructed solely from the knowledge of the two previous sampled outputs and the two previous sampled inputs.

(e) Using the final value theorem given in equation (7.14)

$$c(\infty) = \lim_{z \to 1} \left[ \left( \frac{z-1}{z} \right) \frac{C}{R}(z) R(z) \right] \quad (7.58)$$

$$c(\infty) = \lim_{z \to 1} \left[ \left( \frac{z-1}{z} \right) \left\{ \frac{0.092z + 0.066}{1 - 1.276z + 0.434} \right\} \left( \frac{z}{z-1} \right) \right] \quad (7.59)$$

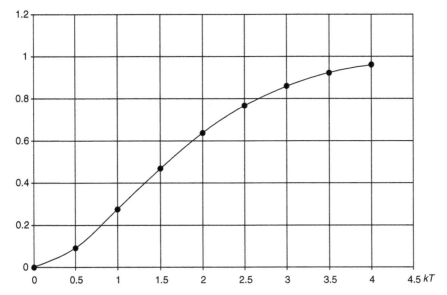

$c(kT)$

**Fig. 7.15** Unit step response for Example 7.4.

$$c(\infty) = \left( \frac{0.092 + 0.066}{1 - 1.276 + 0.434} \right) = 1.0 \tag{7.60}$$

Hence there is no steady-state error.

## 7.6   Stability in the *z*-plane

### 7.6.1   Mapping from the *s*-plane into the *z*-plane

Just as transient analysis of continuous systems may be undertaken in the *s*-plane, stability and transient analysis on discrete systems may be conducted in the *z*-plane.
   It is possible to map from the *s* to the *z*-plane using the relationship

$$z = e^{sT} \tag{7.61}$$

now

$$s = \sigma \pm j\omega$$

therefore

$$z = e^{(\sigma \pm j\omega)T} = e^{\sigma T} e^{j\omega T} \quad \text{(using the positive j}\omega \text{ value)} \tag{7.62}$$

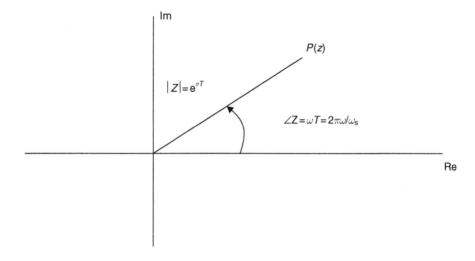

**Fig. 7.16** Mapping from the $s$ to the $z$-plane.

If $e^{\sigma T} = |z|$ and $T = 2\pi/\omega_s$ equation (7.62) can be written

$$z = |z|e^{j(2\pi\omega/\omega_s)} \qquad (7.63)$$

where $\omega_s$ is the sampling frequency.

Equation (7.63) results in a polar diagram in the $z$-plane as shown in Figure 7.16. Figure 7.17 shows mapping of lines of constant $\sigma$ (i.e. constant settling time) from the $s$ to the $z$-plane. From Figure 7.17 it can be seen that the left-hand side (stable) of the $s$-plane corresponds to a region within a circle of unity radius (the unit circle) in the $z$-plane.

Figure 7.18 shows mapping of lines of constant $\omega$ (i.e. constant transient frequency) from the $s$ to the $z$-plane.

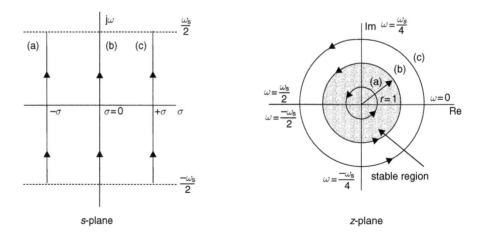

**Fig. 7.17** Mapping constant $\sigma$ from $s$ to $z$-plane.

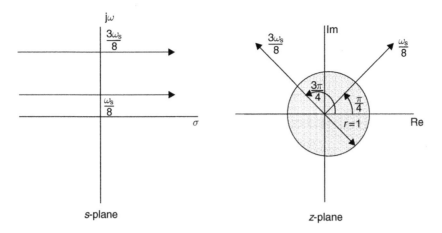

**Fig. 7.18** Mapping constant $\omega$ from $s$ to $z$-plane.

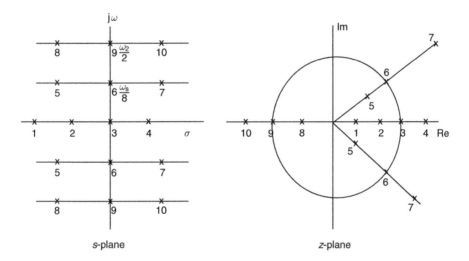

**Fig. 7.19** Corresponding pole locations on both $s$ and $z$-planes.

Figure 7.19 shows corresponding pole locations on both the $s$-plane and $z$-plane.

### 7.6.2 The Jury stability test

In the same way that the Routh–Hurwitz criterion offers a simple method of determining the stability of continuous systems, the Jury (1958) stability test is employed in a similar manner to assess the stability of discrete systems.

Consider the characteristic equation of a sampled-data system

$$Q(z) = a_n z^n + a_{n-1} z^{n-1} + \cdots + a_1 z + a_0 = 0 \qquad (7.64)$$

**Table 7.4** Jury's array

| $z^0$ | $z^1$ | $z^2$ | | $z^{n-1}$ | $z^n$ |
|-------|-------|-------|---|-----------|-------|
| $a_0$ | $a_1$ | $a_2$ | ... | $a_{n-1}$ | $a_n$ |
| $a_n$ | $a_{n-1}$ | $a_{n-2}$ | ... | $a_1$ | $a_0$ |
| $b_0$ | $b_1$ | $b_2$ | ... | $b_{n-1}$ | |
| $b_{n-1}$ | $b_{n-2}$ | $b_{n-3}$ | ... | $b_0$ | |
| . | | | | | |
| . | | | | | |
| . | | | | | |
| $l_0$ | $l_1$ | $l_2$ | ... | $l_3$ | |
| $l_3$ | $l_2$ | $l_1$ | ... | $l_0$ | |
| $m_0$ | $m_1$ | $m_2$ | | | |
| $m_2$ | $m_1$ | $m_0$ | | | |

The array for the Jury stability test is given in Table 7.4 where

$$b_k = \begin{vmatrix} a_0 & a_{n-k} \\ a_n & a_k \end{vmatrix}$$

$$c_k = \begin{vmatrix} b_0 & b_{n-1-k} \\ b_{n-1} & b_k \end{vmatrix} \tag{7.65}$$

$$d_k = \begin{vmatrix} c_0 & c_{n-2-k} \\ c_{n-2} & c_k \end{vmatrix}$$

The necessary and sufficient conditions for the polynomial $Q(z)$ to have no roots outside or on the unit circle are

$$
\begin{aligned}
&\text{Condition 1} \quad Q(1) > 0 \\
&\text{Condition 2} \quad (-1)^n Q(-1) > 0 \\
&\text{Condition 3} \quad |a_0| < a_n \\
&\qquad\qquad\quad\; \vdots \quad\; |b_0| > |b_{n-1}| \\
&\qquad\qquad\quad\; \vdots \quad\; |c_0| > |c_{n-2}| \\
&\qquad\qquad\qquad\qquad\quad \vdots \\
&\qquad\qquad\qquad\qquad\quad \vdots \\
&\text{Condition } n \quad |m_0| > |m_2|
\end{aligned}
\tag{7.66}
$$

*Example 7.5*  (See also Appendix 1, *examp75.m*)
For the system given in Figure 7.14 (i.e. Example 7.4) find the value of the digital compensator gain $K$ to make the system just unstable. For Example 7.4, the characteristic equation is

$$1 + G(z) = 0 \tag{7.67}$$

In Example 7.4, the solution was found assuming that $K = 1$. Therefore, using equation (7.53), the characteristic equation is

$$1 + \frac{K(0.092z + 0.066)}{(z^2 - 1.368z + 0.368)} = 0 \tag{7.68}$$

or

$$Q(z) = z^2 + (0.092K - 1.368)z + (0.368 + 0.066K) = 0 \qquad (7.69)$$

The first row of Jury's array is

$$
\begin{array}{c|ccc}
 & z^0 & z^1 & z^2 \\
\hline
 & (0.368 + 0.066K) & (0.092K - 1.368) & 1
\end{array}
\qquad (7.70)
$$

*Condition 1:* $Q(1) > 0$
From equation (7.69)

$$Q(1) = \{1 + (0.092K - 1.368) + (0.368 + 0.066K)\} > 0 \qquad (7.71)$$

From equation (7.71), $Q(1) > 0$ if $K > 0$.

*Condition 2* $(-1)^n Q(-1) > 0$
From equation (7.69), when $n = 2$

$$(-1)^2 Q(-1) = \{1 - (0.092K - 1.368) + (0.368 + 0.066K)\} > 0 \qquad (7.72)$$

Equation (7.72) simplifies to give

$$2.736 - 0.026K > 0$$

or

$$K < \frac{2.736}{0.026} = 105.23 \qquad (7.73)$$

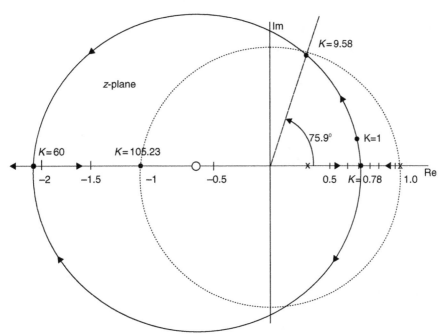

**Fig. 7.20** Root locus diagram for Example 7.4.

*Condition 3:* $|a_0| < a_2$

$$|0.368 + 0.066K| < 1 \qquad (7.74)$$

For marginal stability

$$0.368 + 0.066K = 1$$
$$K = \frac{1 - 0.368}{0.066} = 9.58 \qquad (7.75)$$

Hence the system is marginally stable when $K = 9.58$ and $105.23$ (see also Example 7.6 and Figure 7.20).

### 7.6.3 Root locus analysis in the *z*-plane

As with the continuous systems described in Chapter 5, the root locus of a discrete system is a plot of the locus of the roots of the characteristic equation

$$1 + GH(z) = 0 \qquad (7.76)$$

in the *z*-plane as a function of the open-loop gain constant $K$. The closed-loop system will remain stable providing the loci remain within the unit circle.

### 7.6.4 Root locus construction rules

These are similar to those given in section 5.3.4 for continuous systems.

1. *Starting points* ($K = 0$): The root loci start at the open-loop poles.
2. *Termination points* ($K = \infty$): The root loci terminate at the open-loop zeros when they exist, otherwise at $\infty$.
3. *Number of distinct root loci*: This is equal to the order of the characteristic equation.
4. *Symmetry of root loci*: The root loci are symmetrical about the real axis.
5. *Root locus locations on real axis*: A point on the real axis is part of the loci if the sum of the open-loop poles and zeros to the right of the point concerned is odd.
6. *Breakaway points*: The points at which a locus breaks away from the real axis can be found by obtaining the roots of the equation

$$\frac{\mathrm{d}}{\mathrm{d}z}\{GH(z)\} = 0$$

7. *Unit circle crossover*: This can be obtained by determining the value of $K$ for marginal stability using the Jury test, and substituting it in the characteristic equation (7.76).

*Example 7.6* (See also Appendix 1, *examp76.m*)
Sketch the root locus diagram for Example 7.4, shown in Figure 7.14. Determine the breakaway points, the value of $K$ for marginal stability and the unit circle crossover.

*Solution*
From equation (7.43)

$$G(s) = K\left(1 - \frac{e^{-Ts}}{s}\right)\left\{\frac{1}{s(s+2)}\right\} \tag{7.77}$$

and from equation (7.53), given that $T = 0.5$ seconds

$$G(z) = K\left(\frac{0.092z + 0.066}{z^2 - 1.368z + 0.368}\right) \tag{7.78}$$

Open-loop poles

$$z^2 - 1.368z + 0.368 = 0 \tag{7.79}$$

$$z = 0.684 \pm 0.316$$
$$= 1 \text{ and } 0.368 \tag{7.80}$$

Open-loop zeros

$$0.092z + 0.066 = 0$$

$$z = -0.717 \tag{7.81}$$

From equations (7.67), (7.68) and (7.69) the characteristic equation is

$$z^2 + (0.092K - 1.368)z + (0.368 + 0.066K) = 0 \tag{7.82}$$

*Breakaway points*: Using Rule 6

$$\frac{d}{dz}\{GH(z)\} = 0$$

$$(z^2 - 1.368z + 0.368)K(0.092) - K(0.092z + 0.066)(2z - 1.368) = 0 \tag{7.83}$$

which gives

$$0.092z^2 + 0.132z - 0.1239 = 0$$
$$z = 0.647 \text{ and } -2.084 \tag{7.84}$$

*K for marginal stability*: Using the Jury test, the values of $K$ as the locus crosses the unit circle are given in equations (7.75) and (7.73)

$$K = 9.58 \text{ and } 105.23 \tag{7.85}$$

*Unit circle crossover*: Inserting $K = 9.58$ into the characteristic equation (7.82) gives

$$z^2 - 0.487z + 1 = 0 \tag{7.86}$$

The roots of equation (7.86) are

$$z = 0.244 \pm j0.97 \tag{7.87}$$

or

$$z = 1\angle \pm 75.9° = 1\angle \pm 1.33 \text{ rad} \tag{7.88}$$

Since from equation (7.63) and Figure 7.16

$$z = |z| \angle \omega T \tag{7.89}$$

and $T = 0.5$, then the frequency of oscillation at the onset of instability is

$$0.5\omega = 1.33$$
$$\omega = 2.66 \,\text{rad/s} \tag{7.90}$$

The root locus diagram is shown in Figure 7.20.

It can be seen from Figure 7.20 that the complex loci form a circle. This is usually the case for second-order plant, where

$$\text{Radius} = \sum |\text{open-loop poles}|$$
$$\text{Centre} = (\text{Open-loop zero, } 0) \tag{7.91}$$

The step response shown in Figure 7.15 is for $K = 1$. Inserting $K = 1$ into the characteristic equation gives

$$z^2 - 1.276z + 0.434 = 0$$

or

$$z = 0.638 \pm \text{j}0.164$$

This position is shown in Figure 7.20. The $K$ values at the breakaway points are also shown in Figure 7.20.

## 7.7 Digital compensator design

In sections 5.4 and 6.6, compensator design in the $s$-plane and the frequency domain were discussed for continuous systems. In the same manner, digital compensators may be designed in the $z$-plane for discrete systems.

Figure 7.13 shows the general form of a digital control system. The pulse transfer function of the digital controller/compensator is written

$$\frac{U}{E}(z) = D(z) \tag{7.92}$$

and the closed-loop pulse transfer function become

$$\frac{C}{R}(z) = \frac{D(z)G(z)}{1 + D(z)GH(z)} \tag{7.93}$$

and hence the characteristic equation is

$$1 + D(z)GH(z) = 0 \tag{7.94}$$

## 7.7.1 Digital compensator types

In a continuous system, a differentiation of the error signal $e$ can be represented as

$$u(t) = \frac{de}{dt}$$

Taking Laplace transforms with zero initial conditions

$$\frac{U}{E}(s) = s \tag{7.95}$$

In a discrete system, a differentiation can be approximated to

$$u(kT) = \frac{e(kT) - e(k-1)T}{T}$$

hence

$$\frac{U}{E}(z) = \frac{1 - z^{-1}}{T} \tag{7.96}$$

Hence, the Laplace operator can be approximated to

$$s = \frac{1 - z^{-1}}{T} = \frac{z - 1}{Tz} \tag{7.97}$$

*Digital PID controller*: From equation (4.92), a continuous PID controller can be written as

$$\frac{U}{E}(s) = \frac{K_1(T_i T_d s^2 + T_i s + 1)}{T_i s} \tag{7.98}$$

Inserting equation (7.97) into (7.98) gives

$$\frac{U}{E}(z) = \frac{K_1\left\{T_i T_d\left(\frac{z-1}{Tz}\right)^2 + T_i\left(\frac{z-1}{Tz}\right) + 1\right\}}{T_i\left(\frac{z-1}{Tz}\right)} \tag{7.99}$$

which can be simplified to give

$$\frac{U}{E}(z) = \frac{K_1(b_2 z^2 + b_1 z + b_0)}{z(z - 1)} \tag{7.100}$$

where

$$b_0 = \frac{T_d}{T}$$

$$b_1 = \left(1 - \frac{2T_d}{T}\right) \tag{7.101}$$

$$b_2 = \left(\frac{T_d}{T} + \frac{T}{T_i} + 1\right)$$

*Tustin's Rule:* Tustin's rule, also called the bilinear transformation, gives a better approximation to integration since it is based on a trapizoidal rather than a rectangular area. Tustin's rule approximates the Laplace transform to

$$s = \frac{2(z-1)}{T(z+1)} \tag{7.102}$$

Inserting this value of $s$ into the denominator of equation (7.98), still yields a digital PID controller of the form shown in equation (7.100) where

$$b_0 = \frac{T_d}{T}$$

$$b_1 = \left(\frac{T}{2T_i} - \frac{2T_d}{T} - 1\right) \tag{7.103}$$

$$b_2 = \left(\frac{T}{2T_i} + \frac{T_d}{T} + 1\right)$$

*Example 7.7* (See also Appendix 1, *examp77.m*)
The laser guided missile shown in Figure 5.26 has an open-loop transfer function (combining the fin dynamics and missile dynamics) of

$$G(s)H(s) = \frac{20}{s^2(s+5)} \tag{7.104}$$

A lead compensator, see case study Example 6.6, and equation (6.113) has a transfer function of

$$G(s) = \frac{0.8(1+s)}{(1+0.0625s)} \tag{7.105}$$

(a) Find the $z$-transform of the missile by selecting a sampling frequency of at least 10 times higher than the system bandwidth.
(b) Convert the lead compensator in equation (7.105) into a digital compensator using the simple method, i.e. equation (7.97) and find the step response of the system.
(c) Convert the lead compensator in equation (7.105) into a digital compensator using Tustin's rule, i.e. equation (7.102) and find the step response of the system.
(d) Compare the responses found in (b) and (c) with the continuous step response, and convert the compensator that is closest to this into a difference equation.

*Solution*
(a) From Figure 6.39, lead compensator two, the bandwidth is 5.09 rad/s, or 0.81 Hz. Ten times this is 8.1 Hz, so select a sampling frequency of 10 Hz, i.e.

$T = 0.1$ seconds. For a sample and hold device cascaded with the missile dynamics

$$G(s) = \left(\frac{1 - e^{-Ts}}{s}\right)\left\{\frac{20}{s^2(s+5)}\right\} \tag{7.106}$$

$$G(s) = (1 - e^{-Ts})\left\{\frac{20}{s^3(s+5)}\right\} \tag{7.107}$$

For $T = 0.1$, equation (7.107) has a $z$-transform of

$$G(z) = \frac{0.00296z^2 + 0.01048z + 0.0023}{z^3 - 2.6065z^2 + 2.2131z - 0.6065} \tag{7.108}$$

(b) Substituting

$$s = \frac{z - 1}{Tz}$$

into lead compensator given in equation (7.105) to obtain digital compensator

$$D(z) = 0.8\left\{\frac{\frac{Tz+(z-1)}{Tz}}{\frac{Tz+0.0625(z-1)}{Tz}}\right\}$$

This simplifies to give

$$D(z) = \frac{5.4152z - 4.923}{z - 0.3846} \tag{7.109}$$

(c) Using Tustin's rule

$$s = \frac{2(z - 1)}{T(z + 1)}$$

Substituting into lead compensator

$$D(z) = 0.8\left[\frac{\frac{T(z+1)+2(z-1)}{T(z+1)}}{\frac{T(z+1)+0.0625\{2(z-1)\}}{T(z+1)}}\right]$$

This simplifies to give

$$D(z) = \frac{7.467z - 6.756}{z - 0.111} \tag{7.110}$$

(d) From Figure 7.21, it can be seen that the digital compensator formed using Tustin's rule is closest to the continuous response. From equation (7.110)

$$\frac{U}{E}(z) = \frac{7.467 - 6.756z^{-1}}{1 - 0.111z^{-1}} \tag{7.111}$$

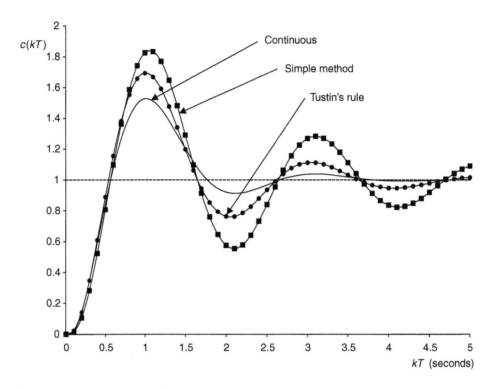

**Fig. 7.21** Comparison between discrete and continuous response.

Hence the difference equation for the digital compensator is

$$u(kT) = 0.111u(k-1)T + 7.467e(kT) - 6.756e(k-1)T \qquad (7.112)$$

## 7.7.2 Digital compensator design using pole placement

### Case study

*Example 7.8* (See also Appendix 1, *examp78.m*)
The continuous control system shown in Figure 7.22(a) is to be replaced by the digital control system shown in Figure 7.22(b).

(a) For the continuous system, find the value of $K$ that gives the system a damping ratio of 0.5. Determine the closed-loop poles in the $s$-plane and hence the values of $\sigma$ and $\omega$.
(b) Find the closed-loop bandwidth $\omega_b$ and make the sampling frequency $\omega_s$ a factor of 10 higher. What is the value of $T$?
(c) For the sampled system shown in Figure 7.22(b), find the open-loop pulse transfer function $G(z)$ when the sample and hold device is in cascade with the plant.
(d) With $D(z)$ set to the value of $K$ found in (a), compare the continuous and discrete step responses.

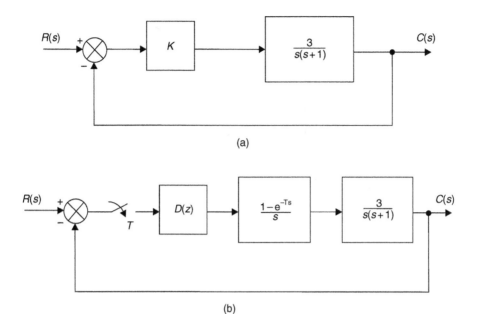

(a)

(b)

**Fig. 7.22** Continuous and digital control systems.

(e) By mapping the closed-loop poles from the $s$ to the $z$-plane, design a compensator $D(z)$ such that both continuous and sampled system have identical closed-loop response, i.e. $\zeta = 0.5$.

*Solution*

(a) The root-locus diagram for the continuous system in shown in Figure 7.23. From Figure 7.23 the closed-loop poles are

$$s = -0.5 \pm j0.866 \qquad (7.113)$$

or

$$\sigma = -0.5, \quad \omega = 0.866\,\text{rad/s}$$

and the value of $K$ is 0.336.

(b) Plotting the closed-loop frequency response for the continuous system gives a bandwidth $\omega_b$ of 1.29 rad/s(0.205 Hz). The sampling frequency should therefore be a factor of 10 higher, i.e. 12.9 rad/s(2.05 Hz). Rounding down to 2.0 Hz gives a sampling time $T$ of 0.5 seconds.

(c)
$$G(z) = (1 - z^{-1})Z\left\{\frac{3}{s^2(s+1)}\right\} \qquad (7.114)$$

Using transform 7 in Table 7.1

$$G(z) = \frac{3\{(e^{-0.5} - 0.5)z + (1 - 1.5e^{-0.5})\}}{(z-1)(z - e^{-0.5})}$$

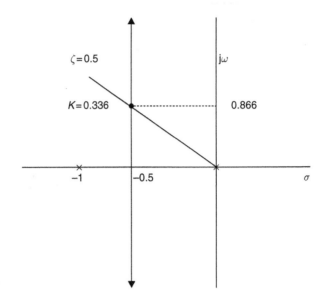

**Fig. 7.23** Root locus diagram for continuous system.

Hence

$$G(z) = \frac{0.3196(z + 0.8467)}{(z - 1)(z - 0.6065)} \quad (7.115)$$

(d) With $D(z) = K = 0.336$, the difference between the continuous and discrete step response can be seen in Figure 7.24.

(e) Mapping closed-loop poles from s to z-plane

$$|z| = e^{\sigma T}$$

inserting values

$$|z| = e^{-0.5 \times 0.5} = 0.779 \quad (7.116)$$

$$\angle z = \omega T$$

$$= 0.866 \times 0.5 = 0.433 \text{ rad} \quad (7.117)$$

$$= 24.8° $$

Converting from polar to cartesian co-ordinates gives the closed-loop poles in the z-plane

$$z = 0.707 \pm \text{j}0.327 \quad (7.118)$$

which provides a z-plane characteristic equation

$$z^2 - 1.414z + 0.607 = 0 \quad (7.119)$$

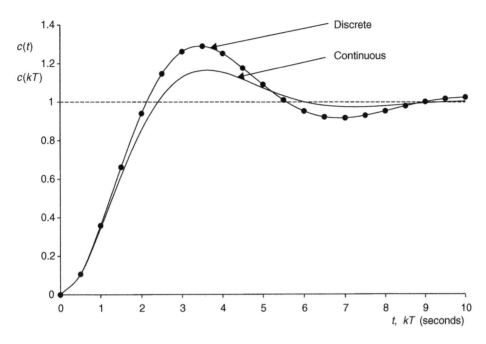

**Fig. 7.24** Continuous and digital controllers set to $K = 0.336$.

The control problem is to design a compensator $D(z)$, which, when cascaded with $G(z)$, provides a characteristic equation

$$1 + D(z)G(z) = 0 \qquad (7.120)$$

such that the equations (7.119) and (7.120) are identical. Let the compensator be of the form

$$D(z) = \frac{K(z - a)}{(z + b)} \qquad (7.121)$$

Select the value of $a$ so that the non-unity pole in $G(z)$ is cancelled

$$D(z)G(z) = \frac{K(z - 0.6065)}{(z + b)} \cdot \frac{0.3196(z + 0.8467)}{(z - 1)(z - 0.6065)} \qquad (7.122)$$

Hence the characteristic equation (7.120) becomes

$$1 + \frac{0.3196K(z + 0.8467)}{(z + b)(z - 1)} = 0$$

which simplifies to give

$$z^2 + (0.3196K + b - 1)z + (0.2706K - b) = 0 \qquad (7.123)$$

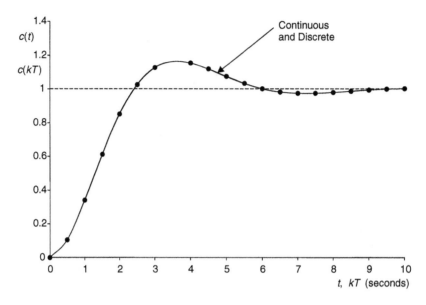

**Fig. 7.25** Identical continuous and discrete step responses as a result of pole placement.

Equating coefficients in equations (7.119) and (7.123) gives

$$0.3196K + b - 1 = -1.414 \qquad (7.124)$$
$$0.2706K - b \quad\;\; = 0.607 \qquad (7.125)$$

Add  $0.5902K - 1 \quad\;\; = -0.807$

or

$$0.5902K = 0.193$$
$$K = 0.327 \qquad (7.126)$$

Inserting equation (7.126) into (7.125)

$$(0.2706 \times 0.327) - 0.607 = b$$
$$b = -0.519 \qquad (7.127)$$

Thus the required compensator is

$$D(z) = \frac{U}{E}(z) = \frac{0.327(z - 0.6065)}{(z - 0.519)} \qquad (7.128)$$

Figure 7.25 shows that the continuous and discrete responses are identical, both with $\zeta = 0.5$. The control algorithm can be implemented as a difference equation

$$\frac{U}{E}(z) = 0.327\frac{(1 - 0.6065z^{-1})}{(1 - 0.519z^{-1})} \qquad (7.129)$$

hence

$$u(kT) = 0.327e(kT) - 0.1983e(k - 1)T + 0.519u(k - 1)T \qquad (7.130)$$

## 7.8 Further problems

*Example 7.9*
Assuming that a sample and hold device is in cascade with the transfer function $G(s)$, determine $G(z)$ for the following

(a) $G(s) = \dfrac{1}{(s+1)}$, $\quad T = 0.1$ seconds

(b) $G(s) = \dfrac{2}{(s+1)(s+2)}$, $\quad T = 0.5$ seconds

(c) $G(s) = \dfrac{1}{s(s+0.5)}$, $\quad T = 1.0$ seconds

*Solution*

(a) $G(z) = \dfrac{0.095}{z - 0.905}$

(b) $G(z) = \dfrac{0.155(z + 0.606)}{z^2 - 0.974z + 0.223}$

(c) $G(z) = \dfrac{0.426(z + 0.847)}{z^2 - 1.607z + 0.607}$

*Example 7.10*
The computer control system shown in Figure 7.26 has a sampling time of 0.5 seconds

(a) Find the open-loop pulse transfer function $G(z)$ and hence determine the open-loop poles and zeros for the combined sample and hold and the plant.
(b) From (a) evaluate the difference equation relating $c(kT)$, $c(k-1)T$, $c(k-2)T$, $u(k-1)T$ and $u(k-2)T$.
(c) If the computer has the control algorithm

$$u(kT) = 1.5e(kT)$$

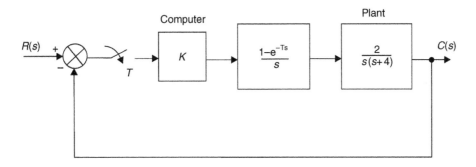

**Fig. 7.26** Computer control system for Example 7.10.

using a tabular approach, calcuate the system response when the input is a unit step applied at $kT = 0$ for the discrete time values of $kT = 0, 0.5, 1.0, 1.5, 2.0$ and 2.5 seconds. Assume that at $kT$ less than zero, all values of input and output are zero.

*Solution*

(a) $G(z) = \dfrac{0.1419(z + 0.523)}{z^2 - 1.135z + 0.135}$

poles   $z = 1, 0.135$

zeros   $z = -0.523$

(b) $c(kT) = 1.135c(k - 1)T - 0.135c(k - 2)T + 0.1419u(k - 1)T + 0.0743u(k - 2)T$

(c)

| $kT$ | 0 | 0.5 | 1.0 | 1.5 | 2.0 | 2.5 |
|------|---|-----|-----|-----|-----|-----|
| $c(kT)$ | 0 | 0.213 | 0.521 | 0.752 | 0.889 | 0.959 |

*Example 7.11*

A unity feedback computer control system, has an open-loop pulse transfer function

$$G(z) = \frac{0.426K(z + 0.847)}{z^2 - 1.607z + 0.607}$$

(a) Determine the open-loop poles and zeros, the characteristic equation and break-away points.
(b) Using the Jury test, determine the value of $K$ at the unit-circle crossover points.
(c) Find the radius and centre of the circular complex loci, and hence sketch the root locus in the $z$-plane.

*Solution*

(a) poles   $z = 1, 0.607$

zeros   $z = -0.847$

$$z^2 + (0.426K - 1.607)z + (0.361K + 0.607) = 0$$

$$\text{breakaway points}\quad z = 0.795, -2.5$$

(b) $K = 1.06, 47.9$
(c) radius $= 1.607$,   centre $= -0.847, 0$

*Example 7.12*

A unity feedback continuous control system has a forward-path transfer function

$$G(s) = \frac{K}{s(s + 5)}$$

(a) Find the value of $K$ to give the closed-loop system a damping ratio of 0.7. The above system is to be replaced by a discrete-time unity feedback control system with a forward-path transfer function

$$G(s) = D(z)\left(\frac{1 - e^{-Ts}}{s}\right)\left(\frac{1}{s(s+5)}\right)$$

(b) If the sampling time is 0.2 seconds, determine the open-loop pulse transfer function.

(c) The discrete-time system is to have the identical time response to the continuous system. What are the desired closed-loop poles and characteristic equations in

    (i) the $s$-plane

    (ii) the $z$-plane

(d) The discrete-time compensator is to take the form

$$D(z) = \frac{K_1(z + a)}{(z + b)}$$

Find the values of $K_1$ and $b$ if $a$ is selected to cancel the non-unity open-loop pole.

*Solution*

(a) $K = 12.8$

(b) $G(z) = \dfrac{0.0147(z + 0.718)}{(z - 1)(z - 0.368)}$

(c) $-2.48 \pm \text{j}2.56, \quad s^2 + 5s + 12.8 = 0$

    $0.531 \pm \text{j}0.298, \quad z^2 - 1.062z + 0.371 = 0$

(d) $D(z) = 12.21\dfrac{(z - 0.368)}{(z - 0.242)}$

# *8*

# State-space methods for control system design

The classical control system design techniques discussed in Chapters 5–7 are generally only applicable to

(a) Single Input, Single Output (SISO) systems
(b) Systems that are linear (or can be linearized) and are time invariant (have parameters that do not vary with time).

The state-space approach is a generalized time-domain method for modelling, analysing and designing a wide range of control systems and is particularly well suited to digital computational techniques. The approach can deal with

(a) Multiple Input, Multiple Output (MIMO) systems, or multivariable systems
(b) Non-linear and time-variant systems
(c) Alternative controller design approaches.

### 8.1.1 The concept of state

The state of a system may be defined as: 'The set of variables (called the state variables) which at some initial time $t_0$, together with the input variables completely determine the behaviour of the system for time $t \geq t_0$'.

The state variables are the smallest number of states that are required to describe the dynamic nature of the system, and it is not a necessary constraint that they are measurable. The manner in which the state variables change as a function of time may be thought of as a trajectory in $n$ dimensional space, called the *state-space*. Two-dimensional state-space is sometimes referred to as the *phase-plane* when one state is the derivative of the other.

## 8.1.2   The state vector differential equation

The state of a system is described by a set of first-order differential equations in terms of the state variables $(x_1, x_2, \ldots, x_n)$ and input variables $(u_1, u_2, \ldots, u_n)$ in the general form

$$\frac{\mathrm{d}x_1}{\mathrm{d}t} = a_{11}x_1 + a_{12}x_2 + \cdots + a_{1n}x_n + b_{11}u_1 + \cdots + b_{1m}u_m$$

$$\frac{\mathrm{d}x_2}{\mathrm{d}t} = a_{21}x_1 + a_{22}x_2 + \cdots + a_{2n}x_n + b_{21}u_1 + \cdots + b_{2m}u_m \tag{8.1}$$

$$\frac{\mathrm{d}x_n}{\mathrm{d}t} = a_{n1}x_1 + a_{n2}x_2 + \cdots + a_{nn}x_n + b_{n1}u_1 + \cdots + b_{nm}u_m$$

The equations set (8.1) may be combined in matrix format. This results in the state vector differential equation

$$\dot{\mathbf{x}} = \mathbf{Ax} + \mathbf{Bu} \tag{8.2}$$

Equation (8.2) is generally called the state equation(s), where lower-case boldface represents vectors and upper-case boldface represents matrices. Thus

$\mathbf{x}$ is the $n$ dimensional state vector

$$\begin{bmatrix} x_1 \\ x_2 \\ \vdots \\ x_n \end{bmatrix} \tag{8.3}$$

$\mathbf{u}$ is the $m$ dimensional input vector

$$\begin{bmatrix} u_1 \\ u_2 \\ \vdots \\ u_m \end{bmatrix} \tag{8.4}$$

$\mathbf{A}$ is the $n \times n$ system matrix

$$\begin{bmatrix} a_{11} & a_{12} & \cdots & a_{1n} \\ a_{21} & a_{22} & \cdots & a_{2n} \\ \vdots & & & \\ a_{n1} & a_{n2} & \cdots & a_{nn} \end{bmatrix} \tag{8.5}$$

$\mathbf{B}$ is the $n \times m$ control matrix

$$\begin{bmatrix} b_{11} & \cdots & b_{1m} \\ b_{21} & \cdots & b_{2m} \\ \vdots & & \\ b_{n1} & \cdots & b_{nm} \end{bmatrix} \tag{8.6}$$

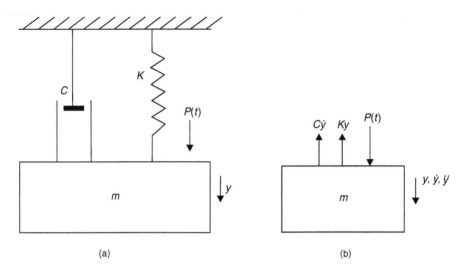

**Fig. 8.1** Spring–mass–damper system and free-body diagram.

In general, the outputs $(y_1, y_2, \ldots, y_n)$ of a linear system can be related to the state variables and the input variables

$$\mathbf{y} = \mathbf{Cx} + \mathbf{Du} \tag{8.7}$$

Equation (8.7) is called the output equation(s).

*Example 8.1*
Write down the state equation and output equation for the spring–mass–damper system shown in Figure 8.1(a).

*Solution*
State variables

$$x_1 = y \tag{8.8}$$

$$x_2 = \frac{dy}{dt} = \dot{x}_1 \tag{8.9}$$

Input variable

$$u = P(t) \tag{8.10}$$

Now

$$\sum F_y = m\ddot{y}$$

From Figure 8.1(b)

$$P(t) - Ky - C\dot{y} = m\ddot{y}$$

or

$$\frac{d^2 y}{dt^2} = -\frac{K}{m} y - \frac{C}{m} \dot{y} + \frac{1}{m} P(t) \tag{8.11}$$

From equations (8.9), (8.10) and (8.11) the set of first-order differential equations are

$$\dot{x}_1 = x_2$$
$$\dot{x}_2 = -\frac{K}{m}x_1 - \frac{C}{m}x_2 + \frac{1}{m}u \qquad (8.12)$$

and the state equations become

$$\begin{bmatrix} \dot{x}_1 \\ \dot{x}_2 \end{bmatrix} = \begin{bmatrix} 0 & 1 \\ -\dfrac{K}{m} & -\dfrac{C}{m} \end{bmatrix} \begin{bmatrix} x_1 \\ x_2 \end{bmatrix} + \begin{bmatrix} 0 \\ \dfrac{1}{m} \end{bmatrix} u \qquad (8.13)$$

From equation (8.8) the output equation is

$$y = \begin{bmatrix} 1 & 0 \end{bmatrix} \begin{bmatrix} x_1 \\ x_2 \end{bmatrix} \qquad (8.14)$$

State variables are not unique, and may be selected to suit the problem being studied.

*Example 8.2*
For the *RCL* network shown in Figure 8.2, write down the state equations when

(a) the state variables are $v_2(t)$ and $\dot{v}_2$
(b) the state variables are $v_2(t)$ and $i(t)$.

*Solution*
(a)

$$x_1 = v_2(t)$$
$$x_2 = \dot{v}_2 = \dot{x}_1 \qquad (8.15)$$

From equation (2.37)

$$LC\frac{d^2 v_2}{dt^2} + RC\frac{dv_2}{dt} + v_2 = v_1(t) \qquad (8.16)$$

From equations (8.15) and (8.16) the set of first-order differential equations are

$$\dot{x}_1 = x_2$$

$$x_2 = -\frac{1}{LC}x_1 - \frac{RC}{LC}x_2 + \frac{1}{LC}u \qquad (8.17)$$

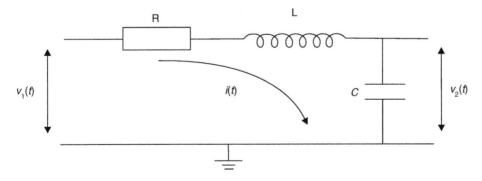

**Fig. 8.2** *RCL* network.

and the state equations are

$$
\begin{bmatrix} \dot{x}_1 \\ \dot{x}_2 \end{bmatrix} = \begin{bmatrix} 0 & 1 \\ -\dfrac{1}{LC} & -\dfrac{R}{L} \end{bmatrix} \begin{bmatrix} x_1 \\ x_2 \end{bmatrix} + \begin{bmatrix} 0 \\ \dfrac{1}{LC} \end{bmatrix} u \tag{8.18}
$$

(b)
$$
\begin{aligned}
x_1 &= v_2(t) \\
x_2 &= i(t)
\end{aligned} \tag{8.19}
$$

From equations (2.34) and (2.35)

$$
L\frac{di}{dt} = -v_2(t) - Ri(t) + v_1(t) \tag{8.20}
$$

$$
C\frac{dv_2}{dt} = i(t) \tag{8.21}
$$

Equations (8.20) and (8.21) are both first-order differential equations, and can be written in the form

$$
\begin{aligned}
\dot{x}_1 &= \frac{1}{C}x_2 \\
\dot{x}_2 &= -\frac{1}{L}x_1 - \frac{R}{L}x_2 + \frac{1}{L}u
\end{aligned} \tag{8.22}
$$

giving the state equations

$$
\begin{bmatrix} \dot{x}_1 \\ \dot{x}_2 \end{bmatrix} = \begin{bmatrix} 0 & 1 \\ -\dfrac{1}{L} & -\dfrac{R}{L} \end{bmatrix} \begin{bmatrix} x_1 \\ x_2 \end{bmatrix} + \begin{bmatrix} 0 \\ \dfrac{1}{L} \end{bmatrix} u \tag{8.23}
$$

*Example 8.3*
For the 2 mass system shown in Figure 8.3, find the state and output equation when the state variables are the position and velocity of each mass.

*Solution*
State variables
$$
\begin{aligned}
x_1 &= y_1 & x_2 &= \dot{y}_1 \\
x_3 &= y_2 & x_4 &= \dot{y}_2
\end{aligned}
$$

System outputs
$$
y_1, y_2
$$

System inputs
$$
u = P(t) \tag{8.24}
$$

For mass $m_1$
$$
\sum F_y = m_1\ddot{y}_1
$$
$$
K_2(y_2 - y_1) - K_1 y_1 + P(t) - C_1\dot{y}_1 = m_1\ddot{y}_1 \tag{8.25}
$$

For mass $m_2$
$$
\sum F_y = m_2\ddot{y}_2
$$
$$
-K_2(y_2 - y_1) = m_2\ddot{y}_2 \tag{8.26}
$$

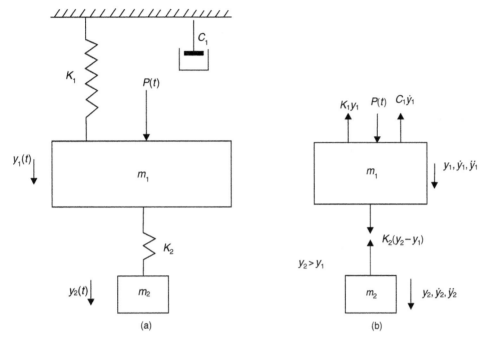

**Fig. 8.3** Two-mass system and free-body diagrams.

From (8.24), (8.25) and (8.26), the four first-order differential equations are

$$\dot{x}_1 = x_2$$

$$\dot{x}_2 = \left(-\frac{K_1}{m_1} - \frac{K_2}{m_1}\right)x_1 - \frac{C_1}{m_1}x_2 + \frac{K_2}{m_1}x_3 + \frac{1}{m_1}u$$

$$\dot{x}_3 = x_4 \tag{8.27}$$

$$\dot{x}_4 = \frac{K_2}{m_2}x_1 - \frac{K_2}{m_2}x_3$$

Hence the state equations are

$$\begin{bmatrix} \dot{x}_1 \\ \dot{x}_2 \\ \dot{x}_3 \\ \dot{x}_4 \end{bmatrix} = \begin{bmatrix} 0 & 1 & 0 & 0 \\ -\left(\dfrac{K_1 + K_2}{m_1}\right) & -\dfrac{C}{m_1} & \dfrac{K_2}{m_1} & 0 \\ 0 & 0 & 0 & 1 \\ \dfrac{K_2}{m_2} & 0 & -\dfrac{K_2}{m_2} & 0 \end{bmatrix} \begin{bmatrix} x_1 \\ x_2 \\ x_3 \\ x_4 \end{bmatrix} + \begin{bmatrix} 0 \\ \dfrac{1}{m_1} \\ 0 \\ 0 \end{bmatrix} u \tag{8.28}$$

and the output equations are

$$\begin{bmatrix} y_1 \\ y_2 \end{bmatrix} = \begin{bmatrix} 1 & 0 & 0 & 0 \\ 0 & 0 & 1 & 0 \end{bmatrix} \begin{bmatrix} x_1 \\ x_2 \\ x_3 \\ x_4 \end{bmatrix} \tag{8.29}$$

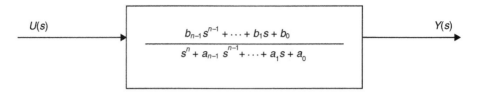

**Fig. 8.4** Generalized transfer function.

### 8.1.3 State equations from transfer functions

Consider the general differential equation

$$\frac{d^n y}{dt^n} + a_{n-1}\frac{d^{n-1}y}{dt^{n-1}} + \cdots + a_1\frac{dy}{dt} + a_0 y = b_{n-1}\frac{d^{n-1}u}{dt^{n-1}} + \cdots + b_1\frac{du}{dt} + b_0 u \quad (8.30)$$

Equation (8.30) can be represented by the transfer function shown in Figure 8.4.
Define a set of state variables such that

$$\begin{aligned}
\dot{x}_1 &= x_2 \\
\dot{x}_2 &= x_3 \\
&\vdots \quad \vdots \\
\dot{x}_n &= -a_0 x_1 - a_1 x_2 - \cdots - a_{n-1}x_n + u
\end{aligned} \quad (8.31)$$

and an output equation

$$y = b_0 x_1 + b_1 x_2 + \cdots + b_{n-1}x_n \quad (8.32)$$

Then the state equation is

$$\begin{bmatrix} \dot{x}_1 \\ \dot{x}_2 \\ \vdots \\ \dot{x}_{n-1} \\ x_n \end{bmatrix} = \begin{bmatrix} 0 & 1 & 0 & \cdots & 0 \\ 0 & 0 & 1 & \cdots & 0 \\ \vdots & & & & \\ 0 & 0 & 0 & \cdots & 1 \\ -a_0 & -a_1 & -a_2 & \cdots & -a_{n-1} \end{bmatrix} \begin{bmatrix} x_1 \\ x_2 \\ \vdots \\ x_{n-1} \\ x_n \end{bmatrix} + \begin{bmatrix} 0 \\ 0 \\ \vdots \\ 0 \\ 1 \end{bmatrix} u \quad (8.33)$$

The state-space representation in equation (8.33) is called the controllable canonical
form and the output equation is

$$y = \begin{bmatrix} b_0 & b_1 & b_2 & \cdots & b_{n-1} \end{bmatrix} \begin{bmatrix} x_1 \\ x_2 \\ x_3 \\ \vdots \\ x_n \end{bmatrix} \quad (8.34)$$

*Example 8.4*   (See also Appendix 1, *examp84.m*)
Find the state and output equations for

$$\frac{Y}{U}(s) = \frac{4}{s^3 + 3s^2 + 6s + 2}$$

*Solution*
State equation

$$\begin{bmatrix} \dot{x}_1 \\ \dot{x}_2 \\ \dot{x}_3 \end{bmatrix} = \begin{bmatrix} 0 & 1 & 0 \\ 0 & 0 & 1 \\ -2 & -6 & -3 \end{bmatrix} \begin{bmatrix} x_1 \\ x_2 \\ x_3 \end{bmatrix} + \begin{bmatrix} 0 \\ 0 \\ 1 \end{bmatrix} u \tag{8.35}$$

Output equation

$$y = \begin{bmatrix} 4 & 0 & 0 \end{bmatrix} \begin{bmatrix} x_1 \\ x_2 \\ x_3 \end{bmatrix} \tag{8.36}$$

*Example 8.5*
Find the state and output equations for

$$\frac{Y}{U}(s) = \frac{5s^2 + 7s + 4}{s^3 + 3s^2 + 6s + 2}$$

*Solution*
The state equation is the same as (8.35). The output equation is

$$y = \begin{bmatrix} 4 & 7 & 5 \end{bmatrix} \begin{bmatrix} x_1 \\ x_2 \\ x_3 \end{bmatrix} \tag{8.37}$$

## 8.2 Solution of the state vector differential equation

Consider the first-order differential equation

$$\frac{\mathrm{d}x}{\mathrm{d}t} = ax(t) + bu(t) \tag{8.38}$$

where $x(t)$ and $u(t)$ are scalar functions of time. Take Laplace transforms

$$sX(s) - x(0) = aX(s) + bU(s) \tag{8.39}$$

where $x(0)$ is the initial condition. From equation (8.39)

$$X(s) = \frac{x(0)}{(s-a)} + \frac{b}{(s-a)} U(s) \tag{8.40}$$

Inverse transform

$$x(t) = \mathrm{e}^{at} x(0) + \int_0^t \mathrm{e}^{a(t-\tau)} bu(\tau)\mathrm{d}\tau \tag{8.41}$$

where the integral term in equation (8.41) is the convolution integral and $\tau$ is a dummy time variable. Note that

$$\mathrm{e}^{at} = 1 + at + \frac{a^2 t^2}{2!} + \cdots + \frac{a^k t^k}{k!} \tag{8.42}$$

Consider now the state vector differential equation

$$\dot{\mathbf{x}} = \mathbf{A}\mathbf{x} + \mathbf{B}\mathbf{u} \tag{8.43}$$

Taking Laplace transforms

$$s\mathbf{X}(s) - \mathbf{x}(0) = \mathbf{A}\mathbf{X}(s) + \mathbf{B}\mathbf{U}(s) \tag{8.44}$$

$$(s\mathbf{I} - \mathbf{A})\mathbf{X}(s) = \mathbf{x}(0) + \mathbf{B}\mathbf{U}(s)$$

Pre-multiplying by $(s\mathbf{I} - \mathbf{A})^{-1}$

$$\mathbf{X}(s) = (s\mathbf{I} - \mathbf{A})^{-1}\mathbf{x}(0) + (s\mathbf{I} - \mathbf{A})^{-1}\mathbf{B}\mathbf{U}(s) \tag{8.45}$$

Inverse transform

$$\mathbf{x}(t) = \mathrm{e}^{\mathbf{A}t}\mathbf{x}(0) + \int_0^t \mathrm{e}^{\mathbf{A}(t-\tau)}\mathbf{B}\mathbf{U}(\tau)\mathrm{d}\tau \tag{8.46}$$

if the initial time is $t_0$, then

$$\mathbf{x}(t) = \mathrm{e}^{\mathbf{A}(t-t_0)}\mathbf{x}(0) + \int_{t_0}^t \mathrm{e}^{\mathbf{A}(t-\tau)}\mathbf{B}\mathbf{u}(\tau)\mathrm{d}\tau \tag{8.47}$$

The exponential matrix $\mathrm{e}^{\mathbf{A}t}$ in equation (8.46) is called the state-transition matrix $\mathbf{\Phi}(t)$ and represents the natural response of the system. Hence

$$\mathbf{\Phi}(s) = (s\mathbf{I} - \mathbf{A})^{-1} \tag{8.48}$$

$$\mathbf{\Phi}(t) = \mathscr{L}^{-1}(s\mathbf{I} - \mathbf{A})^{-1} = \mathrm{e}^{\mathbf{A}t} \tag{8.49}$$

Alternatively

$$\mathbf{\Phi}(t) = \mathbf{I} + \mathbf{A}t + \frac{\mathbf{A}^2 t^2}{2!} + \cdots + \frac{\mathbf{A}^k t^k}{k!} \tag{8.50}$$

Hence equation (8.46) can be written

$$\mathbf{x}(t) = \mathbf{\Phi}(t)\mathbf{x}(0) + \int_0^t \mathbf{\Phi}(t-\tau)\mathbf{B}\mathbf{u}(\tau)\mathrm{d}\tau \tag{8.51}$$

In equation (8.51) the first term represents the response to a set of initial conditions, whilst the integral term represents the response to a forcing function.

### Characteristic equation
Using a state variable representation of a system, the characteristic equation is given by

$$|(s\mathbf{I} - \mathbf{A})| = 0 \tag{8.52}$$

## 8.2.1 Transient solution from a set of initial conditions

*Example 8.6*
For the spring–mass–damper system given in Example 8.1, Figure 8.1, the state equations are shown in equation (8.13)

$$\begin{bmatrix} \dot{x}_1 \\ \dot{x}_2 \end{bmatrix} = \begin{bmatrix} 0 & 1 \\ -\dfrac{K}{m} & -\dfrac{C}{m} \end{bmatrix} \begin{bmatrix} x_1 \\ x_2 \end{bmatrix} + \begin{bmatrix} 0 \\ \dfrac{1}{m} \end{bmatrix} u \tag{8.53}$$

Given: $m = 1\,\text{kg}$, $C = 3\,\text{Ns/m}$, $K = 2\,\text{N/m}$, $u(t) = 0$. Evaluate,

(a) the characteristic equation, its roots, $\omega_n$ and $\zeta$
(b) the transition matrices $\phi(s)$ and $\phi(t)$
(c) the transient response of the state variables from the set of initial conditions

$$y(0) = 1.0,$$
$$\dot{y}(0) = 0$$

*Solution*
Since $x_1 = y$ and $x_2 = \dot{y}$, then $x_1(0) = 1.0$, $x_2(0) = 0$.
Inserting values of system parameters into equation (8.53) gives

$$\begin{bmatrix} \dot{x}_1 \\ \dot{x}_2 \end{bmatrix} = \begin{bmatrix} 0 & 1 \\ -2 & -3 \end{bmatrix} \begin{bmatrix} x_1 \\ x_2 \end{bmatrix} + \begin{bmatrix} 0 \\ 1 \end{bmatrix} u$$

(a) $$(s\mathbf{I} - \mathbf{A}) = \begin{bmatrix} s & 0 \\ 0 & s \end{bmatrix} - \begin{bmatrix} 0 & 1 \\ -2 & -3 \end{bmatrix} = \begin{bmatrix} s & -1 \\ 2 & (s+3) \end{bmatrix} \tag{8.54}$$

From equation (8.52), the characteristic equation is

$$|(s\mathbf{I} - \mathbf{A})| = s(s+3) - (-2) = s^2 + 3s + 2 = 0 \tag{8.55}$$

Roots of characteristic equation

$$s = -1, -2 \tag{8.56}$$

Compare equation (8.55) with the denominator of the standard form in equation (3.43)

$$\omega_n^2 = 2 \quad \text{i.e} \quad \omega_n = 1.414\,\text{rad/s}$$
$$2\zeta\omega_n = 3 \quad \text{i.e} \quad \zeta = 1.061 \tag{8.57}$$

(b) The inverse of any matrix $\mathbf{A}$ (see equation A2.17) is

$$\mathbf{A}^{-1} = \frac{\text{Adjoint } \mathbf{A}}{\det \mathbf{A}} \tag{8.58}$$

From equation (8.48)

$$\Phi(s) = (s\mathbf{I} - \mathbf{A})^{-1}$$

Using the standard matrix operations given in Appendix 2, equation (A2.12)

$$\text{Minors of } \Phi(s) = \begin{bmatrix} (s+3) & 2 \\ -1 & s \end{bmatrix}$$

$$\text{Co-factors of } \Phi(s) = \begin{bmatrix} (s+3) & -2 \\ 1 & s \end{bmatrix}$$

The Adjoint matrix is the transpose of the Co-factor matrix

$$\text{Adjoint of } \Phi(s) = \begin{bmatrix} (s+3) & 1 \\ -2 & s \end{bmatrix} \tag{8.59}$$

Hence, from equations (8.58) and (8.48)

$$\Phi(s) = \begin{bmatrix} \dfrac{(s+3)}{(s+1)(s+2)} & \dfrac{1}{(s+1)(s+2)} \\ \dfrac{-2}{(s+1)(s+2)} & \dfrac{s}{(s+1)(s+2)} \end{bmatrix} \tag{8.60}$$

Using partial fraction expansions

$$\Phi(s) = \begin{bmatrix} \left(\dfrac{2}{s+1} - \dfrac{1}{s+2}\right) & \left(\dfrac{1}{s+1} - \dfrac{1}{s+2}\right) \\ -2\left(\dfrac{1}{s+1} - \dfrac{1}{s+2}\right) & \left(-\dfrac{1}{s+1} + \dfrac{2}{s+2}\right) \end{bmatrix} \tag{8.61}$$

Inverse transform equation (8.61)

$$\Phi(t) = \begin{bmatrix} (2e^{-t} - e^{-2t}) & (e^{-t} - e^{-2t}) \\ -2(e^{-t} - e^{-2t}) & (-e^{-t} + 2e^{-2t}) \end{bmatrix} \tag{8.62}$$

Note that the exponential indices are the roots of the characteristic equation (8.56).

(c) From equation (8.51), the transient response is given by

$$\mathbf{x}(t) = \Phi(t)\mathbf{x}(0) \tag{8.63}$$

Hence

$$\begin{bmatrix} x_1 \\ x_2 \end{bmatrix} = \begin{bmatrix} (2e^{-t} - e^{-2t}) & (e^{-t} - e^{-2t}) \\ -2(e^{-t} - e^{-2t}) & (-e^{-t} + 2e^{-2t}) \end{bmatrix} \begin{bmatrix} 1 \\ 0 \end{bmatrix} \tag{8.64}$$

$$\begin{aligned} x_1(t) &= (2e^{-t} - e^{-2t}) \\ x_2(t) &= -2(e^{-t} - e^{-2t}) \end{aligned} \tag{8.65}$$

The time response of the state variables (i.e. position and velocity) together with the state trajectory is given in Figure 8.5.

*Example 8.7*
For the spring–mass–damper system given in Example 8.6, evaluate the transient response of the state variables to a unit step input using

(a) The convolution integral
(b) Inverse Laplace transforms

Assume zero initial conditions.

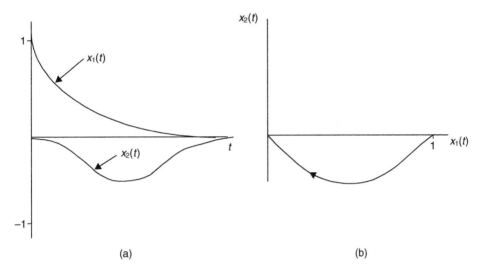

**Fig. 8.5** State variable time response and state trajectory for Example 8.4.

*Solution*

(a) From equation (8.51)

$$\mathbf{x}(t) = \mathbf{\Phi}(t)\begin{bmatrix} 0 \\ 0 \end{bmatrix} + \int_0^t \begin{bmatrix} \phi_{11}(t-\tau) & \phi_{12}(t-\tau) \\ \phi_{21}(t-\tau) & \phi_{22}(t-\tau) \end{bmatrix} \begin{bmatrix} 0 \\ \dfrac{1}{m} \end{bmatrix} \mathbf{u}(\tau)\mathrm{d}\tau \qquad (8.66)$$

Given that $u(t) = 1$ and $1/m = 1$, equation (8.66) reduces to

$$\mathbf{x}(t) = \int_0^t \begin{bmatrix} \phi_{12}(t-\tau) \\ \phi_{22}(t-\tau) \end{bmatrix} \mathrm{d}\tau$$

Inserting values from equation (8.62)

$$\mathbf{x}(t) = \int_0^t \begin{bmatrix} e^{-(t-\tau)} - e^{-2(t-\tau)} \\ e^{-(t-\tau)} + 2e^{-2(t-\tau)} \end{bmatrix} \mathrm{d}\tau \qquad (8.67)$$

Integrating

$$\mathbf{x}(t) = \begin{bmatrix} e^{-(t-\tau)} - \tfrac{1}{2}e^{-2(t-\tau)} \\ e^{-(t-\tau)} + e^{-2(t-\tau)} \end{bmatrix}_0^t \qquad (8.68)$$

Inserting integration limits ($\tau = t$ and $\tau = 0$)

$$\begin{bmatrix} x_1 \\ x_2 \end{bmatrix} = \begin{bmatrix} \tfrac{1}{2} - e^{-t} + \tfrac{1}{2}e^{-2t} \\ e^{-t} - e^{-2t} \end{bmatrix} \qquad (8.69)$$

(b) An alternative method is to inverse transform from an $s$-domain expression. Equation (8.45) may be written

$$\mathbf{X}(s) = \mathbf{\Phi}(s)\mathbf{x}(0) + \mathbf{\Phi}(s)\mathbf{B}\mathbf{U}(s) \qquad (8.70)$$

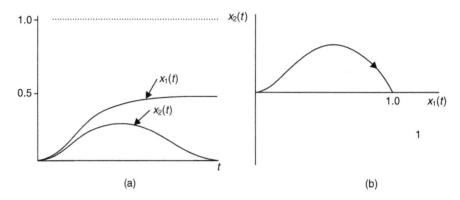

**Fig. 8.6** State variable step response and state trajectory for Example 8.5.

Hence from equation (8.61)

$$\mathbf{X}(s) = \Phi(s)\begin{bmatrix} 0 \\ 0 \end{bmatrix} + \begin{bmatrix} \left(\dfrac{2}{s+1} - \dfrac{1}{s+2}\right) & \left(\dfrac{1}{s+1} - \dfrac{1}{s+2}\right) \\ -2\left(\dfrac{1}{s+1} - \dfrac{1}{s+2}\right) & \left(\dfrac{-1}{s+1} + \dfrac{2}{s+2}\right) \end{bmatrix} \begin{bmatrix} 0 \\ 1 \end{bmatrix}\dfrac{1}{s} \qquad (8.71)$$

Simplifying

$$\mathbf{X}(s) = \begin{bmatrix} \dfrac{1}{s(s+1)} - \dfrac{1}{2}\left\{\dfrac{2}{s(s+2)}\right\} \\ \dfrac{-1}{s(s+1)} + \dfrac{2}{s(s+2)} \end{bmatrix} \qquad (8.72)$$

Inverse transform

$$\mathbf{x}(t) = \begin{bmatrix} (1 - e^{-t}) - \tfrac{1}{2}(1 - e^{-2t}) \\ -(1 - e^{-t}) + (1 - e^{-2t}) \end{bmatrix} \qquad (8.73)$$

which gives

$$\begin{bmatrix} x_1 \\ x_2 \end{bmatrix} = \begin{bmatrix} \tfrac{1}{2} - e^{-t} + \tfrac{1}{2}e^{-2t} \\ e^{-t} - e^{-2t} \end{bmatrix} \qquad (8.74)$$

Equation (8.74) is the same as equation (8.69).

The step response of the state variables, together with the state trajectory, is shown in Figure 8.6.

## 8.3 Discrete-time solution of the state vector differential equation

The discrete-time solution of the state equation may be considered to be the vector equivalent of the scalar difference equation method developed from a $z$-transform approach in Chapter 7.

The continuous-time solution of the state equation is given in equation (8.47). If the time interval $(t - t_0)$ in this equation is $T$, the sampling time of a discrete-time system, then the discrete-time solution of the state equation can be written as

$$\mathbf{x}[(k+1)T] = e^{\mathbf{A}T}\mathbf{x}(kT) + \left\{ \int_0^T e^{\mathbf{A}\tau}\mathbf{B}d\tau \right\} \mathbf{u}(kT) \tag{8.75}$$

Equation (8.75) can be written in the general form

$$\mathbf{x}[(k+1)T] = \mathbf{A}(T)\mathbf{x}(kT) + \mathbf{B}(T)\mathbf{u}(kT) \tag{8.76}$$

Note

$$\mathbf{A}(T) \neq \mathbf{A} \quad \text{and} \quad \mathbf{B}(T) \neq \mathbf{B}$$

Equation (8.76) is called the matrix vector difference equation and can be used for the recursive discrete-time simulation of multivariable systems.

The discrete-time state transition matrix $\mathbf{A}(T)$ may be computed by substituting $T = t$ in equations (8.49) and (8.50), i.e.

$$\mathbf{A}(T) = \mathbf{\Phi}(T) = e^{\mathbf{A}T} \tag{8.77}$$

or

$$\mathbf{A}(T) = \mathbf{I} + \mathbf{A}T + \frac{\mathbf{A}^2 T^2}{2!} + \cdots + \frac{\mathbf{A}^k T^k}{k!} \tag{8.78}$$

Usually sufficient accuracy is obtained with $5 < k < 50$.

The discrete-time control matrix $\mathbf{B}(T)$ from equations (8.75) and (8.76) is

$$\mathbf{B}(T) = \int_0^T e^{\mathbf{A}\tau}\mathbf{B}d\tau \tag{8.79}$$

or

$$\mathbf{B}(T) = \left\{ \sum_{k=0}^{\infty} \frac{\mathbf{A}^k T^k}{(k+1)!} \right\} \mathbf{B}T$$

Put $T$ within the brackets

$$\mathbf{B}(T) = \left\{ \sum_{k=0}^{\infty} \frac{\mathbf{A}^k T^{k+1}}{(k+1)!} \right\} \mathbf{B}$$

Hence

$$\mathbf{B}(T) = \left\{ \mathbf{I}T + \frac{\mathbf{A}T^2}{2!} + \frac{\mathbf{A}^2 T^3}{3!} + \cdots + \frac{\mathbf{A}^k T^{k+1}}{(k+1)!} \right\} \mathbf{B} \tag{8.80}$$

*Example 8.8*   (See also Appendix 1, *examp88.m*)

(a) Calculate the discrete-time transition and control matrices for the spring-mass-damper system in Example 8.6 using a sampling time $T = 0.1$ seconds.
(b) Using the matrix vector difference equation method, determine the unit step response assuming zero initial conditions.

*Solution*

(a) The exact value of $\mathbf{A}(T)$ is found by substituting $T = t$ in equation (8.62)

$$\mathbf{A}(T) = \boldsymbol{\Phi}(T) = \begin{bmatrix} \left(2e^{-0.1} - e^{-0.2}\right) & \left(e^{-0.1} - e^{-0.2}\right) \\ -2\left(e^{-0.1} - e^{-0.2}\right) & \left(-e^{-0.1} + 2e^{-0.2}\right) \end{bmatrix}$$

$$= \begin{bmatrix} 0.991 & 0.086 \\ -0.172 & 0.733 \end{bmatrix} \tag{8.81}$$

An approximate value of $\mathbf{A}(T)$ is found from equation (8.78), taking the series as far as $k = 2$.

$$\mathbf{A}T = \begin{bmatrix} 0 & 0.1 \\ -0.2 & -0.3 \end{bmatrix}$$

$$\frac{\mathbf{A}^2 T^2}{2!} = \begin{bmatrix} 0 & 1 \\ -2 & -3 \end{bmatrix} \begin{bmatrix} 0 & 1 \\ -2 & -3 \end{bmatrix} \frac{0.1^2}{1 \times 2} = \begin{bmatrix} -0.01 & -0.015 \\ 0.03 & 0.035 \end{bmatrix}$$

using the first 3 terms of equation (8.78)

$$\mathbf{A}(T) \approx \begin{bmatrix} 1 & 0 \\ 0 & 1 \end{bmatrix} + \begin{bmatrix} 0 & 0.1 \\ -0.2 & -0.3 \end{bmatrix} + \begin{bmatrix} -0.01 & -0.015 \\ 0.03 & 0.035 \end{bmatrix}$$

$$\approx \begin{bmatrix} 0.99 & 0.085 \\ -0.17 & 0.735 \end{bmatrix} \tag{8.82}$$

Since in equation (8.66), $u(\tau)$ is unity, the exact value of $\mathbf{B}(T)$ can be obtained by substituting $T = t$ in equation (8.69)

$$\mathbf{B}(T) = \begin{bmatrix} \frac{1}{2} - e^{-0.1} + \frac{1}{2}e^{-0.2} \\ e^{-0.1} - e^{-0.2} \end{bmatrix} \tag{8.83}$$

$$\mathbf{B}(T) = \begin{bmatrix} 0.00453 \\ 0.0861 \end{bmatrix} \tag{8.84}$$

An approximate value of $\mathbf{B}(T)$ is found from equation (8.80), taking the series as far as $k = 2$.

$$\mathbf{B}(T) \approx (\mathbf{I}T)\mathbf{B} + \left(\frac{\mathbf{A}T^2}{2!}\right)\mathbf{B} + \left(\frac{\mathbf{A}^2 T^3}{3!}\right)\mathbf{B}$$

$$\approx \begin{bmatrix} 0 \\ 0.1 \end{bmatrix} + \begin{bmatrix} 0.005 \\ -0.015 \end{bmatrix} + \begin{bmatrix} -0.0005 \\ 0.00117 \end{bmatrix}$$

$$\approx \begin{bmatrix} 0.0045 \\ 0.08617 \end{bmatrix} \tag{8.85}$$

(b) Using the values of $\mathbf{A}(T)$ and $\mathbf{B}(T)$ given in equations (8.81) and (8.84), together with the matrix vector difference equation (8.76), the first few recursive steps of the discrete solution to a step input to the system is given in equation (8.86)

$kT = 0$

$$\begin{bmatrix} x_1(0.1) \\ x_2(0.1) \end{bmatrix} = \begin{bmatrix} 0.991 & 0.086 \\ -0.172 & 0.733 \end{bmatrix} \begin{bmatrix} 0 \\ 0 \end{bmatrix} + \begin{bmatrix} 0.00453 \\ 0.0861 \end{bmatrix} 1 = \begin{bmatrix} 0.00453 \\ 0.0861 \end{bmatrix}$$

$kT = 0.1$

$$\begin{bmatrix} x_1(0.2) \\ x_2(0.2) \end{bmatrix} = \begin{bmatrix} 0.991 & 0.086 \\ -0.172 & 0.733 \end{bmatrix} \begin{bmatrix} 0.00453 \\ 0.0861 \end{bmatrix} + \begin{bmatrix} 0.00453 \\ 0.0861 \end{bmatrix} 1 = \begin{bmatrix} 0.016 \\ 0.0148 \end{bmatrix}$$

$kT = 0.2$

$$\begin{bmatrix} x_1(0.3) \\ x_2(0.3) \end{bmatrix} = \begin{bmatrix} 0.991 & 0.086 \\ -0.172 & 0.733 \end{bmatrix} \begin{bmatrix} 0.016 \\ 0.148 \end{bmatrix} + \begin{bmatrix} 0.00453 \\ 0.0861 \end{bmatrix} 1 = \begin{bmatrix} 0.033 \\ 0.192 \end{bmatrix} \quad (8.86)$$

$kT = 0.3$

$$\begin{bmatrix} x_1(0.4) \\ x_2(0.4) \end{bmatrix} = \begin{bmatrix} 0.991 & 0.086 \\ -0.172 & 0.733 \end{bmatrix} \begin{bmatrix} 0.033 \\ 0.192 \end{bmatrix} + \begin{bmatrix} 0.00453 \\ 0.0861 \end{bmatrix} 1 = \begin{bmatrix} 0.054 \\ 0.227 \end{bmatrix}$$

$kT = 0.4$

$$\begin{bmatrix} x_1(0.5) \\ x_2(0.5) \end{bmatrix} = \begin{bmatrix} 0.991 & 0.086 \\ -0.172 & 0.733 \end{bmatrix} \begin{bmatrix} 0.054 \\ 0.227 \end{bmatrix} + \begin{bmatrix} 0.00453 \\ 0.0861 \end{bmatrix} 1 = \begin{bmatrix} 0.078 \\ 0.243 \end{bmatrix}$$

*Example 8.9*

A system has a transfer function

$$\frac{Y}{U}(s) = \frac{1}{s^2 + 2s + 1}$$

The system has an initial condition $y(0) = 1$ and is subject to a unit ramp function $u(t) = t$. Determine

(a) The state and output equations
(b) The transition matrix $\Phi(s)$
(c) Expressions for the time response of the state variables.

*Solution*

(a) $\begin{bmatrix} \dot{x}_1 \\ \dot{x}_2 \end{bmatrix} = \begin{bmatrix} 0 & 1 \\ 1 & -2 \end{bmatrix} \begin{bmatrix} x_1 \\ x_2 \end{bmatrix} + \begin{bmatrix} 0 \\ 1 \end{bmatrix} u$

$y = \begin{bmatrix} 1 & 0 \end{bmatrix} \begin{bmatrix} x_1 \\ x_2 \end{bmatrix}$

(b) $\Phi(s) = \begin{bmatrix} \dfrac{s+2}{(s+1)(s+1)} & \dfrac{1}{(s+1)(s+1)} \\ \dfrac{-1}{(s+1)(s+1)} & \dfrac{s}{(s+1)(s+1)} \end{bmatrix}$

(c) $\begin{bmatrix} x_1 \\ x_2 \end{bmatrix} = \begin{bmatrix} 3e^{-t} + 2te^{-t} - 2 + t \\ 2te^{-t} + 1 - e^{-t} \end{bmatrix}$

## 8.4 Control of multivariable systems

### 8.4.1 Controllability and observability

The concepts of controllability and observability were introduced by Kalman (1960) and play an important role in the control of multivariable systems.

A system is said to be controllable if a control vector $\mathbf{u}(t)$ exists that will transfer the system from any initial state $\mathbf{x}(t_0)$ to some final state $\mathbf{x}(t)$ in a finite time interval.

A system is said to be observable if at time $t_0$, the system state $\mathbf{x}(t_0)$ can be exactly determined from observation of the output $\mathbf{y}(t)$ over a finite time interval.

If a system is described by equations (8.2) and (8.7)

$$\dot{\mathbf{x}} = \mathbf{Ax} + \mathbf{Bu}$$
$$\mathbf{y} = \mathbf{Cx} + \mathbf{Du}$$

(8.87)

then a sufficient condition for complete state controllability is that the $n \times n$ matrix

$$\mathbf{M} = [\mathbf{B} : \mathbf{AB} : \ldots : \mathbf{A}^{n-1}\mathbf{B}]$$

(8.88)

contains $n$ linearly independent row or column vectors, i.e. is of rank $n$ (that is, the matrix is non-singular, i.e. the determinant is non-zero. See Appendix 2). Equation (8.88) is called the controllability matrix.

The system described by equations (8.87) is completely observable if the $n \times n$ matrix

$$\mathbf{N} = \left[\mathbf{C}^{\mathrm{T}} : \mathbf{A}^{\mathrm{T}}\mathbf{C}^{\mathrm{T}} : \ldots : (\mathbf{A}^{\mathrm{T}})^{n-1}\mathbf{C}^{\mathrm{T}}\right]$$

(8.89)

is of rank $n$, i.e. is non-singular having a non-zero determinant. Equation (8.89) is called the observability matrix.

*Example 8.10* (See also Appendix 1, *examp810.m*)
Is the following system completely controllable and observable?

$$\begin{bmatrix} \dot{x}_1 \\ \dot{x}_2 \end{bmatrix} = \begin{bmatrix} -2 & 0 \\ 3 & -5 \end{bmatrix} \begin{bmatrix} x_1 \\ x_2 \end{bmatrix} + \begin{bmatrix} 1 \\ 0 \end{bmatrix} \mathbf{u}$$

$$y = \begin{bmatrix} 1 & -1 \end{bmatrix} \begin{bmatrix} x_1 \\ x_2 \end{bmatrix}$$

*Solution*
From equation (8.88) the controllability matrix is

$$\mathbf{M} = [\mathbf{B} : \mathbf{AB}]$$

where

$$\mathbf{AB} = \begin{bmatrix} -2 & 0 \\ 3 & -5 \end{bmatrix} \begin{bmatrix} 1 \\ 0 \end{bmatrix} = \begin{bmatrix} -2 \\ 3 \end{bmatrix}$$

hence

$$\mathbf{M} = [\mathbf{B} : \mathbf{AB}] = \begin{bmatrix} 0 & -2 \\ 1 & 3 \end{bmatrix}$$

(8.90)

Equation (8.90) is non-singular since it has a non-zero determinant. Also the two row and column vectors can be seen to be linearly independent, so it is of rank 2 and therefore the system is controllable.

From equation (8.89) the observability matrix is

$$\mathbf{N} = \left[\mathbf{C}^T \vdots \mathbf{A}^T\mathbf{C}^T\right]$$

where

$$\mathbf{A}^T\mathbf{C}^T = \begin{bmatrix} -2 & 3 \\ 0 & -5 \end{bmatrix}\begin{bmatrix} 1 \\ -1 \end{bmatrix} = \begin{bmatrix} -5 \\ 5 \end{bmatrix}$$

hence

$$\mathbf{N} = \left[\mathbf{C}^T \vdots \mathbf{A}^T\mathbf{C}^T\right] = \begin{bmatrix} 1 & -5 \\ -1 & 5 \end{bmatrix} \tag{8.91}$$

Equation (8.91) is singular since it has a zero determinant. Also the column vectors are linearly dependent since the second column is $-5$ times the first column and therefore the system is unobservable.

## 8.4.2   State variable feedback design

Consider a system described by the state and output equations

$$\begin{aligned} \dot{\mathbf{x}} &= \mathbf{A}\mathbf{x} + \mathbf{B}\mathbf{u} \\ \mathbf{y} &= \mathbf{C}\mathbf{x} \end{aligned} \tag{8.92}$$

Select a control law of the form

$$\mathbf{u} = (\mathbf{r} - \mathbf{K}\mathbf{x}) \tag{8.93}$$

In equation (8.93), $\mathbf{r}(t)$ is a vector of desired state variables and $\mathbf{K}$ is referred to as the state feedback gain matrix. Equations (8.92) and (8.93) are represented in state variable block diagram form in Figure 8.7.

Substituting equation (8.93) into equation (8.92) gives

$$\dot{\mathbf{x}} = \mathbf{A}\mathbf{x} + \mathbf{B}(\mathbf{r} - \mathbf{K}\mathbf{x})$$

or

$$\dot{\mathbf{x}} = (\mathbf{A} - \mathbf{B}\mathbf{K})\mathbf{x} + \mathbf{B}\mathbf{r} \tag{8.94}$$

In equation (8.94) the matrix $(\mathbf{A} - \mathbf{B}\mathbf{K})$ is the closed-loop system matrix.

For the system described by equation (8.92), and using equation (8.52), the characteristic equation is given by

$$|(s\mathbf{I} - \mathbf{A})| = 0 \tag{8.95}$$

The roots of equation (8.95) are the open-loop poles or eigenvalues. For the closed-loop system described by equation (8.94), the characteristic equation is

$$|(s\mathbf{I} - \mathbf{A} + \mathbf{B}\mathbf{K})| = 0 \tag{8.96}$$

The roots of equation (8.96) are the closed-loop poles or eigenvalues.

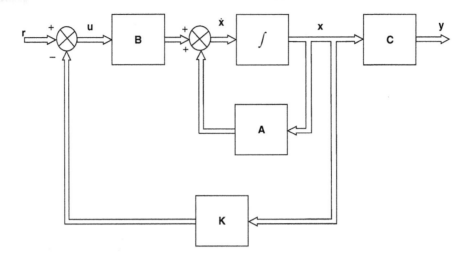

**Fig. 8.7** Control using state variable feedback.

### Regulator design by pole placement

The pole placement control problem is to determine a value of $\mathbf{K}$ that will produce a desired set of closed-loop poles. With a regulator, $\mathbf{r}(t) = \mathbf{0}$ and therefore equation (8.93) becomes

$$\mathbf{u} = -\mathbf{Kx}$$

Thus the control $\mathbf{u}(t)$ will drive the system from a set of initial conditions $\mathbf{x}(0)$ to a set of zero states at time $t_1$, i.e. $\mathbf{x}(t_1) = \mathbf{0}$.

There are several methods that can be used for pole placement.

(a) *Direct comparison method*: If the desired locations of the closed-loop poles (eigenvalues) are

$$s = \mu_1, s = \mu_2, \ldots, s = \mu_n \tag{8.97}$$

then, from equation (8.96)

$$|s\mathbf{I} - \mathbf{A} + \mathbf{BK}| = (s - \mu_1)(s - \mu_2)\ldots(s - \mu_n) \tag{8.98}$$

$$= s^n + \alpha_{n-1}s^{n-1} + \cdots + \alpha_1 s + \alpha_0 \tag{8.99}$$

Solving equation (8.99) will give the elements of the state feedback matrix.

(b) *Controllable canonical form method*: The value of $\mathbf{K}$ can be calculated directly using

$$\mathbf{k} = [\alpha_0 - a_0 : \alpha_1 - a_2 : \ldots : \alpha_{n-2} - a_{n-2} : \alpha_{n-1} - a_{n-1}]\mathbf{T}^{-1} \tag{8.100}$$

where $\mathbf{T}$ is a transformation matrix that transforms the system state equation into the controllable canonical form (see equation (8.33)).

$$\mathbf{T} = \mathbf{MW} \tag{8.101}$$

where **M** is the controllability matrix, equation (8.88)

$$\mathbf{W} = \begin{bmatrix} a_1 & a_2 & \cdots & a_{n-1} & 1 \\ a_2 & a_3 & \cdots & 1 & 0 \\ \vdots & & & & \\ a_{n-1} & 1 & \cdots & 0 & 0 \\ 1 & 0 & \cdots & 0 & 0 \end{bmatrix} \qquad (8.102)$$

Note that $\mathbf{T} = \mathbf{I}$ when the system state equation is already in the controllable canonical form.

(c) *Ackermann's formula*: As with Method 2, Ackermann's formula (1972) is a direct evaluation method. It is only applicable to SISO systems and therefore $u(t)$ and $y(t)$ in equation (8.87) are scalar quantities. Let

$$\mathbf{K} = \begin{bmatrix} 0 & 0 & \cdots & 0 & 1 \end{bmatrix} \mathbf{M}^{-1} \phi(\mathbf{A}) \qquad (8.103)$$

where **M** is the controllability matrix and

$$\phi(\mathbf{A}) = \mathbf{A}^n + \alpha_{n-1}\mathbf{A}^{n-1} + \cdots + \alpha_1\mathbf{A} + \alpha_0\mathbf{I} \qquad (8.104)$$

where **A** is the system matrix and $\alpha_i$ are the coefficients of the desired closed-loop characteristic equation.

*Example 8.11*  (See also Appendix 1, *examp811.m*)
A control system has an open-loop transfer function

$$\frac{Y}{U}(s) = \frac{1}{s(s+4)}$$

When $x_1 = y$ and $x_2 = \dot{x}_1$, express the state equation in the controllable canonical form.
Evaluate the coefficients of the state feedback gain matrix using:

(a) The direct comparison method
(b) The controllable canonical form method
(c) Ackermann's formula

such that the closed-loop poles have the values

$$s = -2, s = -2$$

*Solution*
From the open-loop transfer function

$$\ddot{y} + 4\dot{y} = u \qquad (8.105)$$

Let

$$x_1 = y \qquad (8.106)$$

Then

$$\dot{x}_1 = x_2$$
$$\dot{x}_2 = -4x_2 + u \qquad (8.107)$$

Equation (8.106) provides the output equation and (8.107) the state equation

$$\begin{bmatrix} \dot{x}_1 \\ \dot{x}_2 \end{bmatrix} = \begin{bmatrix} 0 & 1 \\ 0 & -4 \end{bmatrix} \begin{bmatrix} x_1 \\ x_2 \end{bmatrix} + \begin{bmatrix} 0 \\ 1 \end{bmatrix} u \tag{8.108}$$

$$y = \begin{bmatrix} 1 & 0 \end{bmatrix} \begin{bmatrix} x_1 \\ x_2 \end{bmatrix} \tag{8.109}$$

The characteristic equation for the open-loop system is

$$|s\mathbf{I} - \mathbf{A}| = \left| \begin{bmatrix} s & 0 \\ 0 & s \end{bmatrix} - \begin{bmatrix} 0 & 1 \\ 0 & -4 \end{bmatrix} \right| = s^2 + 4s + 0$$

$$= s^2 + a_1 s + a_0 \tag{8.110}$$

Thus

$$a_1 = 4, \quad a_0 = 0$$

The required closed-loop characteristic equation is

$$(s+2)(s+2) = 0$$

or

$$s^2 + 4s + 4 = 0 \tag{8.111}$$

i.e.

$$s^2 + \alpha_1 s + \alpha_0 = 0 \tag{8.112}$$

hence

$$\alpha_1 = 4, \quad \alpha_0 = 4$$

(a) *Direct comparison method*: From equations (8.99) and (8.111)

$$|s\mathbf{I} - \mathbf{A} + \mathbf{BK}| = s^2 + 4s + 4 \tag{8.113}$$

$$\left| \begin{bmatrix} s & 0 \\ 0 & s \end{bmatrix} - \begin{bmatrix} 0 & 1 \\ 0 & -4 \end{bmatrix} + \begin{bmatrix} 0 \\ 1 \end{bmatrix} [k_1 k_2] \right| = s^2 + 4s + 4$$

$$\left| \begin{bmatrix} s & -1 \\ 0 & s+4 \end{bmatrix} + \begin{bmatrix} 0 & 0 \\ k_1 & k_2 \end{bmatrix} \right| = s^2 + 4s + 4$$

$$\left| \begin{matrix} s & -1 \\ k_1 & s+4+k_2 \end{matrix} \right| = s^2 + 4s + 4$$

$$s^2 + (4 + k_2)s + k_1 = s^2 + 4s + 4 \tag{8.114}$$

From equation (8.114)

$$k_1 = 4$$
$$(4 + k_2) = 4 \quad \text{i.e.} \quad k_2 = 0 \tag{8.115}$$

(b) *Controllable canonical form method*: From equation (8.100)

$$\mathbf{K} = [\alpha_0 - a_0 : \alpha_1 - a_1]\mathbf{T}^{-1}$$

$$= [4 - 0 : 4 - 4]\mathbf{T}^{-1}$$

$$= [4 \quad 0]\mathbf{T}^{-1} \tag{8.116}$$

now

$$\mathbf{T} = \mathbf{MW}$$

where

$$\mathbf{M} = [\mathbf{B} : \mathbf{AB}]$$

$$\mathbf{AB} = \begin{bmatrix} 0 & 1 \\ 0 & -4 \end{bmatrix} \begin{bmatrix} 0 \\ 1 \end{bmatrix} = \begin{bmatrix} 1 \\ -4 \end{bmatrix}$$

giving

$$\mathbf{M} = \begin{bmatrix} 0 & 1 \\ 1 & -4 \end{bmatrix} \tag{8.117}$$

Note that the determinant of $\mathbf{M}$ is non-zero, hence the system is controllable.
From equation (8.102)

$$\mathbf{W} = \begin{bmatrix} a_1 & 1 \\ 1 & 0 \end{bmatrix} = \begin{bmatrix} 4 & 1 \\ 1 & 0 \end{bmatrix}$$

Hence

$$\mathbf{T} = \mathbf{MW} = \begin{bmatrix} 0 & 1 \\ 1 & -4 \end{bmatrix} \begin{bmatrix} 4 & 1 \\ 1 & 0 \end{bmatrix} = \begin{bmatrix} 1 & 0 \\ 0 & 1 \end{bmatrix} = \mathbf{I} \tag{8.118}$$

Thus proving that equation (8.108) is already in the controllable canonical form.
Since $\mathbf{T}^{-1}$ is also $\mathbf{I}$, substitute (8.118) into (8.116)

$$\mathbf{K} = [4 \quad 0]\mathbf{I} = [4 \quad 0] \tag{8.119}$$

(c) *Ackermann's formula*: From (8.103)

$$\mathbf{K} = [0 \quad 1]\mathbf{M}^{-1}\phi(\mathbf{A}) \tag{8.120}$$

From (8.117)

$$\mathbf{M}^{-1} = \frac{1}{-1} \begin{bmatrix} -4 & -1 \\ -1 & 0 \end{bmatrix} = \begin{bmatrix} 4 & 1 \\ 1 & 0 \end{bmatrix} \tag{8.121}$$

From (8.104)

$$\phi(\mathbf{A}) = \mathbf{A}^2 + \alpha_1\mathbf{A} + \alpha_0\mathbf{I}$$

inserting values

$$\phi(\mathbf{A}) = \begin{bmatrix} 0 & 1 \\ 0 & -4 \end{bmatrix}^2 + 4\begin{bmatrix} 0 & 1 \\ 0 & -4 \end{bmatrix} + 4\begin{bmatrix} 1 & 0 \\ 0 & 1 \end{bmatrix}$$

$$= \begin{bmatrix} 0 & -4 \\ 0 & 16 \end{bmatrix} + \begin{bmatrix} 0 & 4 \\ 0 & -16 \end{bmatrix} + \begin{bmatrix} 4 & 0 \\ 0 & 4 \end{bmatrix}$$

$$= \begin{bmatrix} 4 & 0 \\ 0 & 4 \end{bmatrix} \tag{8.122}$$

Insert equations (8.121) and (8.122) into (8.120)

$$\mathbf{K} = \begin{bmatrix} 0 & 1 \end{bmatrix}\begin{bmatrix} 4 & 1 \\ 1 & 0 \end{bmatrix}\begin{bmatrix} 4 & 0 \\ 0 & 4 \end{bmatrix}$$

$$= \begin{bmatrix} 0 & 1 \end{bmatrix}\begin{bmatrix} 16 & 4 \\ 4 & 0 \end{bmatrix}$$

$$\mathbf{K} = \begin{bmatrix} 4 & 0 \end{bmatrix} \tag{8.123}$$

These results agree with the root locus diagram in Figure 5.9, where $K = 4$ produces two real roots of $s = -2$, $s = -2$ (i.e. critical damping).

### 8.4.3 State observers

In section 8.4.2 where state feedback design was discussed, it was assumed that all the state variables were available for the control equation (8.93) for a regulator

$$\mathbf{u} = (\mathbf{r} - \mathbf{Kx})$$

when $\mathbf{r} = 0$

$$\mathbf{u} = -\mathbf{Kx} \tag{8.124}$$

Equations (8.124) requires that all state variables must be measured. In practice this may not happen for a number of reasons including cost, or that the state may not physically be measurable. Under these conditions it becomes necessary, if full state feedback is required, to observe, or estimate the state variables.

A full-order state observer estimates all of the system state variables. If, however, some of the state variables are measured, it may only be necessary to estimate a few of them. This is referred to as a reduced-order state observer. All observers use some form of mathematical model to produce an estimate $\hat{\mathbf{x}}$ of the actual state vector $\mathbf{x}$. Figure 8.8 shows a simple arrangement of a full-order state observer.

In Figure 8.8, since the observer dynamics will never exactly equal the system dynamics, this open-loop arrangement means that $\mathbf{x}$ and $\hat{\mathbf{x}}$ will gradually diverge. If however, an output vector $\hat{\mathbf{y}}$ is estimated and subtracted from the actual output vector $\mathbf{y}$, the difference can be used, in a closed-loop sense, to modify the dynamics of the observer so that the output error $(\mathbf{y} - \hat{\mathbf{y}})$ is minimized. This arrangement, sometimes called a Luenberger observer (1964), is shown in Figure 8.9.

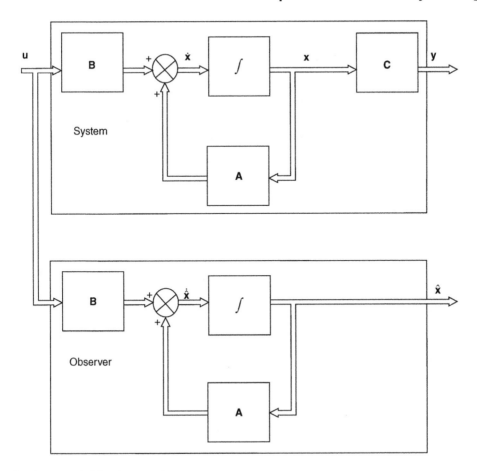

**Fig. 8.8** A simple full-order state observer.

Let the system in Figure 8.9 be defined by

$$\dot{\mathbf{x}} = \mathbf{A}\mathbf{x} + \mathbf{B}\mathbf{u} \tag{8.125}$$

$$\mathbf{y} = \mathbf{C}\mathbf{x} \tag{8.126}$$

Assume that the estimate $\hat{\mathbf{x}}$ of the state vector is

$$\dot{\hat{\mathbf{x}}} = \mathbf{A}\hat{\mathbf{x}} + \mathbf{B}\mathbf{u} + \mathbf{K}_e(\mathbf{y} - \mathbf{C}\hat{\mathbf{x}}) \tag{8.127}$$

where $\mathbf{K}_e$ is the observer gain matrix.

If equation (8.127) is subtracted from (8.125), and $(\mathbf{x} - \hat{\mathbf{x}})$ is the error vector $\mathbf{e}$, then

$$\dot{\mathbf{e}} = (\mathbf{A} - \mathbf{K}_e\mathbf{C})\mathbf{e} \tag{8.128}$$

and, from equation (8.127), the equation for the full-order state observer is

$$\dot{\hat{\mathbf{x}}} = (\mathbf{A} - \mathbf{K}_e\mathbf{C})\hat{\mathbf{x}} + \mathbf{B}\mathbf{u} + \mathbf{K}_e\mathbf{y} \tag{8.129}$$

Thus from equation (8.128) the dynamic behaviour of the error vector depends upon the eigenvalues of $(\mathbf{A} - \mathbf{K}_e\mathbf{C})$. As with any measurement system, these eigenvalues

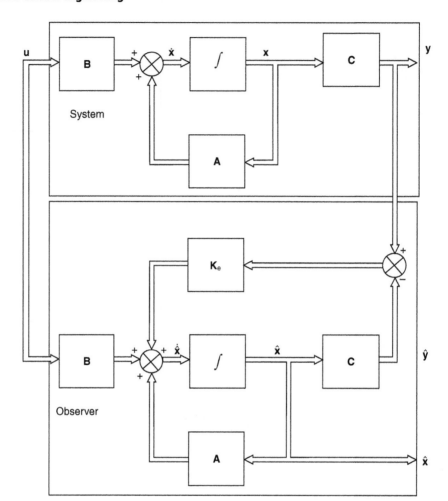

**Fig. 8.9** The Luenberger full-order state observer.

should allow the observer transient response to be more rapid than the system itself (typically a factor of 5), unless a filtering effect is required.

The problem of observer design is essentially the same as the regulator pole placement problem, and similar techniques may be used.

(a) *Direct comparison method*: If the desired locations of the closed-loop poles (eigenvalues) of the observer are

$$s = \mu_1, s = \mu_2, \ldots, s = \mu_n$$

then

$$|s\mathbf{I} - \mathbf{A} + \mathbf{K}_e\mathbf{C}| = (s - \mu_1)(s - \mu_2)\ldots(s - \mu_n)$$
$$= s^n + \alpha_{n-1}s^{n-1} + \cdots + \alpha_1 s + \alpha_0 \quad (8.130)$$

(b) *Observable canonical form method*: For the generalized transfer function shown in Figure 8.4, the observable form of the state equation may be written

$$
\begin{bmatrix} \dot{x}_1 \\ \dot{x}_2 \\ \vdots \\ \dot{x}_n \end{bmatrix} = \begin{bmatrix} 0 & 0 & \dots & 0 & -a_0 \\ 1 & 0 & \dots & 0 & -a_1 \\ \vdots & & & & \vdots \\ 0 & 0 & \dots & 1 & -a_{n-1} \end{bmatrix} \begin{bmatrix} x_1 \\ x_2 \\ \vdots \\ x_n \end{bmatrix} + \begin{bmatrix} b_0 \\ b_1 \\ \vdots \\ b_{n-1} \end{bmatrix} u
$$

$$
y = \begin{bmatrix} 0 & 0 & \dots & 0 & 1 \end{bmatrix} \begin{bmatrix} x_1 \\ x_2 \\ \vdots \\ x_n \end{bmatrix}
$$

(8.131)

Note that the system matrix of the observable canonical form is the transpose of the controllable canonical form given in equation (8.33).

The value of the observer gain matrix $\mathbf{K}_e$ can be calculated directly using

$$
\mathbf{K}_e = \mathbf{Q} \begin{bmatrix} \alpha_0 - a_0 \\ \alpha_1 - a_1 \\ \vdots \\ \alpha_{n-1} - a_{n-1} \end{bmatrix}
$$

(8.132)

$\mathbf{Q}$ is a transformation matrix that transforms the system state equation into the observable canonical form

$$
\mathbf{Q} = (\mathbf{WN}^{\mathrm{T}})^{-1}
$$

(8.133)

where $\mathbf{W}$ is defined in equation (8.102) and $\mathbf{N}$ is the observability matrix given in equation (8.89). If the equation is in the observable canonical form then $\mathbf{Q} = \mathbf{I}$.

(c) *Ackermann's formula*: As with regulator design, this is only applicable to systems where u(t) and y(t) are scalar quantities. It may be used to calculate the observer gain matrix as follows

$$
\mathbf{K}_e = \phi(\mathbf{A})\mathbf{N}^{-1}[0 \quad 0 \quad \dots \quad 0 \quad 1]^{T}
$$

or alternatively

$$
\mathbf{K}_e = \phi(\mathbf{A}) \begin{bmatrix} \mathbf{C} \\ \mathbf{CA} \\ \vdots \\ \mathbf{CA}^{n-1} \end{bmatrix}^{-1} \begin{bmatrix} 0 \\ 0 \\ \vdots \\ 1 \end{bmatrix}
$$

(8.134)

where $\phi(\mathbf{A})$ is defined in equation (8.104).

*Example 8.12* (See also Appendix 1, *examp812.m*)
A system is described by

$$
\begin{bmatrix} \dot{x}_1 \\ \dot{x}_2 \end{bmatrix} = \begin{bmatrix} 0 & 1 \\ -2 & -3 \end{bmatrix} \begin{bmatrix} x_1 \\ x_2 \end{bmatrix} + \begin{bmatrix} 0 \\ 1 \end{bmatrix} u
$$

$$
y = \begin{bmatrix} 1 & 0 \end{bmatrix} \begin{bmatrix} x_1 \\ x_2 \end{bmatrix}
$$

Design a full-order observer that has an undamped natural frequency of 10 rad/s and a damping ratio of 0.5.

*Solution*

From equation (8.89), the observability matrix is

$$\mathbf{N} = \begin{bmatrix} \mathbf{C}^T : \mathbf{A}^T\mathbf{C}^T \end{bmatrix} = \begin{bmatrix} 1 & 0 \\ 0 & 1 \end{bmatrix} \tag{8.135}$$

**N** is of rank 2 and therefore non-singular, hence the system is completely observable and the calculation of an appropriate observer gain matrix $\mathbf{K}_e$ realizable.

Open-loop eigenvalues:

$$|s\mathbf{I} - \mathbf{A}| = s^2 + 3s + 2 = s^2 + a_1 s + a_0 \tag{8.136}$$

Hence

$$a_0 = 2, \quad a_1 = 3$$

And the open-loop eigenvalues are

$$s^2 + 3s + 2 = 0$$
$$(s + 1)(s + 2) = 0$$
$$s = -1, \quad s = -2 \tag{8.137}$$

Desired closed-loop eigenvalues:

$$s^2 + 2\zeta\omega_n s + \omega_n^2 = 0$$
$$s^2 + 10s + 100 = s^2 + \alpha_1 s + \alpha_0 = 0 \tag{8.138}$$

Hence

$$\alpha_0 = 100, \quad \alpha_1 = 10$$

and the desired closed-loop eigenvalues are the roots of equation (8.138)

$$\mu_1 = -5 + j8.66, \quad \mu_2 = -5 - j8.66 \tag{8.139}$$

(a) *Direct comparison method*: From equation (8.130)

$$|s\mathbf{I} - \mathbf{A} + \mathbf{K}_e\mathbf{C}| = s^2 + \alpha_1 + \alpha_0$$

$$\left| \begin{bmatrix} s & 0 \\ 0 & s \end{bmatrix} - \begin{bmatrix} 0 & 1 \\ -2 & -3 \end{bmatrix} + \begin{bmatrix} k_{e1} \\ k_{e2} \end{bmatrix} [1 \ \ 0] \right| = s^2 + 10s + 100$$

$$\left| \begin{bmatrix} s & -1 \\ 2 & s+3 \end{bmatrix} + \begin{bmatrix} k_{e1} & 0 \\ k_{e2} & 0 \end{bmatrix} \right| = s^2 + 10s + 100$$

$$\begin{vmatrix} s + k_{e1} & -1 \\ 2 + k_{e2} & s+3 \end{vmatrix} = s^2 + 10s + 100$$

$$s^2 + (3 + k_{e1})s + (3k_{e1} + 2 + k_{e2}) = s^2 + 10s + 100 \tag{8.140}$$

From equation (8.140)

$$(3 + k_{e1}) = 10, \quad k_{e1} = 7 \tag{8.141}$$

$$(3k_{e1} + 2 + k_{e2}) = 100$$
$$k_{e2} = 100 - 2 - 21 = 77 \tag{8.142}$$

(b) *Observable canonical form method*: From equation (8.132)

$$\mathbf{K}_e = \mathbf{Q} \begin{bmatrix} \alpha_0 - a_0 \\ \alpha_1 - a_1 \end{bmatrix}$$

$$= \mathbf{Q} \begin{bmatrix} 100 - 2 \\ 10 - 3 \end{bmatrix}$$

$$= \mathbf{Q} \begin{bmatrix} 98 \\ 7 \end{bmatrix} \tag{8.143}$$

From equation (8.133)

$$\mathbf{Q} = (\mathbf{W}\mathbf{N}^{\mathrm{T}})^{-1}$$

and from equation (8.102)

$$\mathbf{W} = \begin{bmatrix} a_1 & 1 \\ 1 & 0 \end{bmatrix} = \begin{bmatrix} 3 & 1 \\ 1 & 0 \end{bmatrix} \tag{8.144}$$

Since from equation (8.135)

$$\mathbf{N} = \begin{bmatrix} 1 & 0 \\ 0 & 1 \end{bmatrix}, \quad \mathbf{N}^{\mathrm{T}} = \begin{bmatrix} 1 & 0 \\ 0 & 1 \end{bmatrix} \tag{8.145}$$

Thus

$$\mathbf{W}\mathbf{N}^{\mathrm{T}} = \begin{bmatrix} 3 & 1 \\ 1 & 0 \end{bmatrix} \begin{bmatrix} 1 & 0 \\ 0 & 1 \end{bmatrix} = \begin{bmatrix} 3 & 1 \\ 1 & 0 \end{bmatrix} \tag{8.146}$$

and

$$\mathbf{Q} = \frac{1}{-1} \begin{bmatrix} 0 & -1 \\ -1 & 3 \end{bmatrix} = \begin{bmatrix} 0 & 1 \\ 1 & -3 \end{bmatrix} \tag{8.147}$$

Since $\mathbf{Q} \neq \mathbf{I}$ then $\mathbf{A}$ is not in the observable canonical form.
From equation (8.143)

$$\mathbf{K}_e = \begin{bmatrix} 0 & 1 \\ 1 & -3 \end{bmatrix} \begin{bmatrix} 98 \\ 7 \end{bmatrix} = \begin{bmatrix} 7 \\ 77 \end{bmatrix} \tag{8.148}$$

(c) *Ackermann's Formula*: From (8.134)

$$\mathbf{K}_e = \phi(\mathbf{A}) \begin{bmatrix} \mathbf{C} \\ \mathbf{CA} \end{bmatrix}^{-1} \begin{bmatrix} 0 \\ 1 \end{bmatrix}$$

Using the definition of $\phi(\mathbf{A})$ in equation (8.104)

$$\mathbf{K}_e = (\mathbf{A}^2 + \alpha_1 \mathbf{A} + \alpha_0 \mathbf{I}) \begin{bmatrix} 1 & 0 \\ 0 & 1 \end{bmatrix}^{-1} \begin{bmatrix} 0 \\ 1 \end{bmatrix} \tag{8.149}$$

$$
\begin{aligned}
\mathbf{K}_e &= \left[ \begin{bmatrix} -2 & -3 \\ 6 & 7 \end{bmatrix} + \begin{bmatrix} 0 & 10 \\ -20 & -30 \end{bmatrix} + \begin{bmatrix} 100 & 0 \\ 0 & 100 \end{bmatrix} \right] \begin{bmatrix} 1 & 0 \\ 0 & 1 \end{bmatrix} \begin{bmatrix} 0 \\ 1 \end{bmatrix} \\
&= \begin{bmatrix} 98 & 7 \\ 14 & 77 \end{bmatrix} \begin{bmatrix} 1 & 0 \\ 0 & 1 \end{bmatrix} \begin{bmatrix} 0 \\ 1 \end{bmatrix} \\
&= \begin{bmatrix} 98 & 7 \\ 14 & 77 \end{bmatrix} \begin{bmatrix} 0 \\ 1 \end{bmatrix} = \begin{bmatrix} 7 \\ 77 \end{bmatrix} \tag{8.150}
\end{aligned}
$$

## 8.4.4   Effect of a full-order state observer on a closed-loop system

Figure 8.10 shows a closed-loop system that includes a full-order state observer. In Figure 8.10 the system equations are

$$\dot{\mathbf{x}} = \mathbf{A}\mathbf{x} + \mathbf{B}\mathbf{u}$$
$$\mathbf{y} = \mathbf{C}\mathbf{x} \tag{8.151}$$

The control is implemented using observed state variables

$$\mathbf{u} = -\mathbf{K}\hat{\mathbf{x}} \tag{8.152}$$

If the difference between the actual and observed state variables is

$$\mathbf{e}(t) = \mathbf{x}(t) - \hat{\mathbf{x}}(t)$$

then

$$\hat{\mathbf{x}}(t) = \mathbf{x}(t) - \mathbf{e}(t) \tag{8.153}$$

Combining equations (8.151), (8.152) and (8.153) gives the closed-loop equations

$$\dot{\mathbf{x}} = \mathbf{A}\mathbf{x} - \mathbf{B}\mathbf{K}(\mathbf{x} - \mathbf{e})$$
$$= (\mathbf{A} - \mathbf{B}\mathbf{K})\mathbf{x} + \mathbf{B}\mathbf{K}\mathbf{e} \tag{8.154}$$

The observer error equation from equation (8.128) is

$$\dot{\mathbf{e}} = (\mathbf{A} - \mathbf{K}_e \mathbf{C})\mathbf{e} \tag{8.155}$$

Combining equations (8.154) and (8.155) gives

$$\begin{bmatrix} \dot{\mathbf{x}} \\ \dot{\mathbf{e}} \end{bmatrix} = \begin{bmatrix} \mathbf{A} - \mathbf{B}\mathbf{K} & \mathbf{B}\mathbf{K} \\ 0 & \mathbf{A} - \mathbf{B}_e \mathbf{C} \end{bmatrix} \begin{bmatrix} \mathbf{x} \\ \mathbf{e} \end{bmatrix} \tag{8.156}$$

Equation (8.156) describes the closed-loop dynamics of the observed state feedback control system and the characteristic equation is therefore

$$|s\mathbf{I}\mathbf{A} + \mathbf{B}\mathbf{K}||s\mathbf{I} - \mathbf{A} + \mathbf{K}_e \mathbf{C}| = 0 \tag{8.157}$$

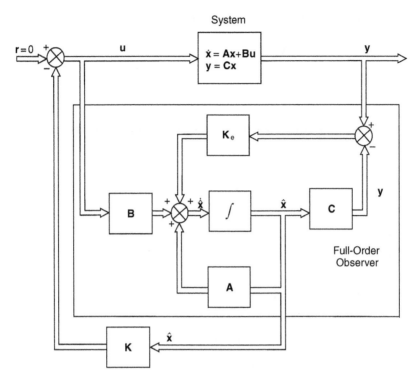

**Fig. 8.10** Closed-loop control system with full-order observer state feedback.

Equation (8.157) shows that the desired closed-loop poles for the control system are not changed by the introduction of the state observer. Since the observer is normally designed to have a more rapid response than the control system with full order observed state feedback, the pole-placement roots will dominate.

Using the state vectors $\mathbf{x}(t)$ and $\hat{\mathbf{x}}(t)$ the state equations for the closed-loop system are

From equations (8.151) and (8.152)

$$\dot{\mathbf{x}} = \mathbf{A}\mathbf{x} - \mathbf{B}\mathbf{K}\hat{\mathbf{x}} \tag{8.158}$$

and from equation (8.129)

$$\begin{aligned}
\dot{\hat{\mathbf{x}}} &= (\mathbf{A} - \mathbf{K_e}\mathbf{C})\hat{\mathbf{x}} - \mathbf{B}\mathbf{K}\hat{\mathbf{x}} + \mathbf{K_e}\mathbf{C}\mathbf{x} \\
&= (\mathbf{A} - \mathbf{K_e}\mathbf{C} - \mathbf{B}\mathbf{K})\hat{\mathbf{x}} + \mathbf{K_e}\mathbf{C}\mathbf{x}
\end{aligned} \tag{8.159}$$

Thus the closed-loop state equations are

$$\begin{bmatrix} \dot{\mathbf{x}} \\ \dot{\hat{\mathbf{x}}} \end{bmatrix} = \begin{bmatrix} \mathbf{A} & -\mathbf{B}\mathbf{K} \\ \mathbf{K_e}\mathbf{C} & \mathbf{A} - \mathbf{K_e}\mathbf{C} - \mathbf{B}\mathbf{K} \end{bmatrix} \begin{bmatrix} \mathbf{x} \\ \hat{\mathbf{x}} \end{bmatrix} \tag{8.160}$$

### 8.4.5 Reduced-order state observers

A full-order state observer estimates all state variables, irrespective of whether they are being measured. In practice, it would appear logical to use a combination of measured states from $y = Cx$ and observed states (for those state variables that are either not being measured, or not being measured with sufficient accuracy).

If the state vector is of $n$th order and the measured output vector is of $m$th order, then it is only necessary to design an $(n - m)$th order state observer.

Consider the case of the measurement of a single state variable $x_1(t)$. The output equation is therefore

$$y = x_1 = \mathbf{Cx} = [1 \quad 0 \ \ldots \ 0]\mathbf{x} \tag{8.161}$$

Partition the state vector

$$\mathbf{x} = \begin{bmatrix} x_1 \\ \mathbf{x}_e \end{bmatrix} \tag{8.162}$$

where $\mathbf{x}_e$ are the state variables to be observed.

Partition the state equations

$$\begin{bmatrix} \dot{x}_1 \\ \dot{\mathbf{x}}_e \end{bmatrix} = \begin{bmatrix} a_{11} & \mathbf{A}_{1e} \\ \mathbf{A}_{e1} & \mathbf{A}_{ee} \end{bmatrix} \begin{bmatrix} x_1 \\ x_2 \end{bmatrix} + \begin{bmatrix} b_1 \\ \mathbf{B}_e \end{bmatrix} u \tag{8.163}$$

If the desired eigenvalues for the reduced-order observer are

$$s = \mu_{1e}, \ s = \mu_{2e}, \ \ldots, \ s = \mu_{(n-1)e}$$

Then it can be shown that the characteristic equation for the reduced-order observer is

$$|s\mathbf{I} - \mathbf{A}_{ee} + \mathbf{K}_e\mathbf{A}_{1e}| = (s - \mu_{1e})\ldots(s - \mu_{(n-1)e})$$
$$= s^{n-1} + \alpha_{(n-2)e}s^{n-2} + \cdots + \alpha_{1e}s + \alpha_{oe} \tag{8.164}$$

In equation (8.164) $\mathbf{A}_{ee}$ replaces $\mathbf{A}$ and $\mathbf{A}_{1e}$ replaces $\mathbf{C}$ in the full-order observer.

The reduced-order observer gain matrix $\mathbf{K}_e$ can also be obtained using appropriate substitutions into equations mentioned earlier. For example, equation (8.132) becomes

$$\mathbf{K}_e = \mathbf{Q}_e \begin{bmatrix} \alpha_{oe} - a_{oe} \\ \alpha_{1e} - a_{1e} \\ \vdots \\ \alpha_{(n-2)e} - a_{(n-2)e} \end{bmatrix} \tag{8.165}$$

where $a_{oe}, \ldots, a_{(n-2)e}$ are the coefficients of the open-loop reduced order characteristics equation

$$|s\mathbf{I} - \mathbf{A}_{ee}| = s^{n-1} + a_{(n-2)e}s^{n-2} + a_{1e}s + a_{oe} \tag{8.166}$$

and

$$\mathbf{Q}_e = (\mathbf{W}_e\mathbf{N}_e^{\mathrm{T}})^{-1} \tag{8.167}$$

where

$$\mathbf{W}_e = \begin{bmatrix} a_1 & a_2 & \ldots & a_{n-2} & 1 \\ a_2 & & a_3 & \ldots & 1 & 0 \\ \vdots & & & & \\ a_{n-2} & & 1 & \ldots & 0 & 0 \\ 1 & & 0 & \ldots & 0 & 0 \end{bmatrix} \tag{8.168}$$

and

$$\mathbf{N}_e = \left[ \mathbf{A}_{1e}^T : \mathbf{A}_{ee} T \mathbf{A}_{1e}^T : \ldots : (\mathbf{A}_{ee} T)^{n-2} \mathbf{A}_{1e}^T \right] \tag{8.169}$$

and Ackermann's formula becomes

$$\mathbf{K}_e = \phi(\mathbf{A}_{ee}) \begin{bmatrix} A_{1e} \\ A_{1e} A_{ee} \\ \vdots \\ A_{1e} A_{ee}^{n-3} \\ A_{1e} A_{ee}^{n-2} \end{bmatrix}^{-1} \begin{bmatrix} 0 \\ 0 \\ \vdots \\ 0 \\ 1 \end{bmatrix} \tag{8.170}$$

where

$$\phi(\mathbf{A}_{ee}) = \mathbf{A}_{ee}^{n-1} + \alpha_{n-2}\mathbf{A}_{ee}^{n-2} + \cdots + \alpha_2\mathbf{A}_{ee} + \alpha_1\mathbf{I} \tag{8.171}$$

Define

$$\dot{\hat{\mathbf{x}}}_{e1} = \hat{\mathbf{x}} - \mathbf{K}_e y$$

Then

$$\hat{\mathbf{x}} = \hat{\mathbf{x}}_{e1} - \mathbf{K}_e y \tag{8.172}$$

The equation for the reduced-order observer can be shown to be

$$\dot{\hat{\mathbf{x}}}_{e1} = (\mathbf{A}_{ee}\mathbf{K}_e\mathbf{A}_{1e})\hat{\mathbf{x}}_{e1} + \{\mathbf{A}_{e1}\mathbf{K}_e a_{11} + (\mathbf{A}_{ee} - \mathbf{K}_e\mathbf{A}_{1e})\mathbf{K}_e\}y + (\mathbf{B}_e - \mathbf{K}_e b_1)u \tag{8.173}$$

Figure 8.11 shows the implementation of a reduced-order state observer.

## Case study

*Example 8.13*   (See also Appendix 1, *examp813.m*)
(a) In case study Example 5.10 a control system has an open-loop transfer function

$$G(s)H(s) = \frac{1}{s(s+2)(s+5)}$$

The controller was a PD compensator of the form

$$G(s) = K_1(s+a)$$

With $K_1 = 15$ and $a = 1$, the system closed-loop poles were

$$s = -3.132 \pm j3.253$$

$$s = -0.736$$

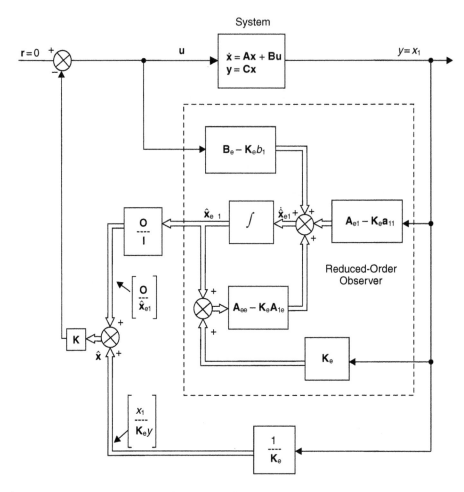

**Fig. 8.11** Implementation of a reduced-order state observer.

with the resulting characteristic equation

$$s^3 + 7s^2 + 25s + 15 = 0$$

Demonstrate that the same result can be achieved using state feedback methods.

(b) Design a reduced second-order state observer for the system such that the poles are a factor of 10 higher than the closed-loop system poles, i.e.

$$s = -31.32 \pm j32.53$$

which correspond to $\omega_n = 45.16\,\text{rad/s}$ and $\zeta = 0.7$.

*Solution*

(a)
$$G(s)H(s) = \frac{1}{s^3 + 7s^2 + 10s + 0}$$

From equations (8.33) and (8.34)

$$\mathbf{A} = \begin{bmatrix} 0 & 1 & 0 \\ 0 & 0 & 1 \\ 0 & -10 & -7 \end{bmatrix} \quad \mathbf{B} = \begin{bmatrix} 0 \\ 0 \\ 1 \end{bmatrix} \quad \mathbf{C} = [1 \quad 0 \quad 0] \tag{8.174}$$

Open-loop characteristic equation

$$s^3 + 7s^2 + 10s + 0 = 0$$
$$s^3 + a_2 s^2 + a_1 s + a_0 = 0 \tag{8.175}$$

Closed-loop characteristic equation

$$s^3 + 7s^2 + 25s + 15 = 0$$
$$s^3 + \alpha_2 s^2 + \alpha_1 s + \alpha_0 = 0 \tag{8.176}$$

Using direct comparison method

$$|s\mathbf{I} - \mathbf{A} + \mathbf{BK}| = s^3 + 7s^2 + 25s + 15$$

$$\left| \begin{bmatrix} s & 0 & 0 \\ 0 & s & 0 \\ 0 & 0 & s \end{bmatrix} - \begin{bmatrix} 0 & 1 & 0 \\ 0 & 0 & 1 \\ 0 & -10 & -7 \end{bmatrix} + \begin{bmatrix} 0 \\ 0 \\ 1 \end{bmatrix}[k_1 \quad k_2 \quad k_3] \right| = s^3 + 7s^2 + 25s + 15$$

$$\left| \begin{bmatrix} s & -1 & 0 \\ 0 & s & -1 \\ 0 & 10 & s+7 \end{bmatrix} + \begin{bmatrix} 0 & 0 & 0 \\ 0 & 0 & 0 \\ k_1 & k_2 & k_3 \end{bmatrix} \right| = s^3 + 7s^2 + 25s + 15$$

$$\begin{vmatrix} s & -1 & 0 \\ 0 & s & -1 \\ k_1 & 10+k_2 & s+7+k_3 \end{vmatrix} = s^3 + 7s^2 + 25s + 15$$

$$\tag{8.177}$$

Expanding the determinant in equation (8.177) gives

$$k_1 = 15, \quad k_2 = 15, \quad k_3 = 0$$

Hence

$$u = -[15 \quad 15 \quad 0] \begin{bmatrix} x_1 \\ x_2 \\ x_3 \end{bmatrix} \tag{8.178}$$

since $x_2 = \dot{x}_1$, this is identical to the original PD controller

$$G(s) = 15(s+1)$$

Although the solution is the same, the important difference is with state feedback, the closed-loop poles are placed at directly the required locations. With root locus, a certain amount of trial and error in placing open-loop zeros was required to achieve the desired closed-loop locations.

(b) *Reduced-order state observer*: Partitioning the system equation $\mathbf{A}$ in (8.174) and inserting in equation (8.164)

$$\left| \begin{bmatrix} s & 0 \\ 0 & s \end{bmatrix} - \begin{bmatrix} 0 & 1 \\ -10 & -7 \end{bmatrix} + \begin{bmatrix} k_{e1} \\ k_{e2} \end{bmatrix} [1 \quad 0] \right| = s^2 + 2\zeta\omega_n s + \omega_n^2$$

$$\left| \begin{bmatrix} s & -1 \\ 10 & s+7 \end{bmatrix} + \begin{bmatrix} k_{e1} & 0 \\ k_{e2} & 0 \end{bmatrix} \right| = s^2 + 63.2s + 2039.4$$

$$\left| \begin{matrix} s+k_{e1} & -1 \\ 10+k_{e2} & s+7 \end{matrix} \right| = s^2 + \alpha_{1e}s + \alpha_{0e}$$

$$s^2 + (7+k_{e1})s + (7k_{e1}+10+k_{e2}) = s^2 + 63.2s + 2039.4 \tag{8.179}$$

Equating coefficients in equation (8.179)

$$(7+k_{e1}) = 63.2 \quad k_{e1} = 56.2 \tag{8.180}$$

$$(7 \times 56.2 + 10 + k_{e2}) = 2039.4$$

$$k_{e2} = 1636 \tag{8.181}$$

Referring to Figure 8.11 and partitioned systems (8.174) and (8.163)

$$
\begin{aligned}
\mathbf{B}_e - \mathbf{K}_e b_1 &= \begin{bmatrix} 0 \\ 1 \end{bmatrix} - \begin{bmatrix} 56.2 \\ 1636 \end{bmatrix} 0 = \begin{bmatrix} 0 \\ 1 \end{bmatrix} \\
\mathbf{A}_{e1} - \mathbf{K}_e a_{11} &= \begin{bmatrix} 0 \\ 0 \end{bmatrix} - \begin{bmatrix} 56.2 \\ 1636 \end{bmatrix} 0 = \begin{bmatrix} 0 \\ 0 \end{bmatrix} \\
\mathbf{A}_{ee} - \mathbf{K}_e \mathbf{A}_{1e} &= \begin{bmatrix} 0 & 1 \\ -10 & -7 \end{bmatrix} - \begin{bmatrix} 56.2 \\ 1636 \end{bmatrix} [1 \quad 0] = \begin{bmatrix} -56.2 & 1 \\ -1646 & -7 \end{bmatrix}
\end{aligned}
\tag{8.182}
$$

Inserting equation (8.182) into Figure 8.11 gives the complete state feedback and reduced observer system shown in Figure 8.12.

Comparing the system shown in Figure 8.12 with the original PD controller given in Example 5.10, the state feedback system may be considered to be a PD controller where the proportional term uses measured output variables and the derivative term uses observed state variables.

## 8.5  Further problems

*Example 8.14*
For the d.c. motor shown in Figure 4.14, the potential difference across the armature winding is given by equation (4.21)

$$e_a(t) - e_b(t) = L_a \frac{di_a}{dt} + R_a i_a(t)$$

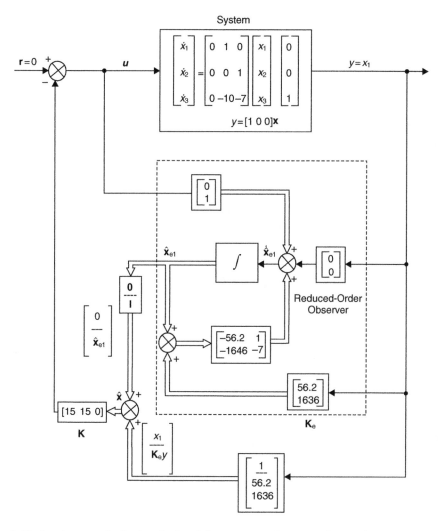

**Fig. 8.12** Complete state feedback and reduced observer system for case study Example 8.11.

where, from equation (4.20)

$$e_b(t) = K_b \frac{d\theta}{dt}$$

and the torque $T_m(t)$ developed by the motor is given by equation (4.18)

$$T_m(t) = K_a i_a(t)$$

If the load consists of a rotor of moment of inertia $I$ and a damping device of damping coefficient $C$, then the load dynamics are

$$T_m(t) - C\frac{d\theta}{dt} = I\frac{d^2\theta}{dt^2}$$

where $\theta$ is the angular displacement of the rotor.

(a) Determine the state and output equations when the state and control variables are

$$x_1 = \theta, \quad x_2 = \dot{x}_1, \quad x_3 = i_a, \quad u = e_a$$

(b) Determine the state and output equations when the state and control variables are

$$x_1 = \theta, \quad x_2 = \dot{x}_1, \quad x_3 = \dot{x}_2, \quad u = e_a$$

*Solution*

(a)

$$\begin{bmatrix} \dot{x}_1 \\ \dot{x}_2 \\ \dot{x}_3 \end{bmatrix} = \begin{bmatrix} 0 & 1 & 0 \\ 0 & -\dfrac{C}{I} & \dfrac{K_a}{I} \\ 0 & -\dfrac{K_b}{L_a} & -\dfrac{R_a}{L_a} \end{bmatrix} \begin{bmatrix} x_1 \\ x_2 \\ x_3 \end{bmatrix} + \begin{bmatrix} 0 \\ 0 \\ \dfrac{1}{L_a} \end{bmatrix} u$$

$$\theta = [1 \quad 0 \quad 0]\mathbf{x}$$

(b)

$$\begin{bmatrix} \dot{x}_1 \\ \dot{x}_2 \\ \dot{x}_3 \end{bmatrix} = \begin{bmatrix} 0 & 1 & 0 \\ 0 & 0 & 1 \\ 0 & -\dfrac{(K_a K_b + R_a C)}{L_a I} & -\left(\dfrac{R_a}{L_a} + \dfrac{C}{I}\right) \end{bmatrix} \begin{bmatrix} x_1 \\ x_2 \\ x_3 \end{bmatrix} + \begin{bmatrix} 0 \\ 0 \\ \dfrac{K_a}{L_a I} \end{bmatrix} u$$

$$\theta = [1 \quad 0 \quad 0]\mathbf{x}$$

*Example 8.15*

Find the state and output equations for the positional servomechanism shown in Figure 8.13 when the state and control variable are

$$x_1 = c(t), \quad x_2 = \dot{x}_1, \quad u = r(t)$$

*Solution*

$$\begin{bmatrix} \dot{x}_1 \\ \dot{x}_2 \end{bmatrix} = \begin{bmatrix} 0 & 1 \\ -\dfrac{K}{m} & -\dfrac{C}{m} \end{bmatrix} \begin{bmatrix} x_1 \\ x_2 \end{bmatrix} + \begin{bmatrix} 0 \\ \dfrac{K}{m} \end{bmatrix} u$$

$$c = [1 \quad 0]\mathbf{x}$$

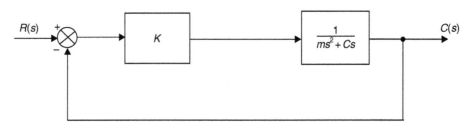

**Fig. 8.13** Block diagram of positional servomechanism.

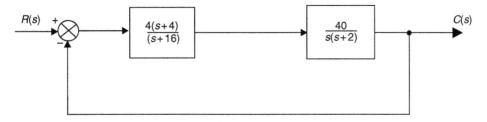

**Fig. 8.14** Closed-loop control system.

*Example 8.16*
Find the state and output equations for the closed-loop control system shown in Figure 8.14 when the state and control variables are

$$x_1 = c(t), \quad x_2 = \dot{x}_1, \quad x_3 = \dot{x}_2, \quad u = r(t)$$

*Solution*

$$\begin{bmatrix} \dot{x}_1 \\ \dot{x}_2 \\ \dot{x}_3 \end{bmatrix} = \begin{bmatrix} 0 & 1 & 0 \\ 0 & 0 & 1 \\ -640 & -192 & 18 \end{bmatrix} \begin{bmatrix} x_1 \\ x_2 \\ x_3 \end{bmatrix} + \begin{bmatrix} 0 \\ 0 \\ 1 \end{bmatrix} u$$

$$c = [\,640 \quad 160 \quad 0\,]\mathbf{x}$$

*Example 8.17*
Figure 8.15 shows the block diagram representation of a car cruise control system where $U(s)$ is the desired speed, $X(s)$ is the accelerator position and $V(s)$ is the actual speed.

(a) Find the state and output equations when the state and control variables are

$$x_1 = x(t), \quad x_2 = v(t), \quad u = u(t)$$

(b) Determine the continuous-time state transition matrix $\boldsymbol{\Phi}(t)$.
(c) For a sampling time of 0.1 seconds, evaluate from $\boldsymbol{\Phi}(t)$ the discrete-time state transition matrix $\mathbf{A}(T)$.

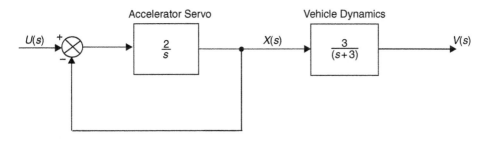

**Fig. 8.15** Car cruise control system.

(d) Using the first three terms of equation (8.80), compute the discrete-time control transition matrix $\mathbf{B}(T)$. Using the difference equations

$$\mathbf{x}(k+1)T = \mathbf{A}(T)\mathbf{x}(kT) + \mathbf{B}\mathbf{u}(kT)$$

determine values for the state variables when $\mathbf{u}(kT)$ is a piece-wise constant function of the form

| $kT$ (sec) | 0 | 0.1 | 0.2 | 0.3 | 0.4 |
|---|---|---|---|---|---|
| $\mathbf{u}(kT)$ | 10 | 15 | 20 | 25 | 30 |

Assume zero initial conditions.

*Solution*

(a) $\begin{bmatrix} \dot{x}_1 \\ \dot{x}_2 \end{bmatrix} = \begin{bmatrix} -2 & 0 \\ 3 & -3 \end{bmatrix} \begin{bmatrix} 2 \\ 0 \end{bmatrix} u$

$v = [0 \quad 1]\mathbf{x}$

(b) $\mathbf{\Phi}(t) = \begin{bmatrix} e^{-2t} & 0 \\ 3(-e^{-3t} + e^{-2t}) & e^{-3t} \end{bmatrix}$

(c) $\mathbf{A}(T) = \begin{bmatrix} 0.819 & 0 \\ 0.234 & 0.741 \end{bmatrix}$

(d) $\mathbf{B}(T) = \begin{bmatrix} 0.181 \\ 0.025 \end{bmatrix}$

| $kT$ (sec) | 0 | 0.1 | 0.2 | 0.3 | 0.4 |
|---|---|---|---|---|---|
| $u(kT)$ | 10 | 15 | 20 | 25 | 30 |
| $x_1$ | 0 | 1.81 | 4.197 | 7.057 | 10.305 |
| $x_2$ | 0 | 0.25 | 0.984 | 2.211 | 3.915 |

*Example 8.18*
The ship roll stabilization system given in case-study Example 5.11 has a forward-path transfer function

$$\frac{\phi_a}{\delta_d}(s) = \frac{K}{(s+1)(s^2 + 0.7s + 2)}$$

(a) For the condition $K = 1$, find the state and output equations when $x_1 = \phi_a(t)$, $x_2 = \dot{x}_1$, $\dot{x}_3 = \dot{x}_2$ and $u = \delta_d(t)$.
(b) Calculate the controllability matrix $\mathbf{M}$ and the observability matrix $\mathbf{N}$ and demonstrate that the system is fully controllable and fully observable.
(c) Determine the state feedback gain matrix $\mathbf{K}$ that produces a set of desired closed-loop poles

$$s = -3.234 \pm j3.3$$
$$s = -3.2$$

(d) Find the observer gain matrix $\mathbf{K}_e$ for a full-order state observer that produces a set of desired closed-loop poles

$$s = -16.15 \pm j16.5$$
$$s = -16$$

(e) If output $\phi_a(t) = x_1$ is measured, design a reduced-order state observer with desired closed-loop poles

$$s = -16.15 \pm j16.5$$

*Solution*

(a) $\begin{bmatrix} \dot{x}_1 \\ \dot{x}_2 \\ \dot{x}_3 \end{bmatrix} = \begin{bmatrix} 0 & 1 & 0 \\ 0 & 0 & 1 \\ -2 & -2.7 & -1.7 \end{bmatrix} \begin{bmatrix} x_1 \\ x_2 \\ x_3 \end{bmatrix} + \begin{bmatrix} 0 \\ 0 \\ 1 \end{bmatrix} u$

$\phi_a = \begin{bmatrix} 1 & 0 & 0 \end{bmatrix} \mathbf{x}$

(b) $\mathbf{M} = \begin{bmatrix} 0 & 0 & 1 \\ 0 & 1 & -1.7 \\ 1 & -1.7 & 0.19 \end{bmatrix}$

$\det(\mathbf{M}) = -1$, $\mathrm{rank}(\mathbf{M}) = 3$

System fully controllable.

$\mathbf{N} = \begin{bmatrix} 1 & 0 & 0 \\ 0 & 1 & 0 \\ 0 & 0 & 1 \end{bmatrix}$

$\det(\mathbf{N}) = 1$, $\mathrm{rank}(\mathbf{N}) = 3$

System fully observable.

(c) $\mathbf{K} = \begin{bmatrix} 66.29 & 39.34 & 7.968 \end{bmatrix}$

(d) $\mathbf{K}_e = \begin{bmatrix} 8527.2 & 1047.2 & 46.6 \end{bmatrix}$

(e) $\mathbf{K}_e = \begin{bmatrix} 530.3 & 30.6 \end{bmatrix}$

# 9

# Optimal and robust control system design

## 9.1 Review of optimal control

An optimal control system seeks to maximize the return from a system for the minimum cost. In general terms, the optimal control problem is to find a control $\mathbf{u}$ which causes the system

$$\dot{\mathbf{x}} = g(\mathbf{x}(t), \mathbf{u}(t), t) \tag{9.1}$$

to follow an optimal trajectory $\mathbf{x}(t)$ that minimizes the performance criterion, or cost function

$$J = \int_{t_0}^{t_1} h(\mathbf{x}(t), \mathbf{u}(t), t)\mathrm{d}t \tag{9.2}$$

The problem is one of constrained functional minimization, and has several approaches.

Variational calculus, Dreyfus (1962), may be employed to obtain a set of differential equations with certain boundary condition properties, known as the Euler–Lagrange equations. The maximum principle of Pontryagin (1962) can also be applied to provide the same boundary conditions by using a Hamiltonian function.

An alternative procedure is the dynamic programming method of Bellman (1957) which is based on the principle of optimality and the imbedding approach. The principle of optimality yields the Hamilton–Jacobi partial differential equation, whose solution results in an optimal control policy. Euler–Lagrange and Pontryagin's equations are applicable to systems with non-linear, time-varying state equations and non-quadratic, time varying performance criteria. The Hamilton–Jacobi equation is usually solved for the important and special case of the linear time-invariant plant with quadratic performance criterion (called the performance index), which takes the form of the matrix Riccati (1724) equation. This produces an optimal control law as a linear function of the state vector components which is always stable, providing the system is controllable.

### 9.1.1 Types of optimal control problems

(a) *The terminal control problem*: This is used to bring the system as close as possible to a given terminal state within a given period of time. An example is an automatic aircraft landing system, whereby the optimum control policy will focus on minimizing errors in the state vector at the point of landing.

(b) *The minimum-time control problem*: This is used to reach the terminal state in the shortest possible time period. This usually results in a 'bang–bang' control policy whereby the control is set to $\mathbf{u}_{max}$ initially, switching to $\mathbf{u}_{min}$ at some specific time. In the case of a car journey, this is the equivalent of the driver keeping his foot flat down on the accelerator for the entire journey, except at the terminal point, when he brakes as hard as possible.

(c) *The minimum energy control problem*: This is used to transfer the system from an initial state to a final state with minimum expenditure of control energy. Used in satellite control.

(d) *The regulator control problem*: With the system initially displaced from equilibrium, will return the system to the equilibrium state in such a manner so as to minimize a given performance index.

(e) *The tracking control problem*: This is used to cause the state of a system to track as close as possible some desired state time history in such a manner so as to minimize a given performance index. This is the generalization of the regulator control problem.

## 9.1.2 Selection of performance index

The decision on the type of performance index to be selected depends upon the nature of the control problem. Consider the design of an autopilot for a racing yacht.

Conventionally, the autopilot is designed for course-keeping, that is to minimise the error $\psi_e(t)$ between that desired course $\psi_d(t)$ and the actual course $\psi_a(t)$ in the presence of disturbances (wind, waves and current). Since $\psi_d(t)$ is fixed for most of the time, this is in essence a regulator problem.

Using classical design techniques, the autopilot will be tuned to return the vessel on the desired course within the minimum transient period. With an optimal control strategy, a wider view is taken. The objective is to win the race, which means completing it in the shortest possible time. This in turn requires:

(a) Minimizing the distance off-track, or cross-track error $y_e(t)$. Wandering off track will increase distance travelled and hence time taken.

(b) Minimizing course or heading error $\psi_e(t)$. It is possible of course to have zero heading error but still be off-track.

(c) Minimizing rudder activity, i.e. actual rudder angle (as distinct from desired rudder angle) $\delta_a(t)$, and hence minimizing the expenditure of control energy.

(d) Minimizing forward speed loss $u_e(t)$. As the vessel yaws as a result of correcting a track or heading error, there is an increased angle of attack of the total velocity vector, which results in increased drag and therefore increased forward speed loss.

From equation (9.2) a general performance index could be written

$$J = \int_{t_0}^{t_1} h(y_e(t), \psi_e(t), u_e(t), \delta_a(t))dt \qquad (9.3)$$

### Quadratic performance indices

If, in the racing yacht example, the following state and control variables are defined

$$x_1 = y_e(t), \quad x_2 = \psi_e(t), \quad x_3 = u_e(t), \quad u = \delta_a(t)$$

then the performance index could be expressed

$$J = \int_{t_0}^{t_1} \{(q_{11}x_1 + q_{22}x_2 + q_{33}x_3) + (r_1 u)\} dt \qquad (9.4)$$

or

$$J = \int_{t_0}^{t_1} (\mathbf{Q}\mathbf{x} + \mathbf{R}\mathbf{u}) dt \qquad (9.5)$$

If the state and control variables in equations (9.4) and (9.5) are squared, then the performance index become quadratic. The advantage of a quadratic performance index is that for a linear system it has a mathematical solution that yields a linear control law of the form

$$\mathbf{u}(t) = -\mathbf{K}\mathbf{x}(t) \qquad (9.6)$$

A quadratic performance index for this example is therefore

$$J = \int_{t_0}^{t_1} \left\{ \left(q_{11}x_1^2 + q_{22}x_2^2 + q_{33}x_3^2\right) + \left(r_1 u^2\right) \right\} dt \qquad (9.7)$$

$$J = \int_{t_0}^{t_1} \left[ \begin{bmatrix} x_1 & x_2 & x_3 \end{bmatrix} \begin{bmatrix} q_{11} & 0 & 0 \\ 0 & q_{22} & 0 \\ 0 & 0 & q_{33} \end{bmatrix} \begin{bmatrix} x_1 \\ x_2 \\ x_3 \end{bmatrix} + [u][r_1][u] \right] dt$$

or, in general

$$J = \int_{t_0}^{t_1} (\mathbf{x}^T\mathbf{Q}\mathbf{x} + \mathbf{u}^T\mathbf{R}\mathbf{u}) dt \qquad (9.8)$$

$\mathbf{Q}$ and $\mathbf{R}$ are the state and control weighting matrices and are always square and symmetric. J is always a scalar quantity.

## 9.2 The Linear Quadratic Regulator

The Linear Quadratic Regulator (LQR) provides an optimal control law for a linear system with a quadratic performance index.

### 9.2.1 Continuous form

Define a functional equation of the form

$$f(\mathbf{x}, t) = \min_{\mathbf{u}} \int_{t_0}^{t_1} h(\mathbf{x}, \mathbf{u}) dt \qquad (9.9)$$

where over the time interval $t_0$ to $t_1$,

$$f(\mathbf{x}, t_0) = f(\mathbf{x}(0))$$
$$f(\mathbf{x}, t_1) = \mathbf{0}$$

From equations (9.1) and (9.2), a Hamilton–Jacobi equation may be expressed as

$$\frac{\partial f}{\partial t} = -\min_{\mathbf{u}} \left[ h(\mathbf{x}, \mathbf{u}) + \left( \frac{\partial f}{\partial \mathbf{x}} \right)^{\mathrm{T}} g(\mathbf{x}, \mathbf{u}) \right] \tag{9.10}$$

For a linear, time invariant plant, equation (9.1) becomes

$$\dot{\mathbf{x}} = \mathbf{A}\mathbf{x} + \mathbf{B}\mathbf{u} \tag{9.11}$$

And if equation (9.2) is a quadratic performance index

$$J = \int_{t_0}^{t_1} (\mathbf{x}^{\mathrm{T}}\mathbf{Q}\mathbf{x} + \mathbf{u}^{\mathrm{T}}\mathbf{R}\mathbf{u})\mathrm{d}t \tag{9.12}$$

Substituting equations (9.11) and (9.12) into (9.10)

$$\frac{\partial f}{\partial t} = -\min_{\mathbf{u}} \left[ \mathbf{x}^{\mathrm{T}}\mathbf{Q}\mathbf{x} + \mathbf{u}^{\mathrm{T}}\mathbf{R}\mathbf{u} + \left( \frac{\partial f}{\partial \mathbf{x}} \right)^{\mathrm{T}} (\mathbf{A}\mathbf{x} + \mathbf{B}\mathbf{u}) \right] \tag{9.13}$$

Introducing a relationship of the form

$$f(\mathbf{x}, t) = \mathbf{x}^{\mathrm{T}}\mathbf{P}\mathbf{x} \tag{9.14}$$

where $\mathbf{P}$ is a square, symmetric matrix, then

$$\frac{\partial f}{\partial t} = \mathbf{x}^{\mathrm{T}} \frac{\partial}{\partial t} \mathbf{P}\mathbf{x} \tag{9.15}$$

and

$$\frac{\partial f}{\partial \mathbf{x}} = 2\mathbf{P}\mathbf{x}$$

$$\left[ \frac{\partial f}{\partial \mathbf{x}} \right]^{\mathrm{T}} = 2\mathbf{x}^{\mathrm{T}}\mathbf{P} \tag{9.16}$$

Inserting equations (9.15) and (9.16) into (9.13) gives

$$\mathbf{x}^{\mathrm{T}} \frac{\partial \mathbf{P}}{\partial t} \mathbf{x} = -\min_{\mathbf{u}} \left[ \mathbf{x}^{\mathrm{T}}\mathbf{Q}\mathbf{x} + \mathbf{u}^{\mathrm{T}}\mathbf{R}\mathbf{u} + 2\mathbf{x}^{\mathrm{T}}\mathbf{P}(\mathbf{A}\mathbf{x} + \mathbf{B}\mathbf{u}) \right] \tag{9.17}$$

To minimize $\mathbf{u}$, from equation (9.17)

$$\frac{\partial [\partial f / \partial t]}{\partial \mathbf{u}} = 2\mathbf{u}^{\mathrm{T}}\mathbf{R} + 2\mathbf{x}^{\mathrm{T}}\mathbf{P}\mathbf{B} = 0 \tag{9.18}$$

Equation (9.18) can be re-arranged to give the optimal control law

$$\mathbf{u}_{\mathrm{opt}} = -\mathbf{R}^{-1}\mathbf{B}^{\mathrm{T}}\mathbf{P}\mathbf{x} \tag{9.19}$$

or

$$\mathbf{u}_{opt} = -\mathbf{Kx} \qquad (9.20)$$

where

$$\mathbf{K} = \mathbf{R}^{-1}\mathbf{B}^{\mathrm{T}}\mathbf{P} \qquad (9.21)$$

Substituting equation (9.19) back into (9.17) gives

$$\mathbf{x}^{\mathrm{T}}\dot{\mathbf{P}}\mathbf{x} = -\mathbf{x}^{\mathrm{T}}(\mathbf{Q} + 2\mathbf{PA} - \mathbf{PBR}^{-1}\mathbf{B}^{\mathrm{T}}\mathbf{P})\mathbf{x} \qquad (9.22)$$

since

$$2\mathbf{x}^{\mathrm{T}}\mathbf{PAx} = \mathbf{x}^{\mathrm{T}}(\mathbf{A}^{\mathrm{T}}\mathbf{P} + \mathbf{PA})\mathbf{x}$$

then

$$\dot{\mathbf{P}} = -\mathbf{PA} - \mathbf{A}^{\mathrm{T}}\mathbf{P} - \mathbf{Q} + \mathbf{PBR}^{-1}\mathbf{B}^{\mathrm{T}}\mathbf{P} \qquad (9.23)$$

Equation (9.23) belongs to a class of non-linear differential equations known as the matrix Riccati equations. The coefficients of $\mathbf{P}(t)$ are found by integration in reverse time starting with the boundary condition

$$\mathbf{x}^{\mathrm{T}}(t_1)\mathbf{P}(t_1)\mathbf{x}(t_1) = \mathbf{0} \qquad (9.24)$$

Kalman demonstrated that as integration in reverse time proceeds, the solutions of $\mathbf{P}(t)$ converge to constant values. Should $t_1$ be infinite, or far removed from $t_0$, the matrix Riccati equations reduce to a set of simultaneous equations

$$\mathbf{PA} + \mathbf{A}^{\mathrm{T}}\mathbf{P} + \mathbf{Q} - \mathbf{PBR}^{-1}\mathbf{B}^{\mathrm{T}}\mathbf{P} = \mathbf{0} \qquad (9.25)$$

Equations (9.23) and (9.25) are the continuous solution of the matrix Riccati equation.

## 9.2.2  Discrete form

From equation (8.76) the discrete solution of the state equation is

$$\mathbf{x}[(k+1)T] = \mathbf{A}(T)\mathbf{x}(kT) + \mathbf{B}(T)\mathbf{u}(kT) \qquad (9.26)$$

For simplicity, if $(kT)$ is written as $(k)$, then

$$\mathbf{x}(k+1) = \mathbf{A}(T)\mathbf{x}(k) + \mathbf{B}(T)\mathbf{u}(k) \qquad (9.27)$$

The discrete quadratic performance index is

$$J = \sum_{k=0}^{N-1}(\mathbf{x}^{\mathrm{T}}(k)\mathbf{Q}\mathbf{x}(k) + \mathbf{u}^{\mathrm{T}}(k)\mathbf{R}\mathbf{u}(k))T \qquad (9.28)$$

The discrete solution of the matrix Riccati equation solves recursively for $\mathbf{K}$ and $\mathbf{P}$ in reverse time, commencing at the terminal time, where

$$\mathbf{K}(N-(k+1)) = [T\mathbf{R} + \mathbf{B}^{\mathrm{T}}(T)\mathbf{P}(N-k)\mathbf{B}(T)]^{-1}\mathbf{B}^{\mathrm{T}}(T)\mathbf{P}(N-k)\mathbf{A}(T) \qquad (9.29)$$

and

$$P(N - (k+1)) = [T\mathbf{Q} + \mathbf{K}^\mathrm{T}(N - (k+1))T\mathbf{R}\mathbf{K}(N - (k+1))] + [\mathbf{A}(T)$$
$$- \mathbf{B}(T)\mathbf{K}(N - (k+1))]^\mathrm{T}\mathbf{P}(N - k)[\mathbf{A}(T) - \mathbf{B}(T)\mathbf{K}(N - (k+1))]$$
$$(9.30)$$

As $k$ is increased from 0 to $N - 1$, the algorithm proceeds in reverse time. When run in forward-time, the optimal control at step $k$ is

$$\mathbf{u}_{\mathrm{opt}}(k) = -\mathbf{K}(k)\mathbf{x}(k) \qquad (9.31)$$

The boundary condition is specified at the terminal time ($k = 0$), where

$$\mathbf{x}^\mathrm{T}(N)\mathbf{P}(N)\mathbf{x}(N) = \mathbf{0} \qquad (9.32)$$

The reverse-time recursive process can commence with $\mathbf{P}(N) = \mathbf{0}$ or alternatively, with $\mathbf{P}(N - 1) = T\mathbf{Q}$.

*Example 9.1*   (See also Appendix 1, *examp91.m*)
The regulator shown in Figure 9.1 contains a plant that is described by

$$\begin{bmatrix} \dot{x}_1 \\ \dot{x}_2 \end{bmatrix} = \begin{bmatrix} 0 & 1 \\ -1 & -2 \end{bmatrix} \begin{bmatrix} x_1 \\ x_2 \end{bmatrix} + \begin{bmatrix} 0 \\ 1 \end{bmatrix} \mathbf{u}$$
$$y = \begin{bmatrix} 1 & 0 \end{bmatrix} \mathbf{x}$$

and has a performance index

$$J = \int_0^\infty \left[ \mathbf{x}^\mathrm{T} \begin{bmatrix} 2 & 0 \\ 0 & 1 \end{bmatrix} \mathbf{x} + \mathbf{u}^2 \right] \mathrm{d}t$$

Determine

(a) the Riccati matrix $\mathbf{P}$
(b) the state feedback matrix $\mathbf{K}$
(c) the closed-loop eigenvalues

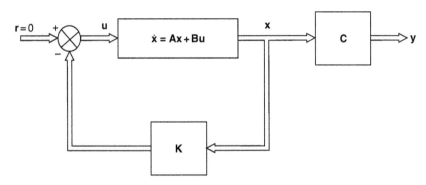

**Fig. 9.1** Optimal regulator.

*Solution*

(a)

$$\mathbf{A} = \begin{bmatrix} 0 & 1 \\ -1 & -2 \end{bmatrix} \quad \mathbf{B} = \begin{bmatrix} 0 \\ 1 \end{bmatrix}$$

$$\mathbf{Q} = \begin{bmatrix} 2 & 0 \\ 0 & 1 \end{bmatrix} \quad \mathbf{R} = \text{scalar} = 1$$

From equation (9.25) the reduced Riccati equation is

$$\mathbf{PA} + \mathbf{A}^{\mathrm{T}}\mathbf{P} + \mathbf{Q} - \mathbf{PBR}^{-1}\mathbf{B}^{\mathrm{T}}\mathbf{P} = 0 \tag{9.33}$$

$$\mathbf{PA} = \begin{bmatrix} p_{11} & p_{12} \\ p_{21} & p_{22} \end{bmatrix} \begin{bmatrix} 0 & 1 \\ -1 & -2 \end{bmatrix} = \begin{bmatrix} -p_{12} & p_{11} - 2p_{12} \\ -p_{22} & p_{21} - 2p_{22} \end{bmatrix} \tag{9.34}$$

$$\mathbf{A}^{\mathrm{T}}\mathbf{P} = \begin{bmatrix} 0 & -1 \\ 1 & -2 \end{bmatrix} \begin{bmatrix} p_{11} & p_{12} \\ p_{21} & p_{22} \end{bmatrix} = \begin{bmatrix} -p_{21} & -p_{22} \\ p_{11} - 2p_{21} & p_{12} - 2p_{22} \end{bmatrix} \tag{9.35}$$

$$\mathbf{PBR}^{-1}\mathbf{B}^{\mathrm{T}}\mathbf{P} = \begin{bmatrix} p_{11} & p_{12} \\ p_{21} & p_{22} \end{bmatrix} \begin{bmatrix} 0 \\ 1 \end{bmatrix} 1[0 \quad 1] \begin{bmatrix} p_{11} & p_{12} \\ p_{21} & p_{22} \end{bmatrix}$$

$$= \begin{bmatrix} p_{12} \\ p_{22} \end{bmatrix} [p_{21} \quad p_{22}]$$

$$= \begin{bmatrix} p_{12}p_{21} & p_{12}p_{22} \\ p_{22}p_{21} & p_{22}^2 \end{bmatrix} \tag{9.36}$$

Combining equations (9.34), (9.35) and (9.36) gives

$$\begin{bmatrix} -p_{12} & p_{11} - 2p_{12} \\ -p_{22} & p_{21} - 2p_{22} \end{bmatrix} + \begin{bmatrix} -p_{21} & -p_{22} \\ p_{11} - 2p_{21} & p_{12} - 2p_{22} \end{bmatrix}$$
$$+ \begin{bmatrix} 2 & 0 \\ 0 & 1 \end{bmatrix} - \begin{bmatrix} p_{12}p_{21} & p_{12}p_{22} \\ p_{22}p_{21} & p_{22}^2 \end{bmatrix} = \mathbf{0} \tag{9.37}$$

Since **P** is symmetric, $p_{21} = p_{12}$. Equation (9.37) can be expressed as four simultaneous equations

$$-p_{12} - p_{12} + 2 - p_{12}^2 = 0 \tag{9.38}$$

$$p_{11} - 2p_{12} - p_{22} - p_{12}p_{22} = 0 \tag{9.39}$$

$$-p_{22} + p_{11} - 2p_{12} - p_{12}p_{22} = 0 \tag{9.40}$$

$$p_{12} - 2p_{22} + p_{12} - 2p_{22} + 1 - p_{22}^2 = 0 \tag{9.41}$$

Note that equations (9.39) and (9.40) are the same. From equation (9.38)

$$p_{12}^2 + 2p_{12} - 2 = 0$$

solving

$$p_{12} = p_{21} = 0.732 \quad \text{and} \quad -2.732$$

Using positive value

$$p_{12} = p_{21} = 0.732 \tag{9.42}$$

From equation (9.41)

$$2p_{12} - 4p_{22} + 1 - p_{22}^2 = 0$$
$$p_{22}^2 + 4p_{22} - 2.464 = 0$$

solving

$$p_{22} = 0.542 \text{ and } -4.542$$

Using positive value

$$p_{22} = 0.542 \tag{9.43}$$

From equation (9.39)

$$p_{11} - (2 \times 0.732) - 0.542 - (0.732 \times 0.542) = 0$$
$$p_{11} = 2.403 \tag{9.44}$$

From equations (9.42), (9.43) and (9.44) the Riccati matrix is

$$\mathbf{P} = \begin{bmatrix} 2.403 & 0.732 \\ 0.732 & 0.542 \end{bmatrix} \tag{9.45}$$

(b) Equation (9.21) gives the state feedback matrix

$$\mathbf{K} = \mathbf{R}^{-1}\mathbf{B}^{\mathsf{T}}\mathbf{P} = 1[0 \quad 1]\begin{bmatrix} 2.403 & 0.732 \\ 0.732 & 0.542 \end{bmatrix} \tag{9.46}$$

Hence

$$\mathbf{K} = [0.732 \quad 0.542]$$

(c) From equation (8.96), the closed-loop eigenvalues are

$$|s\mathbf{I} - \mathbf{A} + \mathbf{B}\mathbf{K}| = 0$$

$$\left| \begin{bmatrix} s & 0 \\ 0 & s \end{bmatrix} - \begin{bmatrix} 0 & 1 \\ -1 & -2 \end{bmatrix} + \begin{bmatrix} 0 \\ 1 \end{bmatrix} [0.732 \quad 0.542] \right| = 0$$

$$\left| \begin{bmatrix} s & -1 \\ 1 & s+2 \end{bmatrix} + \begin{bmatrix} 0 & 0 \\ 0.732 & 0.542 \end{bmatrix} \right| = 0$$

$$\left| \begin{matrix} s & -1 \\ 1.732 & s+2.542 \end{matrix} \right| = 0$$

$$s^2 + 2.542s + 1.732 = 0$$

$$s_1, s_2 = -1.271 \pm \text{j}0.341$$

In general, for a second-order system, when $\mathbf{Q}$ is a diagonal matrix and $\mathbf{R}$ is a scalar quantity, the elements of the Riccati matrix $\mathbf{P}$ are

$$p_{11} = \left[ \frac{b_2^2}{r} \right] p_{12}p_{22} - p_{22}a_{21} - p_{12}a_{22}$$

$$p_{12} = p_{21} = \frac{r}{b_2^2} \left[ a_{21} \pm \sqrt{a_{21}^2 + \frac{q_{11}b_2^2}{r}} \right] \tag{9.47}$$

$$p_{22} = \frac{r}{b_2^2} \left[ a_{22} \pm \sqrt{a_{22}^2 + \frac{(2p_{12} + q_{22})}{r}} \right]$$

## 9.3 The linear quadratic tracking problem

The tracking or servomechanism problem is defined in section 9.1.1(e), and is directed at applying a control $\mathbf{u}(t)$ to drive a plant so that the state vector $\mathbf{x}(t)$ follows a desired state trajectory $\mathbf{r}(t)$ in some optimal manner.

### 9.3.1 Continuous form

The quadratic performance index to be minimized is

$$J = \int_{t_0}^{t_1} \left[ (\mathbf{r} - \mathbf{x})^{\mathrm{T}} \mathbf{Q} (\mathbf{r} - \mathbf{x}) + \mathbf{u}^{\mathrm{T}} \mathbf{R} \mathbf{u} \right] \mathrm{d}t \tag{9.48}$$

It can be shown that the constrained functional minimization of equation (9.48) yields again the matrix Riccati equations (9.23) and (9.25) obtained for the LQR, combined with the additional set of reverse-time state tracking equations

$$\dot{\mathbf{s}} = (\mathbf{A} - \mathbf{B}\mathbf{R}^{-1}\mathbf{B}^{\mathrm{T}}\mathbf{P})^{\mathrm{T}}\mathbf{s} - \mathbf{Q}\mathbf{r} \tag{9.49}$$

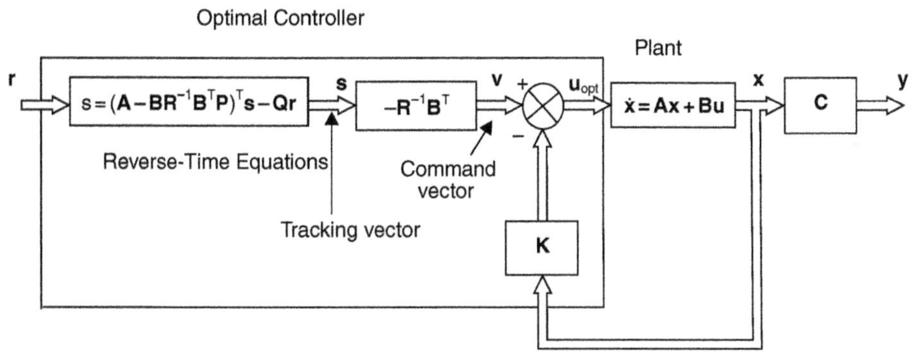

Optimal Controller

Plant

**Fig. 9.2** Optimal tracking system.

where **s** is a tracking vector, whose boundary condition is

$$\mathbf{s}(t_1) = 0 \tag{9.50}$$

and the optimal control law is given by

$$\mathbf{u}_{\text{opt}} = -\mathbf{R}^{-1}\mathbf{B}^T\mathbf{P}\mathbf{x} - \mathbf{R}^{-1}\mathbf{B}^T\mathbf{s}$$

If

$$\mathbf{v} = -\mathbf{R}^{-1}\mathbf{B}^T\mathbf{s}$$

and

$$\mathbf{K} = \mathbf{R}^{-1}\mathbf{B}^T\mathbf{P}$$

Then

$$\mathbf{u}_{\text{opt}} = \mathbf{v} - \mathbf{K}\mathbf{x} \tag{9.51}$$

Hence, if the desired state vector $\mathbf{r}(t)$ is known in advance, tracking errors may be reduced by allowing the system to follow a command vector $\mathbf{v}(t)$ computed in advance using the reverse-time equation (9.49). An optimal controller for a tracking system is shown in Figure 9.2.

## 9.3.2 Discrete form

The discrete quadratic performance index, writing $(kT)$ as $(k)$, is

$$J = \sum_{k=0}^{N-1} [(\mathbf{r}(k) - \mathbf{x}(k))^T\mathbf{Q}(\mathbf{r}(k) - \mathbf{x}(k)) + \mathbf{u}^T(k)\mathbf{R}\mathbf{u}(k)]T \tag{9.52}$$

Discrete minimization gives the recursive Riccati equations (9.29) and (9.30). These are run in reverse-time together with the discrete reverse-time state tracking equation

$$\mathbf{s}(N - (k + 1)) = \mathbf{F}(T)\mathbf{s}(N - k) + \mathbf{G}(T)\mathbf{r}(N - k) \tag{9.53}$$

having the boundary condition

$$\mathbf{s}(N) = 0$$

$\mathbf{F}(T)$ and $\mathbf{G}(T)$ are the discrete transition and control matrices and are obtained by converting the matrices in the continuous equation (9.49) into discrete form using equations (8.78) and (8.80).

The command vector $\mathbf{v}$ is given by

$$\mathbf{v}(N - k) = -\mathbf{R}^{-1}\mathbf{B}^{\mathrm{T}}\mathbf{s}(N - k) \tag{9.54}$$

When run in forward-time, the optimal control at time $(kT)$ is

$$\mathbf{u}_{\mathrm{opt}}(kT) = \mathbf{v}(kT) - \mathbf{K}(kT)\mathbf{x}(kT) \tag{9.55}$$

The values of $\mathbf{x}(kT)$ are calculated using the plant state transition equation

$$\mathbf{x}(k + 1)T = \mathbf{A}(T)\mathbf{x}(kT) + \mathbf{B}(T)\mathbf{u}_{\mathrm{opt}}(kT) \tag{9.56}$$

*Example 9.2* (See also Appendix 1, *examp92.m*)
The optimal tracking system shown in Figure 9.2 contains a plant that is described by

$$\begin{bmatrix} \dot{x}_1 \\ \dot{x}_2 \end{bmatrix} = \begin{bmatrix} 0 & 1 \\ -1 & -1 \end{bmatrix} \begin{bmatrix} x_1 \\ x_2 \end{bmatrix} + \begin{bmatrix} 0 \\ 1 \end{bmatrix} u$$

$$\mathbf{y} = \begin{bmatrix} 1 & 0 \\ 0 & 1 \end{bmatrix} \begin{bmatrix} x_1 \\ x_2 \end{bmatrix}$$

The discrete performance index is

$$J = \sum_{k=0}^{200} \left[ (r_1(kT) - x_1(kT))(r_2(kT) - x_2(kT)) \begin{bmatrix} 10 & 0 \\ 0 & 1 \end{bmatrix} \begin{bmatrix} r_1(kT) - x_1(kT) \\ r_2(kT) - x_2(kT) \end{bmatrix} + u(kT)^2 \right] T$$

It is required that the system tracks the following desired state vector

$$\begin{bmatrix} r_1(kT) \\ r_2(kT) \end{bmatrix} = \begin{bmatrix} 1.0 \sin (0.6284kT) \\ 0.6 \cos (0.6284kT) \end{bmatrix}$$

over a period of 0–20 seconds. The sampling time $T$ is 0.1 seconds.

In reverse-time, starting with $\mathbf{P}(N) = \mathbf{0}$ at $NT = 20$ seconds, compute the state feedback gain matrix $\mathbf{K}(kT)$ and Riccati matrix $\mathbf{P}(kT)$ using equations (9.29) and (9.30). Also in reverse time, use the desired state vector $\mathbf{r}(kT)$ to drive the tracking equation (9.53) with the boundary condition $\mathbf{s}(N) = 0$ and hence compute the command vector $\mathbf{v}(kT)$.

In forward-time, use the command vector $\mathbf{v}(kT)$ and state vector $\mathbf{x}(kT)$ to calculate $\mathbf{u}_{\mathrm{opt}}(kT)$ in equation (9.55) and hence, using the plant state transition equation (9.56) calculate the state trajectories.

*Solution*

The reverse-time calculations are shown in Figure 9.3. Using equations (9.29) and (9.30) and commencing with $\mathbf{P}(N) = 0$, it can be seen that the solution for $\mathbf{K}$ (and also $\mathbf{P}$) settle down after about 2 seconds to give steady-state values of

$$\mathbf{K}(kT) = [2.0658 \quad 1.4880]$$

$$\mathbf{P}(kT) = \begin{bmatrix} 8.0518 & 2.3145 \\ 2.3145 & 1.6310 \end{bmatrix} \tag{9.57}$$

Using equation (9.49), together with equations (8.78) and (8.80), to calculate $\mathbf{F}(T)$ and $\mathbf{G}(T)$ in equation (9.53), for $T = 0.1$ seconds, the discrete reverse-time state tracking equation is

$$\begin{bmatrix} s_1(N - (k+1)) \\ s_2(N - (k+1)) \end{bmatrix} = \begin{bmatrix} 0.9859 & -0.2700 \\ 0.0881 & 0.7668 \end{bmatrix} \begin{bmatrix} s_1(N - k) \\ s_2(N - k) \end{bmatrix}$$
$$+ \begin{bmatrix} -0.9952 & 0.0141 \\ -0.0460 & -0.0881 \end{bmatrix} \begin{bmatrix} r_1(N - k) \\ r_2(N - k) \end{bmatrix}$$

and

$$\mathbf{v}(N - k) = -1[0 \quad 1] \begin{bmatrix} s_1(N - k) \\ s_2(N - k) \end{bmatrix} \tag{9.58}$$

Then the command vector $\mathbf{v}$ (in this case a scalar) is generated in reverse-time as shown in Figure 9.3. The forward-time response is shown in Figure 9.4.

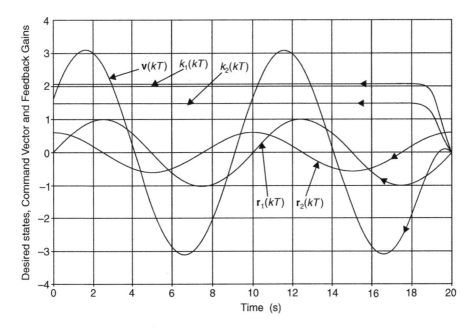

**Fig. 9.3** Reverse-time solutions for a tracking system.

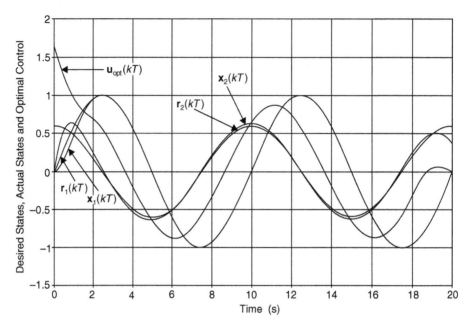

**Fig. 9.4** Forward-time response of a tracking system.

The optimal control is calculated using equation (9.55) and the values of the state variables are found using the plant state transition equation (9.56)

$$\begin{bmatrix} x_1(k+1)T \\ x_2(k+1)T \end{bmatrix} = \begin{bmatrix} 0.9952 & 0.0950 \\ -0.0950 & 0.9002 \end{bmatrix} \begin{bmatrix} x_1(kT) \\ x_2(kT) \end{bmatrix} + \begin{bmatrix} 0.0048 \\ 0.0950 \end{bmatrix} u_{\text{opt}}(kT) \qquad (9.59)$$

where $\mathbf{A}(T)$ and $\mathbf{B}(T)$ are calculated from equations (8.78) and (8.80). From Figure 9.4 it can be seen that after an initial transient period, the optimal control law takes the form of a phase lead compensator. Because of the weighting of the $\mathbf{Q}$ matrix in the performance index, $x_1(kT)$ tracks $r_1(kT)$ more closely than $x_2(kT)$ tracks $r_2(kT)$.

## 9.4 The Kalman filter

In the design of state observers in section 8.4.3, it was assumed that the measurements $\mathbf{y} = \mathbf{Cx}$ were noise free. In practice, this is not usually the case and therefore the observed state vector $\hat{\mathbf{x}}$ may also be contaminated with noise.

### 9.4.1 The state estimation process

State estimation is the process of extracting a best estimate of a variable from a number of measurements that contain noise.

The classical problem of obtaining a best estimate of a signal by combining two noisy continuous measurements of the same signal was first solved by Weiner (1949).

His solution required that both the signal and noise be modelled as random process with known statistical properties.

This work was extended by Kalman and Bucy (1961) who designed a state estimation process based upon an optimal minimum variance filter, generally referred to as a Kalman filter.

## 9.4.2   The Kalman filter single variable estimation problem

The Kalman filter is a complementary form of the Weiner filter. Let $A_x$ be a measurement of a parameter $x$ and let its variance $P_a$ be given by

$$P_a = E\left\{(A_x - \bar{A}_x)^2\right\}$$ (9.60)

where $\bar{A}_x$ is the mean and $E\{\ \}$ is the expected value.

Let $B_x$ be a measurement from another system of the same parameter and the variance $P_b$ is

$$P_b = E\left\{(B_x - \bar{B}_x)^2\right\}$$ (9.61)

Assume that $x$ can be expressed by the parametric relationship

$$x = A_x K + B_x(1 - K)$$ (9.62)

where $K$ is any weighting factor between 0 and 1. The problem is to derive a value of $K$ which gives an optimal combination of $A_x$ and $B_x$ and hence the best estimate of measured variable $x$, which is given the symbol $\hat{x}$ (pronounced x hat).

Let $P$ be the variance of the weighted mean

$$P = E\left\{(x - \bar{x})^2\right\}$$ (9.63)

The optimal value of $K$ is the one that yields the minimum variance, i.e.

$$\frac{dP}{dK} = 0$$ (9.64)

Substitution of equation (9.62) into (9.63) gives

$$P = K^2 P_A + (1 - K)^2 P_B$$ (9.65)

Hence $K$ is given by

$$\frac{d}{dK}\left\{K^2 P_A + (1 - K)^2 P_B\right\} = 0$$

From which

$$K = \frac{P_B}{P_A + P_B}$$ (9.66)

Substitution of equation (9.66) into (9.62) provides

$$\hat{x} = A_x - \left\{\frac{P_A}{P_A + P_B}\right\}(A_x - B_x)$$ (9.67)

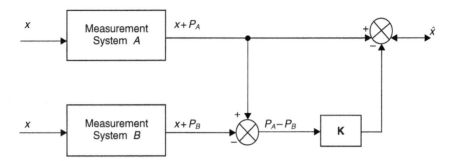

**Fig. 9.5** Integration of two measurement systems to obtain optimal estimate.

or

$$\hat{x} = A_x - K(A_x - B_x) \tag{9.68}$$

$K$ is the Kalman gain and the total error variance expected is

$$P = P_A - K(P_A - P_B) \tag{9.69}$$

so that

$$\hat{x} = x + P_A - K(P_A - P_B) \tag{9.70}$$

Equation (9.70) is illustrated in Figure 9.5.

## 9.4.3 The Kalman filter multivariable state estimation problem

Consider a plant that is subject to a Gaussian sequence of disturbances $\mathbf{w}(kT)$ with disturbance transition matrix $\mathbf{C_d}(T)$. Measurements $\mathbf{z}(k+1)T$ contain a Gaussian noise sequence $\mathbf{v}(k+1)T$ as shown in Figure 9.6.

The general form of the Kalman filter usually contains a discrete model of the system together with a set of recursive equations that continuously update the Kalman gain matrix $\mathbf{K}$ and the system covariance matrix $\mathbf{P}$.

The state estimate $\hat{\mathbf{x}}(k+1/k+1)$ is obtained by calculating the predicted state $\hat{\mathbf{x}}(k+1/k)$ from

$$\hat{\mathbf{x}}(k+1/k)T = \mathbf{A}(T)\hat{\mathbf{x}}(k/k)T + \mathbf{B}(T)\mathbf{u}(kT) \tag{9.71}$$

and then determining the estimated state at time $(k+1)T$ using

$$\hat{\mathbf{x}}(k+1/k+1)T = \hat{\mathbf{x}}(k+1/k)T + \mathbf{K}(k+1)\{\mathbf{z}(k+1)T - \mathbf{C}(T)\hat{\mathbf{x}}(k+1/k)T\} \tag{9.72}$$

The term $(k/k)$ means data at time $k$ based on information available at time $k$. The term $(k+1/k)$ means data used at time $k+1$ based on information available at time $k$. Similarly $(k+1/k+1)$ means data at time $k+1$ based on information available at time $k+1$.

The vector of measurements is given by

$$\mathbf{z}(k+1)T = \mathbf{C}(T)\mathbf{x}(k+1)T + \mathbf{v}(k+1)T \tag{9.73}$$

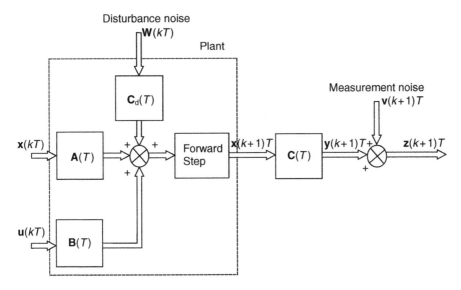

**Fig. 9.6** Plant with disturbance and measurement noise.

where

  $z(k + 1)T$ is the measurement vector

  $C(T)$ is the measurement matrix

  $v(k + 1)T$ is a Gaussian noise sequence

The Kalman gain matrix **K** is obtained from a set of recursive equations that commence from some initial covariance matrix $P(k/k)$

$$P(k + 1/k) = A(T)P(k/k)A^T(T) + C_d(T)QC_d^T(T) \tag{9.74}$$

$$K(k + 1) = P(k + 1/k)C^T(T)\{C(T)P(k + 1/k)C^T(T) + R\}^{-1} \tag{9.75}$$

$$P(k + 1/k + 1) = \{I - K(k + 1)C(T)\}P(k + 1/k) \tag{9.76}$$

The recursive process continues by substituting the covariance matrix $P(k + 1/k + 1)$ computed in equation (9.76) back into (9.74) as $P(k/k)$ until $K(k + 1)$ settles to a steady value, see Appendix 1, script files *kalfilc.m* for the continuous solution and *kalfild.m* for the above discrete solution. In equations (9.74)–(9.76)

$C_d(T)$ is the disturbance transition matrix

**Q** is the disturbance noise covariance matrix

**R** is the measurement noise covariance matrix

Equations (9.71)–(9.76) are illustrated in Figure 9.7 which shows the block diagram of the Kalman filter.

  The recursive equations (9.74)–(9.76) that calculate the Kalman gain matrix and covariance matrix for a Kalman filter are similar to equations (9.29) and (9.30) that

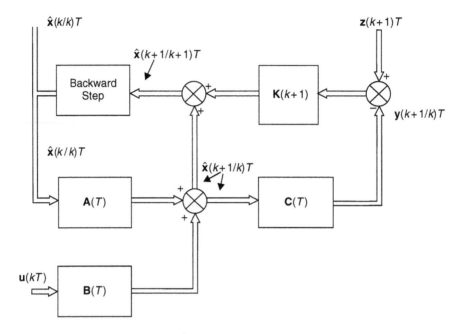

**Fig. 9.7** The Kalman filter.

calculate the feedback matrix and Riccati matrix for a linear quadratic regulator. The difference is that the Kalman filter is computed in forward-time, the LQR being computed in reverse-time.

## 9.5 Linear Quadratic Gaussian control system design

A control system that contains a LQ Regulator/Tracking controller together with a Kalman filter state estimator as shown in Figure 9.8 is called a Linear Quadratic Gaussian (LQG) control system.

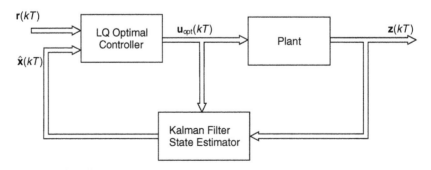

**Fig. 9.8** Linear Quadratic Gaussian (LQG) control system.

## Case study

*Example 9.3*   (See also Appendix 1, *examp93.m*)
China clay is used in the paper, ceramics and fertilizer industries, and is washed from
quarry faces, by high pressure hoses. A pressing operation reduces the moisture
content in the clay to about 30%, and then the clay is extruded into small cylindrical
shaped noodles. The clay noodles are then passed through the band drying oven
shown in Figure 9.9 at rates varying between 2 and 15 tonnes/hour. Upon exit, the
moisture content of the clay should be controlled to a desired level of between 4 and
12%, with a deviation of no more than ±1%. The process air is heated by mixing the
exhaust gas from a gas burner with a large quantity of dilution air to meet the
specified air flow-rate into the dryer.

An existing control arrangement uses a PID controller to control the temperature
of the process air (measured by thermocouples) and the dry clay moisture content
measured by samples taken by the works laboratory. If this is out of specification,
then the process air temperature is raised or lowered. The dry clay moisture content
can be measured by an infrared absorption analyser, but on its own, this is considered to be too noisy and unreliable.

The important process parameters are

(a)  Burner exhaust temperature $t_b(t)$ (°C)
(b)  Dryer outlet temperature $t_d(t)$ (°C)
(c)  Dryer outlet clay moisture content $m_f(t)$ (%)

The important control parameters are

 (i)  Burner gas supply valve angle $v_a(t)$ (rad)
(ii)  Dryer clay feed-rate $f_i(t)$ (tonnes/hour)

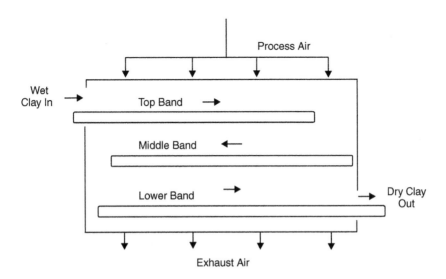

**Fig. 9.9** Band drying oven.

A proposed control scheme by Drew and Burns (1992) uses an LQG design, whereby the three process parameters are controlled in an optimal manner, their values (particularly the moisture control) being estimated.

*System model*: The dynamic characteristics of the dryer were measured experimentally. This yielded the following transfer functions

Burner model

$$G_1(s) = \frac{T_b}{V_a}(s) = \frac{420}{1 + 47s} \tag{9.77}$$

Dryer model

$$G_2(s) = \frac{T_d}{T_b}(s) = \frac{0.119}{1 + 200s} \tag{9.78}$$

Moisture models

$$G_{31}(s) = \frac{M_f}{T_d}(s) = \frac{-0.167}{1 + 440s} \tag{9.79}$$

$$G_{32}(s) = \frac{M_f}{F_i}(s) = \frac{0.582}{1 + 440s} \tag{9.80}$$

Equations (9.79) and (9.80) can be combined to give

$$M_f(s) = \frac{-0.167T_d(s) + 0.582F_i(s)}{1 + 440s} \tag{9.81}$$

The block diagram of the system is shown in Figure 9.10.

Continuous state-space model: From equations (9.77)–(9.81)

$$47\dot{t}_b + t_b(t) = 420v_a(t)$$
$$200\dot{t}_d + t_d(t) = 0.119t_d(t) \tag{9.82}$$
$$440\dot{m}_f + m_f(t) = -0.167t_d(t) + 0.582f_i(t)$$

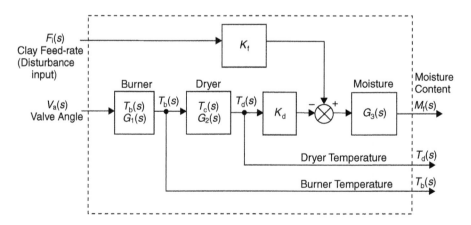

**Fig. 9.10** Model of band dryer system.

Define the state variables

$$x_1 = t_b, \quad x_2 = t_d, \quad x_3 = m_f$$

and the control variables

$$u_1 = v_a$$

and the disturbance variables

$$w_1 = 0, \quad w_2 = 0, \quad w_3 = f_i$$

then equations (9.82) can now be written as

$$\dot{x}_1 = -0.02128x_1 + 8.93617u_1 + c_{d11}w_1$$
$$\dot{x}_2 = 0.00060x_1 - 0.00500x_2 + c_{d22}w_2 \tag{9.83}$$
$$\dot{x}_3 = -0.00038x_2 - 0.00227x_3 + 0.00132w_3$$

where $c_{d11}$ and $c_{d22}$ are unknown burner and dryer disturbance coefficients, and are given an arbitrary value of 0.1. The state equations are written in the form

$$\dot{\mathbf{x}} = \mathbf{A}\mathbf{x} + \mathbf{B}\mathbf{u} + \mathbf{C}_d\mathbf{w}$$

$$\begin{bmatrix} \dot{x}_1 \\ \dot{x}_2 \\ \dot{x}_3 \end{bmatrix} = \begin{bmatrix} -0.0213 & 0 & 0 \\ 0.0006 & -0.005 & 0 \\ 0 & -0.00038 & -0.0023 \end{bmatrix} \begin{bmatrix} x_1 \\ x_2 \\ x_3 \end{bmatrix} + \begin{bmatrix} 8.9362 \\ 0 \\ 0 \end{bmatrix} u_1$$

$$+ \begin{bmatrix} 0.1 & 0 & 0 \\ 0 & 0.1 & 0 \\ 0 & 0 & 0.00132 \end{bmatrix} \begin{bmatrix} w_1 \\ w_2 \\ w_3 \end{bmatrix}$$

and the output equation is

$$\begin{bmatrix} t_b \\ t_d \\ m_f \end{bmatrix} = \begin{bmatrix} 1 & 0 & 0 \\ 0 & 1 & 0 \\ 0 & 0 & 1 \end{bmatrix} \begin{bmatrix} x_1 \\ x_2 \\ x_3 \end{bmatrix} \tag{9.84}$$

Discrete system model: The discrete system model (without disturbances) is given by

$$\mathbf{x}(k+1)T = \mathbf{A}(T)\mathbf{x}(kT) + \mathbf{B}(T)\mathbf{u}(kT) \tag{9.85}$$

For a sampling time of 2 seconds, from equations (8.78) and (8.80)

$$\mathbf{A}(T) = \begin{bmatrix} 0.9583 & 0 & 0 \\ 0.0012 & 0.9900 & 0 \\ 0 & -0.0008 & 0.9955 \end{bmatrix}; \quad \mathbf{B}(T) = \begin{bmatrix} 17.4974 \\ 0.0105 \\ 0 \end{bmatrix} \tag{9.86}$$

LQR Design: Using the quadratic performance index

$$J = \int_0^\infty (\mathbf{x}^T\mathbf{Q}\mathbf{x} + \mathbf{u}^T\mathbf{R}\mathbf{u})dt$$

where $\mathbf{Q}$ and $\mathbf{R}$ are diagonal matrices of the form

$$\mathbf{Q} = \begin{bmatrix} q_{11} & 0 & 0 \\ 0 & q_{22} & 0 \\ 0 & 0 & q_{33} \end{bmatrix}; \quad \mathbf{R} = \begin{bmatrix} r_{11} & 0 \\ 0 & r_{22} \end{bmatrix} \tag{9.87}$$

From equations (9.20) and (9.21), the optimal control law is

$$\mathbf{u}_{opt} = -\mathbf{Kx}$$

where

$$\mathbf{K} = \mathbf{R}^{-1}\mathbf{B}^{T}\mathbf{P} \tag{9.88}$$

The design procedure employed was to maintain $\mathbf{R}$ as unity scalar, and systematically vary the diagonal elements of $\mathbf{Q}$ to achieve the performance specification. This was to maintain a dry clay moisture content of 6%, ±1%, as the clay feed-rate varied from 6 to 10 tonnes/hour. Also the drying oven temperature $t_d$ should not vary more than ±3 °C from the set point of 50 °C. At each design setting, the clay feed-rate was varied according to

$$w_3(t) = 8 + 2\sin(0.00154t) \tag{9.89}$$

Some results are presented in Table 9.1.

It was found that $q_{11}$ had little effect, and was set to zero. From Table 9.1, the settings that meet the performance specification are

$$\mathbf{Q} = \begin{bmatrix} 0 & 0 & 0 \\ 0 & 0.5 & 0 \\ 0 & 0 & 20 \end{bmatrix} \quad r = 1 \tag{9.90}$$

From equation (9.25), the Riccati matrix is

$$\mathbf{P} = \begin{bmatrix} 0 & 0.1 & -0.2 \\ 0.1 & 10.8 & -30 \\ -0.2 & -30 & 3670.4 \end{bmatrix} \tag{9.91}$$

which gives, from equation (9.88), the feedback gain matrix

$$\mathbf{K} = [0.0072 \quad 0.6442 \quad -1.8265] \tag{9.92}$$

The same results are also obtained from the discrete equations (9.29) and (9.30).

Table 9.1 Variations in dryer temperature and moisture content

| $q_{22}$ | $q_{33}$ | Variation in temperature $t_d$ (°C) | | Variation in moisture content (%) | |
|---|---|---|---|---|---|
| | | Max | Min | Max | Min |
| 3 | 1 | 0.17 | 0 | 2.09 | −2.11 |
| 1 | 3 | 0.99 | 0 | 1.74 | −2.13 |
| 0.5 | 3 | 1.524 | 0 | 1.5 | −2.15 |
| 0.5 | 6 | 2.05 | −2.05 | 1.27 | −1.27 |
| 0.5 | 10 | 2.42 | −2.42 | 1.1 | −1.1 |
| 0.5 | 20 | 2.86 | −2.86 | 0.89 | −0.89 |

The closed-loop eigenvalues are

$$s = -0.0449 \pm j0.0422$$
$$s = -0.0033$$

(9.93)

Implementation: The optimal control law was implemented by using

$$\mathbf{u}_1 = \mathbf{Ke}$$

where

$$\mathbf{e} = (\mathbf{r} - \mathbf{x})$$

(9.94)

This is shown in Figure 9.11.

A discrete simulation was undertaken using equations (9.85) and (9.86) together with a disturbance transition matrix $\mathbf{C_d}(T)$, which was calculated using $\mathbf{C_d}$ in equation (9.84) and equation (8.80) for $\mathbf{B}(T)$, with a sampling time of 2 seconds.

$$\mathbf{C_d}(T) = \begin{bmatrix} 0.1958 & 0 & 0 \\ 0.0001 & 0.199 & 0 \\ 0 & -0.0001 & 0.0026 \end{bmatrix}$$

(9.95)

The desired state vector was

$$\mathbf{r} = \begin{bmatrix} 450 \\ 50 \\ -6 \end{bmatrix}$$

(9.96)

Note that the moisture content $r_3$ is negative because of the moisture model in equation (9.79). The initial conditions were

$$\mathbf{x}(0) = \begin{bmatrix} 200 \\ 30 \\ -30 \end{bmatrix}$$

(9.97)

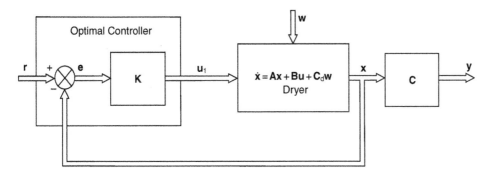

**Fig. 9.11** Optimal control of band dryer.

and the disturbance vector

$$\mathbf{w} = \begin{bmatrix} 0 \\ 0 \\ w_3 \end{bmatrix}$$ (9.98)

where $w_3$, the clay feed-rate was set at a value between 6 and 10 tonnes/hour. Figure 9.12 shows the time response of $u_1(t)$, the gas-valve angle in radians. The valve angle was not allowed to have a negative value, so remained closed for the first 80 seconds of the simulation, when the dryer was cooling. The steady-state angle was 0.95 radians, or 54°.

Figure 9.13 indicates the burner temperature time response $t_b(t)$. The temperature falls from its initial value, since the gas valve is closed, and then climbs with a response indicated by the eigenvalues in equation (9.93) to a steady-state value of 400 °C, or a steady-state error of 50 °C.

Figure 9.14 shows the combined response of the dryer temperature $t_d(t)$ and the moisture content $m_f(t)$, the latter being shown as a positive number. The dryer temperature climbs to 48 °C (steady-state error of 2 °C) and the moisture falls to 6%, with no steady-state error. In this simulation the clay feed-rate $w_3(t)$ was constant at 8 tonnes/hour.

As the band dryer is a type zero system, and there are no integrators in the controller, steady-state errors must be expected. However, the selection of the elements in the $\mathbf{Q}$ matrix, equation (9.90), focuses the control effort on control-

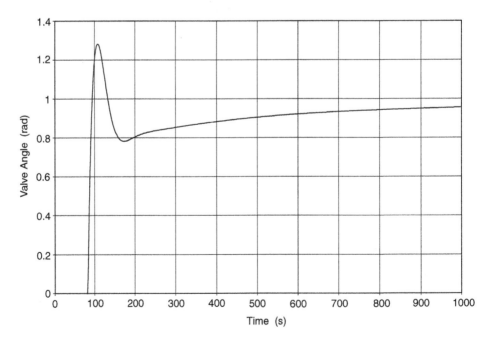

**Fig. 9.12** Time response of gas-valve angle $u_1(t)$.

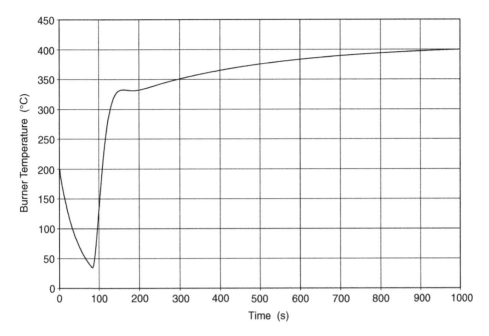

**Fig. 9.13** Time response of burner temperature $t_b(t)$.

ling the moisture content, at the expense of, in particular, the burner temperature $t_b(t)$.

Figure 9.15 shows the final 500 seconds of the moisture content simulation as the clay feed-rate is varied between 6 and 10 tonnes/hour. After 1 000 seconds, as the clay leaves the dryer, the moisture content is between 5.2% and 6.8%, which is within the specification of ±1% of the set point of 6%.

*Kalman filter design*: If the three stages of the covariance matrix $\mathbf{P}$ are written as $\mathbf{P}(k/k) = \mathbf{P}_1$; $\mathbf{P}(k+1/k) = \mathbf{P}_2$ and $\mathbf{P}(k+1/k+1) = \mathbf{P}_3$, then recursive equations (9.74), (9.75) and (9.76) become

$$\mathbf{P}_2 = \mathbf{A}\mathbf{P}_1\mathbf{A}^{\mathsf{T}} + \mathbf{C}_d\mathbf{Q}\mathbf{C}_d^{\mathsf{T}}$$
$$\mathbf{K} = \mathbf{P}_2\mathbf{C}^{\mathsf{T}}\{\mathbf{C}\mathbf{P}_2\mathbf{C}^{\mathsf{T}} + \mathbf{R}\}^{-1} \tag{9.99}$$
$$\mathbf{P}_3 = \{\mathbf{I} - \mathbf{K}\mathbf{C}\}\mathbf{P}_2$$

Equation set (9.99) is simpler to visualize, but remember the system matrices are the transition matrices $\mathbf{A}(T)$, $\mathbf{B}(T)$ and $\mathbf{C}_d(T)$. Before recursion can start, values for $\mathbf{R}$, the measurement noise covariance matrix, and $\mathbf{Q}$, the disturbance noise covariance matrix must be selected.

*Measurement noise covariance matrix* $\mathbf{R}$: The main problem with the instrumentation system was the randomness of the infrared absorption moisture content analyser. A number of measurements were taken from the analyser and compared with samples taken simultaneously by work laboratory staff. The errors could be approximated to a normal distribution with a standard deviation of 2.73%, or a variance of 7.46.

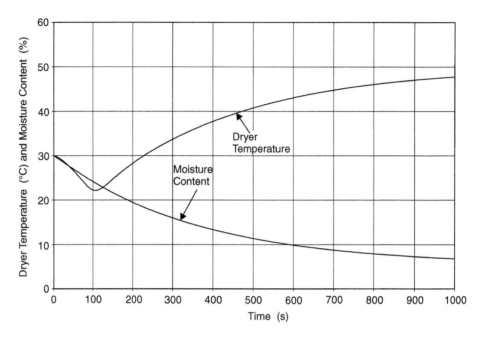

**Fig. 9.14** Combined response of dryer temperature $t_d(t)$ and moisture content $m_f(t)$.

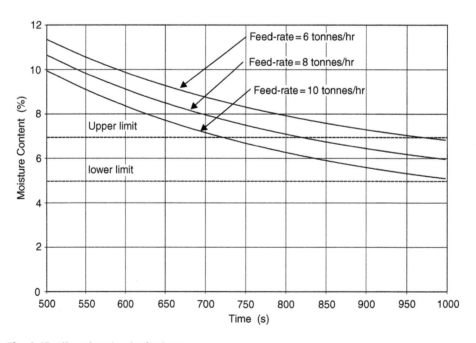

**Fig. 9.15** Effect of varying clay feed-rate.

The thermocouples measuring the burner and dryer temperatures were relatively noise-free, with standard deviations in the order of 0.1 °C. The variance was therefore set at 0.01, giving

$$\mathbf{R} = \begin{bmatrix} 0.01 & 0 & 0 \\ 0 & 0.01 & 0 \\ 0 & 0 & 7.46 \end{bmatrix} \tag{9.100}$$

*Disturbance noise covariance matrix* **Q**: This was set as a diagonal matrix, where $q_{11}$ and $q_{22}$ represent changes in the burner and dryer temperatures as a result of changing heat transfer through the walls of the dryer, due to wind and variations in external temperature.

On the other hand, $q_{33}$ is a measure of clay feed-rate variations, and a standard deviation of 0.3 tonnes/hour seemed appropriate. In the absence of any other information, standard deviations of the burner and dryer temperatures was also thought to be in the order of 0.3 °C. Thus, when these values are squared, the **Q** matrix becomes

$$\mathbf{Q} = \begin{bmatrix} 0.1 & 0 & 0 \\ 0 & 0.1 & 0 \\ 0 & 0 & 0.1 \end{bmatrix} \tag{9.101}$$

Before equations (9.99) can be run, and initial value of $\mathbf{P}(k/k)$ is required. Ideally, they should not be close to the final value, so that convergence can be seen to have taken place. In this instance, $\mathbf{P}(k/k)$ was set to an identity matrix. Figure 9.16 shows the diagonal elements of the Kalman gain matrix during the first 20 steps of the recursive equation (9.99).

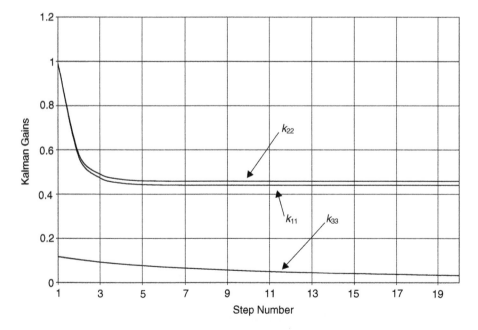

**Fig. 9.16** Convergence of diagonal elements of Kalman gain matrix.

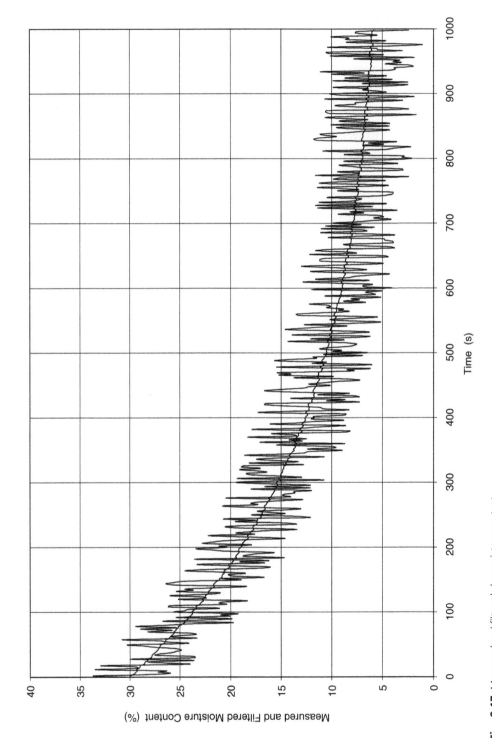

**Fig. 9.17** Measured and filtered clay moisture content.

The final values of the Kalman Gain matrix $\mathbf{K}$ and covariance matrix $\mathbf{P}$ were

$$\mathbf{K} = \begin{bmatrix} 0.4408 & 0.0003 & 0 \\ 0.0003 & 0.4579 & 0 \\ 0 & -0.0006 & 0.0325 \end{bmatrix}; \quad \mathbf{P} = \begin{bmatrix} 0.0044 & 0 & 0 \\ 0 & 0.0046 & 0 \\ 0 & 0 & 0.2426 \end{bmatrix} \quad (9.102)$$

The full LQG system, comprising of the LQ optimal controller and Kalman filter was then constructed. Figure 9.17 shows a set of moisture content measurements $z_3(kT)$ together with the estimated moisture content $\hat{\mathbf{x}}_3(kT)$.

## 9.6 Robust control

### 9.6.1 Introduction

The robust control problem is to find a control law which maintains system response and error signals within prescribed tolerances despite the effects of uncertainty on the system. Forms of uncertainty include

- Disturbance effects on the plant
- Measurement noise
- Modelling errors due to nonlinearities
- Modelling errors due to time-varying parameters

In previous chapters, linear models have been used to describe plant dynamics. However, in section 2.7.1 it was demonstrated that nonlinear functions could be linearized for small perturbations about an operating point. It is therefore possible to describe a nonlinear system by a series of linear models each constructed about a known operating point. If the operating point can be linked to a measurement, then a simple robust system may be constructed using an LQG approach. The feedback and Kalman gain matrices are calculated in advance for each operating point and some form of interpolation used to provide a 'Gain Scheduling Controller.'

The disturbance and measurement noise is taken into account by the Kalman filter. In the following example, undertaken by the author (1984), a non-linear simulation of a ship of length 161 m and displacement 17 000 tonnes was given a series of step changes in demanded rudder-angle at forward speeds of 2.572 m/s (5 knots), 5.145 m/s (10 knots) and 7.717 m/s (15 knots). At each forward speed a linear model was constructed and the $\mathbf{Q}$ and $\mathbf{R}$ matrices in an LQG implementation selected to return the closed-loop eigenvalues back to some desired value (Ackermann's formula could not be used since $\mathbf{y}(t)$ and $\mathbf{u}(t)$ were vector, not scalar quantities).

A subset of the state error variables is

$$e_1(t) = \text{cross-track position error}$$
$$e_2(t) = \text{cross-track velocity error}$$
$$e_3(t) = \text{heading error}$$
$$e_4(t) = \text{heading-rate error}$$

The feedback control is of the form

$$\mathbf{u}_{opt} = \mathbf{Ke}$$

where the values of $\mathbf{K}$ for the three forward speeds are

$$\begin{aligned}
\mathbf{K}_{2.572} &= [0.0121 \quad 1.035 \quad 7.596 \quad 160.26] \\
\mathbf{K}_{5.145} &= [0.0029 \quad 0.3292 \quad 1.81 \quad 25.963] \\
\mathbf{K}_{7.717} &= [0.0013 \quad 0.1532 \quad 0.8419 \quad 8.047]
\end{aligned} \tag{9.103}$$

If the forward velocity of the ship is the state variable $u_s$, a best estimate of which is given by the Kalman filter, the gain scheduling controller can be expressed as

$$\begin{aligned}
k_1 &= 0.08 u_s^{-2.0} \\
k_2 &= 6.0 u_s^{-1.8} \\
k_3 &= 50.0 u_s^{-2.0} \\
k_4 &= 2090.0 u_s^{-2.72}
\end{aligned} \tag{9.104}$$

Equation set (9.104) approximates to an inverse square law, and increases the controller gains at low speeds, where the control surfaces are at their most insensitive.

In general, however, robust control system design uses an idealized, or nominal model of the plant $G_m(s)$. Uncertainty in the nominal model is taken into account by considering a family of models that include all possible variations. The control system is said to have *robust stability* if a controller can be found that will stabilize all plants within the family. However, on its own, robust stability is not enough, since there may be certain plants within the family that are on the verge of instability. A controller is said to have *robust performance* if all the plants within the family meet a given performance specification.

## 9.6.2 Classical feedback control

Figure 9.18 shows a classical feedback control system $D(s)$ is a disturbance input, $N(s)$ is measurement noise, and therefore

$$\begin{aligned}
Y(s) &= G(s)U(s) + D(s) \\
B(s) &= Y(s) + N(s) \\
U(s) &= C(s)(R(s) - B(s))
\end{aligned} \tag{9.105}$$

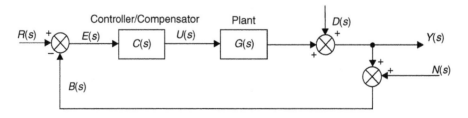

**Fig. 9.18** Classical feedback control system.

Eliminating $U(s)$ and $B(s)$ from equations (9.105) gives

$$Y(s) = \frac{G(s)C(s)R(s)}{1 + G(s)C(s)} + \frac{D(s)}{1 + G(s)C(s)} - \frac{G(s)C(s)N(s)}{1 + G(s)C(s)} \quad (9.106)$$

Define a sensitivity function $S(s)$ that relates $Y(s)$ and $D(s)$ when $R(s) = N(s) = 0$

$$\frac{Y}{D}(s) = S(s) = \frac{1}{1 + G(s)C(s)} \quad (9.107)$$

and define a complementary sensitivity function

$$T(s) = 1 - S(s) = \frac{G(s)C(s)}{1 + G(s)C(s)} \quad (9.108)$$

Thus, when $N(s) = 0$, equation (9.106) may be written

$$Y(s) = T(s)R(s) + S(s)D(s) \quad (9.109)$$

If $T(s) = 1$ and $S(s) = 0$ there is perfect set-point tracking and disturbance rejection. This requires that $G(s)C(s)$ is strictly proper (has more poles than zeros), so that

$$\lim_{s \to \infty} G(s)C(s) = 0 \quad (9.110)$$

However, if $N(s) \neq 0$, then equation (9.106) becomes

$$Y(s) = T(s)R(s) + S(s)D(s) - T(s)N(s) \quad (9.111)$$

Hence, if $T(s) = 1$, there will be both perfect set-point tracking and noise acceptance. Considering the problem in the frequency domain however, it may be possible that at low frequencies $T(j\omega) \to 1$ (good set-point tracking) and at high frequencies $T(j\omega) \to 0$ (good noise rejection).

## 9.6.3 Internal Model Control (IMC)

Consider the system shown in Figure 9.19 $G(s)$ is the plant, $G_m(s)$ is the nominal model, $R(s)$ is the desired value, $U(s)$ is the control, $D(s)$ is a disturbance input, $Y(s)$ is the output and $N(s)$ is the measurement noise. $C(s)$ is called the IMC controller and is to be designed so that $y(t)$ is kept as close as possible to $r(t)$ at all times.

From Figure 9.19, the feedback signal $B(s)$ is

$$B(s) + G(s)U(s) + D(s) + N(s) - G_m(s)U(s)$$

or

$$B(s) = (G(s) - G_m(s))U(s) + D(s) + N(s) \quad (9.112)$$

If, in equation (9.112) the model is exact, i.e. $G_m(s) = G(s)$ and the disturbance $D(s)$ and noise $N(s)$ are both zero, then $B(s)$ is also zero and the control system is effectively open-loop. This is the condition when there is no uncertainty. However, if $G_m(s) \neq G(s)$, and $D(s)$ and $N(s)$ are not zero, then $B(s)$ expresses the uncertainty of the process.

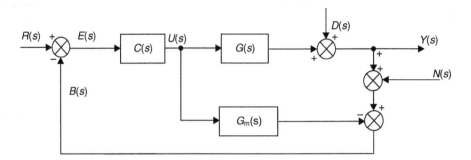

**Fig. 9.19** Block diagram of an IMC system.

## 9.6.4  IMC performance

From Figure 9.19

$$Y(s) = G(s)U(s) + D(s)$$

$$B(s) = Y(s) + N(s) - G_m(s)U(s) \quad\quad\quad (9.113)$$

$$U(s) = C(s)(R(s) - B(s))$$

Eliminating $U(s)$ and $B(s)$ from equations (9.113) gives

$$Y(s) = \frac{G(s)C(s)R(s)}{1 + C(s)(G(s) - G_m(s))} + \frac{(1 - C(s)G_m(s))D(s)}{1 + C(s)(G(s) - G_m(s))} - \frac{G(s)C(s)N(s)}{1 + C(s)(G(s) - G_m(s))}$$
$$(9.114)$$

The sensitivity function $S(s)$ that relates $Y(s)$ and $D(s)$ when $R(s) = N(s) = 0$ is

$$\frac{Y}{D}(s) = S(s) = \frac{1 - C(s)G_m(s)}{1 + C(s)(G(s) - G_m(s))} \quad\quad\quad (9.115)$$

and the complementary sensitivity function

$$T(s) = 1 - S(s) = \frac{C(s)G(s)}{1 + C(s)(G(s) - G_m(s))} \qu\quad\quad (9.116)$$

Thus, when $N(s) = 0$, equation (9.114) may be written

$$Y(s) = T(s)R(s) + S(s)D(s) \qu\quad\quad (9.117)$$

If $T(s) = 1$ there is perfect set-point tracking. This will occur if $G_m(s) = G(s)$ and $C(s) = 1/G(s)$. If $S(s) = 0$ there is perfect disturbance rejection. Again, this will occur if $G_m(s) = G(s)$ and $C(s) = 1/G_m(s)$.

### Two degree-of-freedom IMC system
If good set-point tracking and good disturbance rejection is required when the dynamic characteristics of $R(s)$ and $D(s)$ are substantially different, then it may be

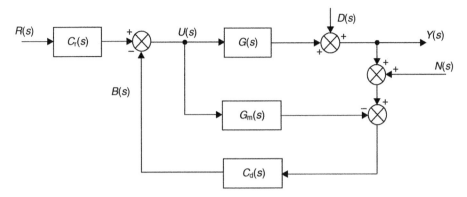

**Fig. 9.20** Two degree-of-freedom IMC system.

necessary to introduce a second controller, which provides a second degree-of-freedom of control action. A two degree-of-freedom IMC system is shown in Figure 9.20.

With a two degree-of-freedom IMC system, equation (9.114) becomes

$$Y(s) = \frac{G(s)C_r(s)R(s)}{1 + C_d(s)(G(s) - G_m(s))} + \frac{(1 - C_d(s)G_m(s))D(s)}{1 + C_d(s)(G(s) - G_m(s))}$$
$$- \frac{G(s)C_d(s)N(s)}{1 + C_d(s)(G(s) - G_m(s))}$$

(9.118)

In equation (9.118) $C_r(s)$ is designed for set-point tracking and $C_d(s)$ for disturbance rejection.

### 9.6.5 Structured and unstructured model uncertainty

Unstructured model uncertainty relates to unmodelled effects such as plant disturbances and are related to the nominal plant $G_m(s)$ as either additive uncertainty $\ell_a(s)$

$$G(s) = G_m(s) + \ell_a(s)$$

(9.119)

or multiplicative uncertainty $\ell_m(s)$

$$G(s) = (1 + \ell_m(s))G_m(s)$$

(9.120)

Equating (9.119) and (9.120) gives

$$\ell_a(s) = \ell_m(s)G_m(s)$$

(9.121)

Block diagram representations of additive and multiplicative model uncertainly are shown in Figure 9.21.

Structured uncertainty relates to parametric variations in the plant dynamics, i.e. uncertain variations in coefficients in plant differential equations.

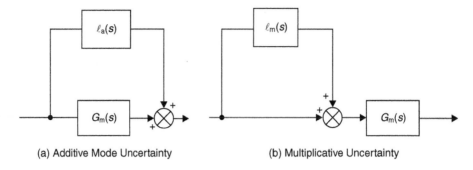

(a) Additive Mode Uncertainty       (b) Multiplicative Uncertainty

**Fig. 9.21** Additive and multiplicative model uncertainty.

## 9.6.6 Normalized system inputs

All inputs to the control loop (changes in set-point or disturbances) are generically represented by $V(s)$. The input $V(s)$ is found by passing a mathematically bounded normalized input $V^1(s)$ through a transfer function block $W(s)$, called the input weight, as shown in Figure 9.22.

### Specific inputs

$$\begin{aligned} \text{Impulse} \quad & V^1(s) = 1 \quad W(s) = 1 \\ \text{Step} \quad & V^1(s) = 1 \quad W(s) = 1/s \end{aligned} \tag{9.122}$$

Thus for specific inputs

$$V(s) = W(s)V^1(s) = W(s) \tag{9.123}$$

Sets of bounded inputs may be represented by

$$\left\| v^1(t) \right\|_2^2 = \int_0^\infty (v^1(t))^2 \mathrm{d}t \leq 1 \tag{9.124}$$

The left-hand side of equation (9.124) is called 'the 2-norm of the input signal $v^1(t)$ squared'. Norms are mathematical measures that enable objects belonging to the

**Fig. 9.22** Transformation of a normalized bounded input $V^1(s)$ into an actual input $V(s)$.

same set to be compared. Using Parseval's theorem, equation (9.124) may be transformed into the frequency domain

$$\left\|v^1(t)\right\|_2^2 = \frac{1}{2\pi}\int_{-\infty}^{\infty}\left|\frac{V(j\omega)}{W(j\omega)}\right|^2 d\omega \le 1 \tag{9.125}$$

For a realizable controller to exist, all external signals that enter the control loop must be bounded.

## 9.7 $H_2$- and $H_\infty$-optimal control

### 9.7.1 Linear quadratic $H_2$-optimal control

The scalar version of equation (9.48), when $\mathbf{u}(t)$ is not constrained, and $Q$ is unity, is called the Integral Squared Error (ISE), i.e.

$$\text{ISE} = \int_{t_0}^{t_1} e^2(t)dt \tag{9.126}$$

The $H_2$-optimal control problem is to find a contoller $c(t)$ such that the 2-norm of the ISE (written $\|e(t)\|_2^2$) is minimized for just one specific input $v(t)$.

If, in equation set (9.105) $B(s)$ and $Y(s)$ are eliminated and $U(s)$ is written as $C(s)E(s)$, then

$$\begin{aligned}E(s) &= \frac{1}{1 + G(s)C(s)}\{R(s) - D(s) - N(s)\}\\ &= S(s)\{R(s) - D(s) - N(s)\}\end{aligned} \tag{9.127}$$

Also, from equation (9.123), for a specific input

$$V(s) = W(s) \tag{9.128}$$

Using Parseval's theorem, from equation (9.126) the $H_2$-optimal control problem can be expressed in the frequency domain as

$$\min_c\|e(t)\|_2^2 = \min_c\frac{1}{2\pi}\int_{-\infty}^{\infty}|E(j\omega)|^2 d\omega \tag{9.129}$$

Substituting equations (9.127) and (9.128) into (9.129) gives

$$\min_c\|e(t)\|_2^2 = \min_c\frac{1}{2\pi}\int_{-\infty}^{\infty}|S(j\omega)W(j\omega)|^2 d\omega \tag{9.130}$$

Thus the $H_2$-optimal controller minimizes the average magnitude of the sensitivity function weighted by $W(j\omega)$, where $W(j\omega)$ depends upon the specific input $V(j\omega)$. In mathematical terms, the controller minimizes the 2-norm of the sensitivity function weighted by $W(j\omega)$.

### 9.7.2 $H_\infty$-optimal control

With $H_\infty$-optimal control the inputs $V(j\omega)$ are assumed to belong to a set of norm-bounded functions with weight $W(j\omega)$ as given by equation (9.125). Each input $V(j\omega)$ in the set will result in a corresponding error $E(j\omega)$. The $H_\infty$-optimal controller is designed to minimise the worst error that can arise from any input in the set, and can be expressed as

$$\min_c \|e(t)\|_\infty = \min_c \ \sup_\omega |S(j\omega)W(j\omega)| \qquad (9.131)$$

In equation (9.131), sup is short for supremum, which means the final result is the least upper bound. Thus the $H_\infty$-optimal controller minimizes the maximum magnitude of the weighted sensitivity function over frequency range $\omega$, or in mathematical terms, minimizes the $\infty$-norm of the sensitivity function weighted by $W(j\omega)$.

## 9.8 Robust stability and robust performance

### 9.8.1 Robust stability

Robust stability can be investigated in the frequency domain, using the Nyquist stability criterion, defined in section 6.4.2.

Consider a Nyquist contour for the nominal open-loop system $G_m(j\omega)C(j\omega)$ with the model uncertainty given by equation (9.119). Let $\bar{\ell}_a(\omega)$ be the bound of additive uncertainty and therefore be the radius of a disk superimposed upon the nominal Nyquist contour. This means that $G(j\omega)$ lies within a family of plants $\pi\,(G(j\omega) \in \pi)$ described by the disk, defined mathematically as

$$\pi = \{G: |G(j\omega) - G_m(j\omega)| \le \bar{\ell}_a(\omega)\} \qquad (9.132)$$

and therefore

$$|\ell_a(j\omega)| \le \bar{\ell}_a(\omega) \qquad (9.133)$$

If the multiplicative uncertainty in equations (9.120) and (9.121) is defined as

$$\ell_m(j\omega) = \frac{\ell_a(j\omega)}{G_m(j\omega)C(j\omega)} \qquad (9.134)$$

and the bound of multiplicative uncertainty

$$\bar{\ell}_m(\omega) = \frac{\bar{\ell}_a(\omega)}{|G_m(j\omega)C(j\omega)|} \qquad (9.135)$$

From equation (9.135) the disk radius (bound of uncertainty) is

$$\bar{\ell}_a(\omega) = |G_m(j\omega)C(j\omega)|\bar{\ell}_m(\omega) \qquad (9.136)$$

From the Nyquist stability criterion, let $N(k, G(j\omega))$ be the net number of clockwise encirclements of a point $(k, 0)$ of the Nyquist contour. Assume that all plants in the family $\pi$, expressed in equation (9.132) have the same number $(n)$ of right-hand plane (RHP) poles.

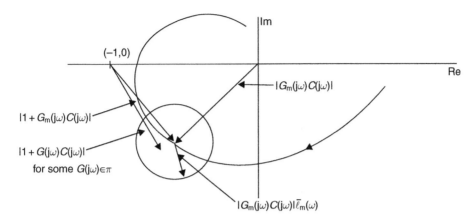

**Fig. 9.23** Robust stability.

There will be robust stability of a specific controller $C(j\omega)$ if and only if

$$N(-1, G(j\omega)C(j\omega)) = -n \quad \text{for all } G(j\omega) \in \pi \tag{9.137}$$

It is also necessary for the nominal plant $G_m(j\omega)$ to be stable

$$N(-1, G_m(j\omega)C(j\omega)) = -n \tag{9.138}$$

From Figure 9.23 robust stability occurs when the vector magnitude $|1 + G_m(j\omega)C(j\omega)|$ (see also Figure 6.25) exceeds the disk radius $|G_m(j\omega)C(j\omega)|\bar{\ell}_m(\omega)$

$$|1 + G_m(j\omega)C(j\omega)| > |G_m(j\omega)C(j\omega)|\bar{\ell}_m(\omega) \quad \text{for all } \omega$$

or

$$\left| \frac{G_m(j\omega)C(j\omega)}{1 + G_m(j\omega)C(j\omega)} \right| \bar{\ell}_m(\omega) < 1 \tag{9.139}$$

Equation (9.139) uses the magnitude of the complementary sensitivity function $T(j\omega)$ as defined in equation (9.108). Thus

$$|T(j\omega)|\bar{\ell}_m(\omega) < 1 \quad \text{for all } \omega \tag{9.140}$$

Robust stability can therefore be stated as: 'If all plants $G(s)$ in the family $\pi$ have the same number of RHP poles and that a particular controller $C(s)$ stabilizes the nominal plant $G_m(s)$, then the system is robustly stable with the controller $C(s)$ if and only if the complementary sensitivity function $T(s)$ for the nominal plant $G_m(s)$ satisfies the following bound

$$\|T(j\omega)\bar{\ell}_m(\omega)\|\infty = \sup_\omega |T(j\omega)\bar{\ell}_m(\omega)| < 1 \tag{9.141}$$

where the LHS of equation (9.141) is the infinity norm of $T(j\omega)\bar{\ell}_m(j\omega)$. This means that robust stability imposes a bound on the $\infty$ norm of the complementary sensitivity function $T(j\omega)$ weighted by $\bar{\ell}_m(\omega)$'.

## 9.8.2  Robust performance

Robust stability provides a minimum requirement in an environment where there is plant model uncertainty. For a control system to have robust performance it should be capable of minimizing the error for the *worst* plant (i.e. the one giving the largest error) in the family $G(j\omega) \in \pi$.

For the $H_\infty$-control problem, from equation (9.131), the $\infty$-norm of the weighted sensitivity function can be written

$$\|SW\|_\infty = \sup_\omega |S(j\omega)W(j\omega)| \tag{9.142}$$

If, as part of the design process, a bound is placed upon the sensitivity function

$$|S(j\omega)| < |W(j\omega)|^{-1} \tag{9.143}$$

Should an $H_\infty$ controller be found such that

$$\|SW\|_\infty < 1 \tag{9.144}$$

then the bound in equation (9.143) is met. Hence, for robust performance

$$\|SW\|_\infty = \sup_\omega |S(j\omega)W(j\omega)| < 1 \quad \text{for all } G(j\omega) \in \pi \tag{9.145}$$

From Figure 9.23 representing robust stability, the actual frequency response $G(j\omega)C(j\omega)$ will always lie inside the region of uncertainty denoted by the disk, or

$$|1 + G(j\omega)C(j\omega)| \geq |1 + G_m(j\omega)C(j\omega)| - |G_m(j\omega)C(j\omega)|\bar{\ell}_m(\omega) \quad \text{for all } G(j\omega) \in \pi \tag{9.146}$$

giving

$$|S(j\omega)| = \left| \frac{1}{1 + G(j\omega)C(j\omega)} \right| \leq \frac{|S_m(j\omega)|}{1 - |T_m(j\omega)|\bar{\ell}_m(\omega)} \quad \text{for all } G(j\omega) \in \pi \tag{9.147}$$

where $S_m(j\omega)$ is the sensitivity function for the nominal plant

$$S_m(j\omega) = \frac{1}{1 + G_m(j\omega)C(j\omega)} \tag{9.148}$$

Using equation (9.147), equation (9.145) can be expressed as

$$\frac{|S_m(j\omega)W(j\omega)|}{1 - |T_m(j\omega)|\bar{\ell}_m(\omega)} < 1 \quad \text{for all } \omega \tag{9.149}$$

or

$$|T_m(j\omega)\bar{\ell}_m(\omega)| + |S_m(j\omega)W(j\omega)| < 1 \quad \text{for all } \omega \tag{9.150}$$

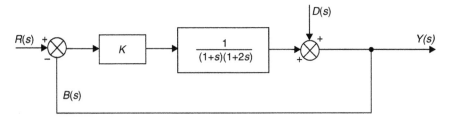

**Fig. 9.24** Control system.

Robust performance then means that the closed-loop system will meet the performance specification given in equation (9.145) if and only if the nominal system is
closed-loop stable (equation (9.141)) and that the sensitivity function $S_m(j\omega)$ and
complementary sensitivity function $T_m(j\omega)$ for the nominal system satisfy the relationship given in equation (9.150).

*Example 9.4*
(a) For the control system shown in Figure 9.24 produce the Bode magnitude plots
   for the sensitivity function $|S(j\omega)|$ and the complementary sensitivity function
   $|T(j\omega)|$ when $K = 10$. Comment on their values.
(b) For a step input, let $W(s) = 1/s$. Produce Bode magnitude plots for $|S(j\omega)W(j\omega)|$
   when $K = 10$, 50 and 100 and identify the optimal value using both $H_2$ and $H_\infty$
   criteria.

Use a frequency range of 0.01–100 rad/s for both (a) and (b).

*Solution*
(a) From equation (9.107)

$$S(s) = \frac{1}{1 + G(s)C(s)} = \frac{1}{1 + \frac{K}{(1+s)(1+2s)}}$$

$$= \frac{2s^2 + 3s + 1}{2s^2 + 3s + (1 + K)} \tag{9.151}$$

From equation (9.108)

$$T(s) = 1 - S(s) = 1 - \left\{ \frac{2s^2 + 3s + 1}{2s^2 + 3s + (1 + K)} \right\}$$

$$= \frac{K}{2s^2 + 3s + (1 + K)} \tag{9.152}$$

The Bode magnitude plots for $|S(j\omega)|$ and $|T(j\omega)|$ are shown in Figure 9.25 for
$K = 10$. From Figure 9.25 it can be seen that up to 1 rad/s, the system has a set-
point tracking error of $-0.8$ dB ($|T(j\omega)|$) and a disturbance rejection of
$-20$ dB ($|S(j\omega)|$).

(b) For a specific input of a unit step, let $W(s) = 1/s$. Hence the weighted sensitivity function is

$$S(s)W(s) = \frac{2s^2 + 3s + 1}{s\{2s^2 + 3s + (1 + K)\}} \tag{9.153}$$

The Bode magnitude plots for $|S(j\omega)W(j\omega)|$ for $K = 10$, 50 and 100 are shown in Figure 9.26.

From Figure 9.26 it can be seen that the $H_2$-norm, or average value of the weighted sensitivity function (equation (9.130)) reduces as $K$ increases and hence, using this criteria, $K = 100$ is the best value. Using the $H_\infty$-norm as defined in equation (9.131), the maximum magnitude of the weighted sensitivity function occurs at the lowest frequency. The least upper bound therefore is 0 dB, occurring at 0.01 rad/s when $K = 100$, so this again is the best value.

*Example 9.5*

A closed-loop control system has a nominal forward-path transfer function equal to that given in Example 6.4, i.e.

$$G_m(s)C(s) = \frac{K}{s(s^2 + 2s + 4)}$$

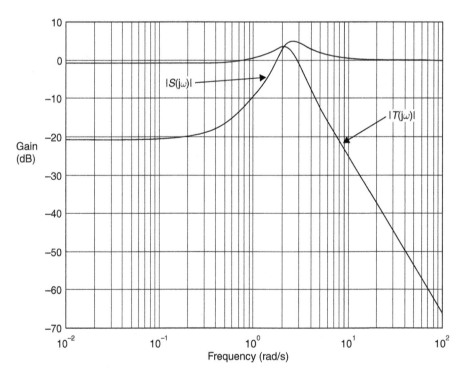

**Fig. 9.25** Bode magnitude plots for $|S(j\omega)|$ and $|T(j\omega)|$.

Let the bound of the multiplicative model uncertainty be

$$\bar{\ell}_m(s) = \frac{0.5(1+s)}{(1+0.25s)}$$

What is the maximum value that $K$ can have for robust stability?

*Solution*
At frequencies below $1\,\text{rad/s}$, $\bar{\ell}_m(\omega) \rightarrow 0.5$ and at frequencies above $4\,\text{rad/s}$ $\bar{\ell}_m(\omega) \rightarrow 2.0$. From equation (9.141), for robust stability

$$|T(j\omega)\bar{\ell}_m(\omega)| < 1 \qquad (9.154)$$

now

$$T(s) = \frac{G_m(s)C(s)}{1 + G_m(s)C(s)}$$

therefore

$$T(s) = \frac{\frac{K}{s^3+2s^2+4s}}{1 + \frac{K}{s^3+2s^2+4s}}$$

$$T(s) = \frac{K}{s^3 + 2s^2 + 4s + K}$$

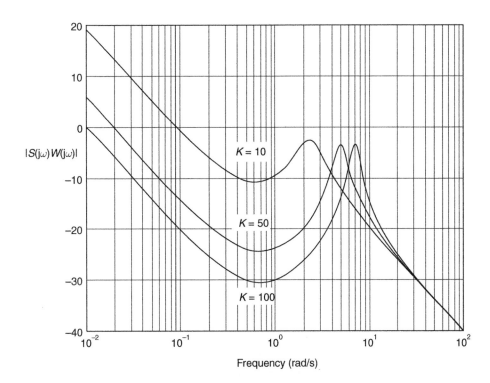

**Fig. 9.26** Bode magnitude plot of weighted sensitivity function for Example 9.4.

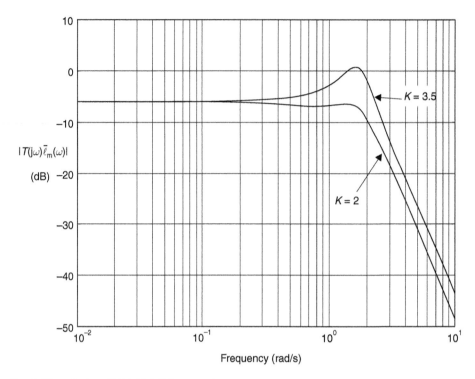

**Fig. 9.27** Bode plot of $|T(j\omega)\bar{\ell}_m(\omega)|$ for Example 9.5.

thus

$$T(s)\bar{\ell}_m(s) = \frac{0.5K(1+s)}{(1+0.25s)(s^3 + 2s^2 + 4s + K)} \qquad (9.155)$$

The Bode magnitude plot for equation (9.155) is shown in Figure 9.27 when $K = 2$ and 3.5.

In Example 6.4, when there was no model uncertainty, $K$ for marginal stability was 8, and for a gain margin of 6 dB, $K$ was 4. In this example with model uncertainty, from equation (9.154) marginal stability occurs with $K = 3.5$, so this is the maximum value for robust stability. For robust performance, equation (9.150) applies. For a specific step input let $W(s) = 1/s$ now

$$S_m(s) = \frac{s^3 + 2s^2 + 4s}{s^3 + 2s^2 + 4s + K} \qquad (9.156)$$

and

$$S_m(s)W(s) = \frac{s(s^2 + 2s + 4)}{s(s^3 + 2s^2 + 4s + K)}$$

hence

$$S_m(s)W(s) = \frac{s^2 + 2s + 4}{s^3 + 2s^2 + 4s + K} \qquad (9.157)$$

The Bode magnitude plot of the weighted sensitivity function is shown in Figure 9.28 for $K = 2$, 2.5 and 3.5.

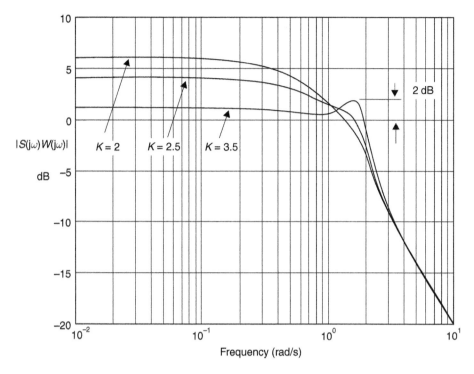

**Fig. 9.28** Bode magnitude plot of weighted sensitivity function for Example 9.5.

For robust performance

$$\left|T_{\mathrm{m}}(\mathrm{j}\omega) + \bar{\ell}_{\mathrm{m}}(\omega)\right| + \left|S_{\mathrm{m}}(\mathrm{j}\omega)W(\mathrm{j}\omega)\right| < 1 \quad \text{for all } \omega \qquad (9.158)$$

From Table 9.2 it can be concluded that:

(a) The control system has robust stability up to $K = 3.5$.
(b) The $H_\infty$-norm is $>1$ for all values of $K$ considered. Therefore equations (9.145) and (9.150) are not met and the system cannot be considered to have robust performance.

From (b) above, it must be concluded that the controller $C(s)$ must be something more sophisticated than a simple gain constant $K$.

**Table 9.2** Robust performance for Example 9.5

| $\omega$ (rad/s) | 0.01 | | | 1.5 | | |
|---|---|---|---|---|---|---|
| $K$ | 2 | 2.5 | 3.5 | 2 | 2.5 | 3.5 |
| $\left|T_{\mathrm{m}}(\mathrm{j}\omega)\bar{\ell}_{\mathrm{m}}(\omega)\right|$ | 0.5 (−6 dB) | 0.5 (not shown in Figure 9.27) | 0.5 | 0.5 | 0.63 (not shown in Figure 9.27) | 1.0 |
| $\left|S_{\mathrm{m}}(\mathrm{j}\omega)W(\mathrm{j}\omega)\right|$ | 2.0 (6 dB) | 1.58 | 1.12 | 0.96 | 1.05 | 1.26 |
| Sum | 2.5 | 2.08 | 1.62 | 1.46 | 1.68 | 2.26 |

## 9.9 Multivariable robust control

### 9.9.1 Plant equations

The canonical robust control problem is shown in Figure 9.29.

In Figure 9.29, $\mathbf{u}_2$ are the inputs to the plant $\mathbf{P}_m$ from the controller and $\mathbf{u}_1$ are the disturbance and noise inputs. Also, $\mathbf{y}_1$ are the outputs to be controlled and $\mathbf{y}_2$ are the outputs that are fed back to the controller.

If $\mathbf{P}_m(s)$ and the plant uncertainty $\Delta(s)$ are combined to give $\mathbf{P}(s)$, then Figure 9.29 can be simplified as shown in Figure 9.30, also referred to as the two-port state-space representation.

The state and output equations are

$$\dot{\mathbf{x}} = \mathbf{A}\mathbf{x} + \mathbf{B}_1\mathbf{u}_1 + \mathbf{B}_2\mathbf{u}_2$$
$$\mathbf{y}_1 = \mathbf{C}_1\mathbf{x} + \mathbf{D}_{11}\mathbf{u}_1 + \mathbf{D}_{12}\mathbf{u}_2$$
$$\mathbf{y}_2 = \mathbf{C}_2\mathbf{x} + \mathbf{D}_{21}\mathbf{u}_1 + \mathbf{D}_{22}\mathbf{u}_2 \qquad (9.159)$$

Equation (9.159) can be combined

$$\begin{bmatrix} \dot{\mathbf{x}} \\ \mathbf{y}_1 \\ \mathbf{y}_2 \end{bmatrix} = \begin{bmatrix} \mathbf{A} & \mathbf{B}_1 & \mathbf{B}_2 \\ \mathbf{C}_1 & \mathbf{D}_{11} & \mathbf{D}_{12} \\ \mathbf{C}_2 & \mathbf{D}_{21} & \mathbf{D}_{22} \end{bmatrix} \begin{bmatrix} \mathbf{x} \\ \mathbf{u}_1 \\ \mathbf{u}_2 \end{bmatrix} \qquad (9.160)$$

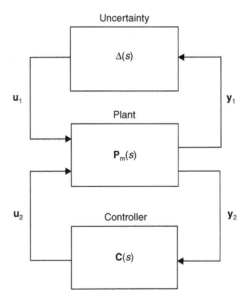

**Fig. 9.29** The canonical robust control problem.

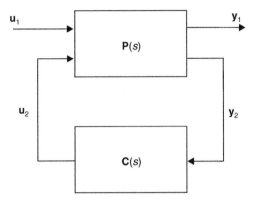

**Fig. 9.30** Two-port state-space augmented plant and controller.

Hence the augmented plant matrix $\mathbf{P}(s)$ in Figure 9.30 is

$$\mathbf{P}(s) = \left[\begin{array}{c:cc} \mathbf{A} & \mathbf{B}_1 & \mathbf{B}_2 \\ \hdashline \mathbf{C}_1 & \mathbf{D}_{11} & \mathbf{D}_{12} \\ \mathbf{C}_2 & \mathbf{D}_{21} & \mathbf{D}_{22} \end{array}\right] = \left[\begin{array}{c:c} \mathbf{P}_{11} & \mathbf{P}_{12} \\ \hdashline \mathbf{P}_{21} & \mathbf{P}_{22} \end{array}\right] \qquad (9.161)$$

From the partitioned matrix in equation (9.161), the closed-loop transfer function matrix relating $\mathbf{y}_1$ and $\mathbf{u}_1$ is

$$\mathbf{T}\mathbf{y}_1\mathbf{u}_1 = \mathbf{P}_{11}(s) + \mathbf{P}_{12}(s)\left(\mathbf{I} - \mathbf{C}(s)\,\mathbf{P}_{22}(s)\right)^{-1}\mathbf{C}(s)\,\mathbf{P}_{21}(s), \qquad (9.162)$$

where

$$\mathbf{u}_2(s) = \mathbf{C}(s)\mathbf{y}_2(s) \qquad (9.163)$$

## 9.9.2   Singular value loop shaping

The singular values of a complex $n \times m$ matrix $\mathbf{A}$, denoted by $\sigma_i(\mathbf{A})$ are the non-negative square-roots of the eigenvalues of $\mathbf{A}^T\mathbf{A}$ ordered such that

$$\sigma_1 \geq \sigma_2 \geq \cdots \geq \sigma_\mathrm{p} \quad p = \min\{n, m\} \qquad (9.164)$$

The maximum singular value $\bar{\sigma}$ of $\mathbf{A}$ and the minimum singular value $\underline{\sigma}$ of $\mathbf{A}$ are defined by

$$\begin{aligned} \bar{\sigma}(\mathbf{A}) &= \|\mathbf{A}\|_2 \\ \underline{\sigma}(\mathbf{A}) &= \|\mathbf{A}^{-1}\|_2^{-1} \quad \text{if } \mathbf{A}^{-1} \text{ exists} \end{aligned} \qquad (9.165)$$

As with a SISO system, a sensitivity function may be defined

$$\mathbf{S}(s) = (\mathbf{I} + \mathbf{G}(s)\mathbf{C}(s))^{-1} \qquad (9.166)$$

where $G(s)$ is the non-augmented plant matrix. For good performance $S(s)$ should be as small as possible. The complementary sensitivity function is

$$T(s) = G(s)\,C(s)\,(I + G(s)C(s))^{-1} \tag{9.167}$$

where

$$S(s) + T(s) = I \tag{9.168}$$

The singular value of the sensitivity function $\bar{\sigma}(S(j\omega))$ and of the complementary sensitivity function $\bar{\sigma}(T(j\omega))$ can be displayed as Bode plots and play an important role in robust multivariable control system design.

The singular values of $S$ determine the disturbance attenuation, and thus a performance specification may be written

$$\bar{\sigma}(S(j\omega)) \le \left| W_s^{-1}(j\omega) \right| \tag{9.169}$$

where $\left| W_s^{-1}(j\omega) \right|$ is a desired disturbance attenuation factor. If $\Delta_m(s)$ is a diagonal matrix of multiplicative plant uncertainty as illustrated in Figure 9.29, it can be shown that the size of the smallest stable $\Delta_m(s)$ for which the system becomes unstable is

$$\bar{\sigma}(\Delta_m(j\omega)) = 1/\bar{\sigma}(T(j\omega)) \tag{9.170}$$

or alternatively

$$\bar{\sigma}(T(j\omega)) \le \left| W_T^{-1}(j\omega) \right| \tag{9.171}$$

where $|W_T(j\omega)|$ is the size of the largest anticipated multiplicative plant uncertainty.

### 9.9.3   Multivariable $H_2$ and $H_\infty$ robust control

The $H_2$-optimal control problem is to find a stabilizing controller $C(s)$ in equation (9.163) for an augmented plant $P(s)$ in equation (9.161), such that the closed-loop transfer function matrix $Ty_1u_1$ in equation (9.162) is minimized.

Thus

$$\min_{C(s)} \|Ty_1u_1\|_2 = \min_{C(s)} \left\{ \frac{1}{\pi} \int_0^\infty \text{trace}(Ty_1u_1(j\omega)^T Ty_1u_1(j\omega)) d\omega \right\}^{1/2} \tag{9.172}$$

where $T$ is the complex conjugate transpose, and trace is the sum of the diagonal elements. The $H_\infty$ robust control problem is to find a stabilizing controller $C(s)$ for an augmented plant $P(s)$, such that the closed-loop transfer function matrix $Ty_1u_1$ satisfies the infinity-norm inequality

$$\|Ty_1u_1\|_\infty = \sup_\omega \sigma_{max}(Ty_1u_1(j\omega)) < 1 \tag{9.173}$$

Equation (9.173) is also called the 'small gain' infinity-norm control problem.

## 9.9.4   The weighted mixed-sensitivity approach

Multivariable loop shaping in robust control system design may be achieved using a weighted mixed sensitivity approach. As with the SISO systems described in section 9.8.2, the sensitivity function $\mathbf{S}(s)$ given in equation (9.166) and the complementary sensitivity function $\mathbf{T}(s)$ given in equation (9.167) may be combined with weights $\mathbf{W}_s(s)$ and $\mathbf{W}_T(s)$ to give

$$\mathbf{Ty}_1\mathbf{u}_1 = \begin{bmatrix} \mathbf{W}_s(s) & \mathbf{S}(s) \\ \mathbf{W}_T(s) & \mathbf{T}(s) \end{bmatrix} \tag{9.174}$$

where the infinity norm of $\mathbf{Ty}_1\mathbf{u}_1$ is $<1$ as given in equation (9.173). Equation (9.174) defines a mixed-sensitivity cost function since both $\mathbf{S}(s)$ and $\mathbf{T}(s)$ are penalized. Note that if $\mathbf{W}_s(s)$ weights the error and $\mathbf{W}_T(s)$ the output, the two-port augmented plant given in Figure 9.30 may be represented by Figure 9.31.

*Example 9.6*   (See also Appendix 1, *examp96.m*)
A plant has a transfer function

$$G(s) = \frac{200}{s^3 + 3s^2 + 102s + 200} \tag{9.175}$$

given the sensitivity and complementary weighting functions

$$W_s(s) = \gamma\left(\frac{100 + s}{1 + 100s}\right)$$

$$W_T(s) = \left(\frac{1 + 100s}{100 + s}\right) \tag{9.176}$$

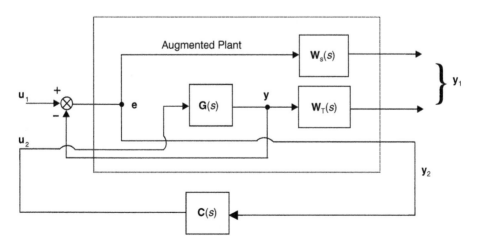

**Fig. 9.31** Weighted mixed-sensitivity approach.

determine the singular value Bode magnitude plots for

(a) the plant $G(j\omega)$
(b) the weighting functions $W_s^{-1}(j\omega)$ and $W_T^{-1}(j\omega)$
(c) the cost function $\mathbf{Ty_1u_1}(j\omega)$ at its optimal value of $\gamma$ (given in $W_s(s)$)
(d) the $H_\infty$-optimal controller $C(j\omega)$

Find also the state-space and transfer function expressions for the controller.

*Solution*
The state-space representation of the plant $G(s)$ is

$$
\begin{bmatrix} \mathbf{Ag} & \mathbf{Bg} \\ \mathbf{Cd} & \mathbf{Dg} \end{bmatrix} = \left[ \begin{array}{cccc} -3 & -102 & -200 & \vdots & 1 \\ 1 & 0 & 0 & \vdots & 0 \\ 0 & 1 & 0 & \vdots & 0 \\ \hdashline 0 & 0 & 200 & \vdots & 0 \end{array} \right] \tag{9.177}
$$

The singular value frequency response $G(j\omega)$ is shown in Figure 9.32.

The frequency response of the reciprocal of the weighting functions $W_s^{-1}(j\omega)$ ($\gamma = 1$) and $W_T^{-1}(j\omega)$ are given in Figure 9.33.

The optimal value of $\mathbf{Ty_1u_1}$ is achieved when ($\gamma = 0.13$) and its singular value frequency response is shown in Figure 9.34.

The controller single value frequency response $C(j\omega)$ is illustrated in Figure 9.35.

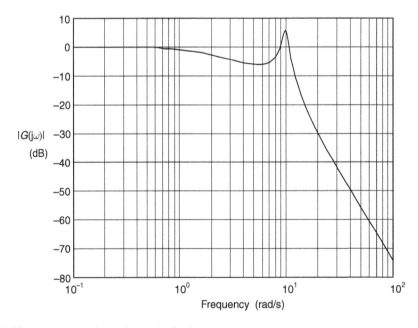

**Fig. 9.32** Plant singular value Bode magnitude plot.

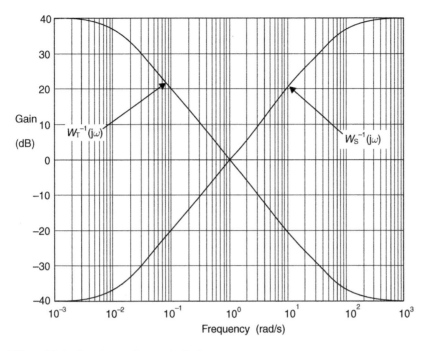

**Fig. 9.33** Weighting functions Bode magnitude plots.

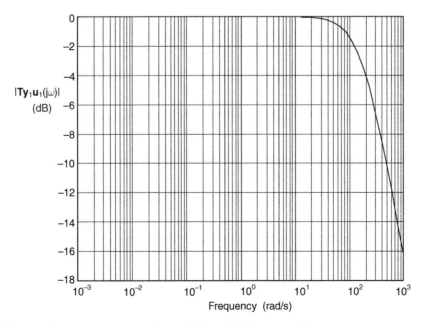

**Fig. 9.34** Singular value Bode magnitude plot of $|Ty_1u_1(j\omega)|$ when $\gamma = 0.13$.

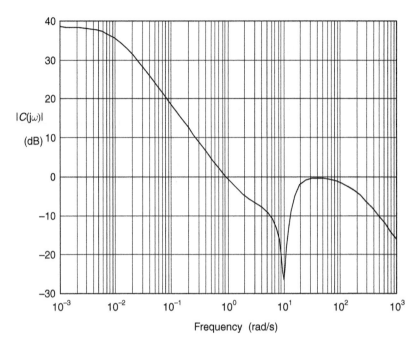

**Fig. 9.35** Controller single value Bode magnitude plot $|C(j\omega)|$.

The state-space representation of the controller $C(s)$ is

$$
\begin{bmatrix} \mathbf{A_c} & \mathbf{B_c} \\ \mathbf{C_c} & \mathbf{D_c} \end{bmatrix} =
\left[
\begin{array}{ccccc:c}
-0.01 & -0.002 & 0.004 & 0.015 & 0.137 & -7.976 \\
-0.009 & -7.763 & 21.653 & 37.164 & 621.89 & 0.091 \\
-0.046 & -2.032 & -3.148 & -2.234 & -81.88 & 0.582 \\
0.435 & 0.107 & 8.807 & -102.25 & -37.78 & -8.8 \\
0.676 & 0.197 & 13.558 & -3.728 & -162.51 & -13.66 \\
\hdashline
-0.086 & 0.169 & 0.121 & -0.654 & -11.17 & 0
\end{array}
\right]
$$

(9.178)

and the controller transfer function is

$$
C(s) = \frac{159s^4 + 16.4 \times 10^3 s^3 + 63.9 \times 10^3 s^2 + 1.6 \times 10^6 s + 3.18 \times 10^6}{s^5 + 275s^4 + 20.4 \times 10^3 s^3 + 324.8 \times 10^3 s^2 + 3.78 \times 10^6 s + 37.8 \times 10^3}
$$

(9.179)

The results in this example were obtained using the MATLAB Robust Control Toolbox.

## 9.10 Further problems

*Example 9.7*
In a multivariable optimal regulator system, the plant state equations are

$$\begin{bmatrix} \dot{x}_1 \\ \dot{x}_2 \end{bmatrix} = \begin{bmatrix} 0 & 1 \\ -4 & -2 \end{bmatrix} \begin{bmatrix} x_1 \\ x_2 \end{bmatrix} + \begin{bmatrix} 0 \\ 4 \end{bmatrix} u \qquad (9.180)$$
$$y = \begin{bmatrix} 1 & 0 \end{bmatrix} \mathbf{x}$$

If the performance index to be minimized is

$$J = \int_0^\infty (\mathbf{x}^T \mathbf{Q} \mathbf{x} + u^2) dt \qquad (9.181)$$

(a) Determine, by hand, the elements of the Riccati matrix $\mathbf{P}$ in the reduced Riccati equation

$$\mathbf{P}\mathbf{A} + \mathbf{A}^T\mathbf{P} + \mathbf{Q} - \mathbf{P}\mathbf{B}\mathbf{R}^{-1}\mathbf{B}^T\mathbf{P} = 0 \qquad (9.182)$$

given that

$$\mathbf{Q} = \begin{bmatrix} 2 & 0 \\ 0 & 1 \end{bmatrix} \qquad (9.183)$$

(b) Find the optimal feedback matrix $\mathbf{K}$ so that

$$\mathbf{K} = \mathbf{R}^{-1}\mathbf{B}^T\mathbf{P} \qquad (9.184)$$

and hence calculate the closed-loop eigenvalues.

*Solutions*
(a) $\mathbf{P} = \begin{bmatrix} 1.703 & 0.183 \\ 0.183 & 0.193 \end{bmatrix}$

(b) $\mathbf{K} = \begin{bmatrix} 0.732 & 0.772 \end{bmatrix}$

$s = -2.544 \pm j0.675$

*Example 9.8*
A plant and measurement system are described by

$$\dot{\mathbf{x}} = \mathbf{A}\mathbf{x} + \mathbf{B}\mathbf{u} + \mathbf{C}_d\mathbf{w}$$
$$\mathbf{y} = \mathbf{C}\mathbf{x} \qquad (9.185)$$
$$\mathbf{z} = \mathbf{y} + \mathbf{v}$$

where $\mathbf{w}(t)$ is a Gaussian sequence of disturbances and $\mathbf{v}(t)$ is a Gaussian sequence of measurement noise. $\mathbf{z}(t)$ is the measured value of $\mathbf{y}(t)$ that is contaminated with measurement noise $\mathbf{v}(t)$. The plant parameters are

$$\mathbf{A} = \begin{bmatrix} 0 & 1 & 0 \\ 0 & 0 & 1 \\ -50 & -102 & -4.5 \end{bmatrix} \quad \mathbf{B} = \begin{bmatrix} 0 \\ 0 \\ 100 \end{bmatrix}$$

$$\mathbf{C_d} = \begin{bmatrix} 0.5 & 0 & 0 \\ 0 & 0.5 & 0 \\ 0 & 0 & 10 \end{bmatrix} \quad \mathbf{C} = \begin{bmatrix} 1 & 0 & 0 \\ 0 & 1 & 0 \\ 0 & 0 & 1 \end{bmatrix}$$

(9.186)

The measurement noise and disturbance covariance matrices are

$$\mathbf{R} = \begin{bmatrix} 0.1 & 0 & 0 \\ 0 & 0.1 & 0 \\ 0 & 0 & 6 \end{bmatrix} \quad \mathbf{Q} = \begin{bmatrix} 0.1 & 0 & 0 \\ 0 & 0.1 & 0 \\ 0 & 0 & 2 \end{bmatrix}$$

(9.187)

(a) For a sampling time of 0.1 seconds, using equations (8.78) and (8.80) calculate the discrete-time state transition, control and disturbance matrices $\mathbf{A}(T)$, $\mathbf{B}(T)$ and $\mathbf{C_d}(T)$.

(b) Starting with an initial covariance matrix $\mathbf{P}(k/k)$ equal to the identity matrix, perform 20 recursions of equations (9.74), (9.75) and (9.76) to compute the Kalman gain matrix $\mathbf{K}(k+1)$ and covariance matrix $\mathbf{P}(k+1/k+1)$.

*Solutions*

(a) $\quad \mathbf{A}(T) = \begin{bmatrix} 0.993 & 0.085 & 0.004 \\ -0.199 & 0.587 & 0.067 \\ -3.370 & -7.074 & 0.284 \end{bmatrix} \quad \mathbf{B}(T) = \begin{bmatrix} 0.014 \\ 0.398 \\ 6.740 \end{bmatrix}$

$\mathbf{C_d}(T) = \begin{bmatrix} 0.050 & 0.002 & 0.001 \\ -0.004 & 0.043 & 0.040 \\ -0.100 & -0.207 & 0.674 \end{bmatrix}$

(b) $\quad \mathbf{K}(k+1) = \begin{bmatrix} 0.040 & -0.006 & -0.002 \\ -0.006 & 0.137 & 0.003 \\ -0.133 & 0.155 & 0.218 \end{bmatrix}$

$\mathbf{P}(k+1/k+1) = \begin{bmatrix} 0.004 & -0.001 & -0.013 \\ -0.001 & 0.014 & 0.016 \\ -0.013 & 0.016 & 1.310 \end{bmatrix}$

*Example 9.9*
The plant described in Example 9.8 by equations (9.185) and (9.186) is to be controlled by a Linear Quadratic Gaussian (LQG) control scheme that consists of a LQ Regulator combined with the Kalman filter designed in Example 9.8. The

quadratic performance index to be minimized for the LQ regulator is of the form given in equation (9.181) where

$$Q = \begin{bmatrix} 10 & 0 & 0 \\ 0 & 5 & 0 \\ 0 & 0 & 1 \end{bmatrix} \quad (R = 1) \tag{9.188}$$

Using the recursive equations (9.29) and (9.30), solve, in reverse time, the Riccati equation commencing with $P(N) = 0$.

If the sampling time is 0.1 seconds, the values of the discrete-time state transition and control matrices $A(T)$ and $B(T)$ calculated in Example 9.8 may be used in the recursive solution.

Continue the recursive steps until the solution settles down (when $k = 50$, or $kT = 5$ seconds) and hence determine the steady-state value of the feedback matrix $K(0)$ and Riccati matrix $P(0)$. What are the closed-loop eigenvalues?

*Solutions*

$$K(0) = [\,-0.106 \quad -0.581 \quad 0.064\,]$$

$$P(0) = \begin{bmatrix} 11.474 & 3.406 & 0.153 \\ 3.406 & 3.952 & 0.163 \\ 0.153 & 0.163 & 0.1086 \end{bmatrix}$$

closed-loop eigenvalues $= -1.230$

$$-4.816 \pm j2.974$$

*Example 9.10*
A unity-feedback control system has a nominal plant transfer function

$$G_m(s) = \frac{1}{(s+2)(s+5)} \tag{9.189}$$

and an integral controller in the forward path

$$C(s) = \frac{K}{s} \tag{9.190}$$

If the bound of the multiplicative model uncertainty is

$$\bar{\ell}_m(s) = \frac{0.25(1+4s)}{(1+0.25s)} \tag{9.191}$$

determine:

(a) Expressions for the sensitivity and complementary sensitivity function $S(s)$ and $T(s)$ for the nominal plant.
(b) The maximum value that $K$ can have for robust stability.

*Solutions*

(a) $S(s) = \dfrac{s^3 + 7s^2 + 10s}{s^3 + 7s^2 + 10s + K}$

$T(s) = \dfrac{K}{s^3 + 7s^2 + 10s + K}$

(b) $K_{max} = 4.5$

*Example 9.11*
A plant has a transfer function

$$G(s) = \frac{100}{s^2 + 2s + 100}$$

and sensitivity and complementary weighting functions

$$W_s(s) = \gamma\left(\frac{s + 100}{s + 1}\right)$$

$$W_T(s) = \left(\frac{s + 1}{s + 100}\right)$$

Find the optimal value for $\gamma$ and hence the state-space and transfer functions for the $H_\infty$-optimal controller $C(s)$.

*Solutions*

$\gamma_{opt} = 0.0576$

$$
\begin{bmatrix} \mathbf{A_c} & \mathbf{B_c} \\ \mathbf{C_c} & \mathbf{D_c} \end{bmatrix} =
\left[
\begin{array}{cccc:c}
-2.8 & 3.0 & -1.4 & 150.9 & -4.55 \\
-3.0 & -22.2 & 50.1 & 258.0 & 3.10 \\
-2.7 & 39.6 & -74.5 & 196.1 & 2.99 \\
-3.2 & -36.3 & -23.2 & -1871.1 & 7.44 \\
\hdashline
-0.33 & 1.68 & 1.02 & -49.42 & 0
\end{array}
\right]
$$

$$C(s) = \frac{1.86 \times 10^3 s^3 + 0.1898 \times 10^6 s^2 + 0.5581 \times 10^6 s + 18.6031 \times 10^6}{s^4 + 1.97 \times 10^3 s^3 + 0.2005 \times 10^6 s^2 + 1.3531 \times 10^6 s + 1.1546 \times 10^6}$$

# Intelligent control system design

### 10.1.1 Intelligence in machines

According to the Oxford dictionary, the word intelligence is derived from intellect, which is the faculty of knowing, reasoning and understanding. Intelligent behaviour is therefore the ability to reason, plan and learn, which in turn requires access to knowledge.

Artificial Intelligence (AI) is a by-product of the Information Technology (IT) revolution, and is an attempt to replace human intelligence with machine intelligence. An intelligent control system combines the techniques from the fields of AI with those of control engineering to design autonomous systems that can sense, reason, plan, learn and act in an intelligent manner. Such a system should be able to achieve sustained desired behaviour under conditions of uncertainty, which include:

(a) uncertainty in plant models
(b) unpredictable environmental changes
(c) incomplete, inconsistent or unreliable sensor information
(d) actuator malfunction.

### 10.1.2 Control system structure

An intelligent control system, as considered by Johnson and Picton (1995), comprises of a number of subsystems as shown in Figure 10.1.

#### The perception subsystem
This collects information from the plant and the environment, and processes it into a form suitable for the cognition subsystem. The essential elements are:

(a) *Sensor array* which provides raw data about the plant and the environment
(b) *Signal processing* which transforms information into a suitable form
(c) *Data fusion* which uses multidimensional data spaces to build representations of the plant and its environment. A key technology here is pattern recognition.

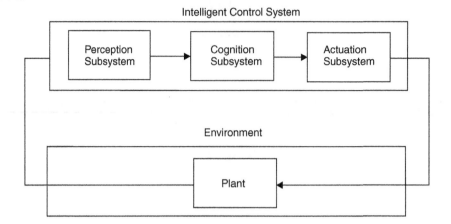

**Fig. 10.1** Intelligent control system structure (adapted from Johnson and Picton).

### The cognition subsystem

Cognition in an intelligent control system is concerned with the decision making process under conditions of uncertainty. Key activities include:

(a) Reasoning, using
    (i) knowledge-based systems
    (ii) fuzzy logic
(b) Strategic planning, using
    (i) optimum policy evaluation
    (ii) adaptive search and genetic algorithms
    (iii) path planning
(c) Learning, using
    (i) supervised learning in neural networks
    (ii) unsupervised learning in neural networks
    (iii) adaptive learning

### The actuation subsystem

The actuators operate using signals from the cognition subsystem in order to drive the plant to some desired states. In the event of actuator (or sensor) failure, an intelligent control system should be capable of being able to re-configure its control strategy.

This chapter is mainly concerned with some of the processes that are contained within the cognition subsystem.

## 10.2 Fuzzy logic control systems

### 10.2.1 Fuzzy set theory

Fuzzy logic was first proposed by Zadeh (1965) and is based on the concept of fuzzy sets. Fuzzy set theory provides a means for representing uncertainty. In general, probability theory is the primary tool for analysing uncertainty, and assumes that the uncertainty is a random process. However, not all uncertainty is random, and fuzzy set theory is used to model the kind of uncertainty associated with imprecision, vagueness and lack of information.

Conventional set theory distinguishes between those elements that are members of a set and those that are not, there being very clear, or crisp boundaries. Figure 10.2 shows the crisp set 'medium temperature'. Temperatures between 20 and 30 °C lie within the crisp set, and have a membership value of one.

The central concept of fuzzy set theory is that the membership function $\mu$, like probability theory, can have a value of between 0 and 1. In Figure 10.3, the membership function $\mu$ has a linear relationship with the $x$-axis, called the universe of discourse U. This produces a triangular shaped fuzzy set.

Fuzzy sets represented by symmetrical triangles are commonly used because they give good results and computation is simple. Other arrangements include non-symmetrical triangles, trapezoids, Gaussian and bell shaped curves.

Let the fuzzy set 'medium temperature' be called fuzzy set M. If an element $u$ of the universe of discourse U lies within fuzzy set M, it will have a value of between 0 and 1. This is expressed mathematically as

$$\mu_M(u) \in [0, 1] \tag{10.1}$$

When the universe of discourse is discrete and finite, fuzzy set M may be expressed as

$$M = \sum_{i=1}^{n} \mu_M(u_i)/u_i \tag{10.2}$$

In equation (10.2) '/' is a delimiter. Hence the numerator of each term is the membership value in fuzzy set M associated with the element of the universe indicated in the denominator. When $n = 11$, equation (10.2) can be written as

$$M = 0/0 + 0/5 + 0/10 + 0.33/15 + 0.67/20 + 1/25 + 0.67/30 + 0.33/35 \\ + 0/40 + 0/45 + 0/50 \tag{10.3}$$

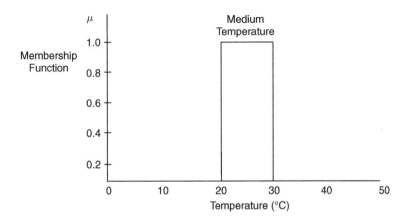

**Fig. 10.2** Crisp set 'medium temperature'.

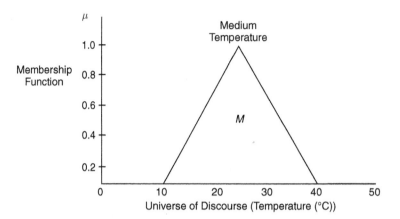

**Fig. 10.3** Fuzzy set 'medium temperature'.

Note the symbol '+' is not an addition in the normal algebraic sense, but in fuzzy arithmetic denotes a union operation.

## 10.2.2   Basic fuzzy set operations

Let A and B be two fuzzy sets within a universe of discourse U with membership functions $\mu_A$ and $\mu_B$ respectively. The following fuzzy set operations can be defined as

*Equality*: Two fuzzy sets A and B are equal if they have the same membership function within a universe of discourse U.

$$\mu_A(u) = \mu_B(u) \quad \text{for all } u \in U \tag{10.4}$$

*Union*: The union of two fuzzy sets A and B corresponds to the Boolean OR function and is given by

$$\mu_{A\cup B}(u) = \mu_{A+B}(u) = \max\{\mu_A(u), \mu_B(u)\} \quad \text{for all } u \in U \tag{10.5}$$

*Intersection*: The intersection of two fuzzy sets A and B corresponds to the Boolean AND function and is given by

$$\mu_{A\cap B}(u) = \min\{\mu_A(u), \mu_B(u)\} \quad \text{for all } u \in U \tag{10.6}$$

*Complement*: The complement of fuzzy set A corresponds to the Boolean NOT function and is given by

$$\mu_{\neg A}(u) = 1 - \mu_A(u) \quad \text{for all } u \in U \tag{10.7}$$

*Example 10.1*
Find the union and intersection of fuzzy set low temperature L and medium temperature M shown in Figure 10.4. Find also the complement of fuzzy set M. Using equation (10.2) the fuzzy sets for $n = 11$ are

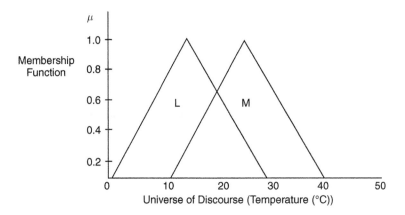

**Fig. 10.4** Overlapping sets 'low' and 'medium temperature'.

$$L = 0/0 + 0.33/5 + 0.67/10 + 1/15 + 0.67/20 + 0.33/25$$
$$+ 0/30 + 0/35 + \cdots + 0/50$$
$$M = 0/0 + 0/5 + 0/10 + 0.33/15 + 0.67/20 + 1/25 + 0.67/30 \quad (10.8)$$
$$+ 0.33/35 + 0/40 + \cdots + 0/50$$

(a) *Union*: Using equation (10.5)

$$\mu_{L+M}(u) = \max(0, 0)/0 + \max(0.33, 0)/5 + \max(0.67, 0)/10$$
$$+ \max(1, 0.33)/15 + \max(0.67, 0.67)/20 + \max(0.33, 1)/25$$
$$+ \max(0, 0.67)/30 + \max(0, 0.33)/35 + \max(0, 0)/40 + \cdots$$
$$+ \max(0, 0)/50 \quad (10.9)$$

$$\mu_{L+M}(u) = 0/0 + 0.33/5 + 0.67/10 + 1/15 + 0.67/20 + 1/25 + 0.67/30$$
$$+ 0.33/35 + 0/40 + \cdots + 0/50 \quad (10.10)$$

(b) *Intersection*: Using equation (10.6) and replacing 'max' by 'min' in equation (10.9) gives

$$\mu_{L \cap M}(u) = 0/0 + 0/5 + 0/10 + 0.33/15 + 0.67/20 + 0.33/25$$
$$+ 0/30 + \cdots + 0/50 \quad (10.11)$$

Equations (10.10) and (10.11) are shown in Figure 10.5.

(c) *Complement*: Using equation (10.7)

$$\mu_{\neg M}(u) = (1 - 0)/0 + (1 - 0)/5 + (1 - 0)/10 + (1 - 0.33)/15$$
$$+ (1 - 0.67)/20 + (1 - 1)/25 + (1 - 0.67)/30 + (1 - 0.33)/35$$
$$+ (1 - 0)/40 + \cdots + (1 - 0)/50 \quad (10.12)$$

Equation (10.12) is illustrated in Figure 10.6.

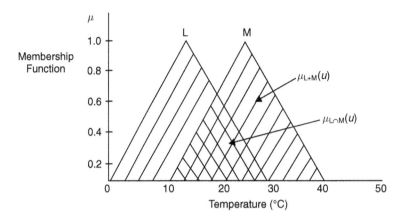

**Fig. 10.5** 'Union' and 'intersection' functions.

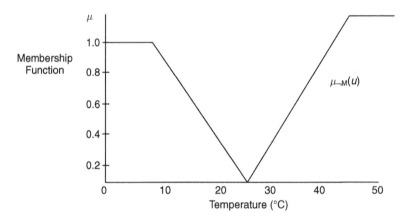

**Fig. 10.6** The complement of fuzzy set M.

## 10.2.3 Fuzzy relations

An important aspect of fuzzy logic is the ability to relate sets with different universes of discourse. Consider the relationship

$$\text{IF L THEN M} \tag{10.13}$$

In equation (10.13) L is known as the *antecedent* and M as the *consequent*. The relationship is denoted by

$$A = L \times M \tag{10.14}$$

or

$$L \times M = \begin{bmatrix} \min\{\mu_L(u_1), \mu_M(v_1)\} \dots \min\{\mu_L(u_1), \mu_M(v_k)\} \\ \min\{\mu_L(u_j), \mu_M(v_1)\} \dots \min\{\mu_L(u_j), \mu_M(v_k)\} \end{bmatrix} \tag{10.15}$$

where $u_1 \rightarrow u_j$ and $v_1 \rightarrow v_k$ are the discretized universe of discourse. Consider the statement

$$\text{IF L is low THEN M is medium} \tag{10.16}$$

Then for the fuzzy sets L and M defined by equation (10.8), for U from 5 to 35 in steps of 5

$$
L \times M =
\begin{bmatrix}
\min(0.33, 0) & \cdots & \min(0.33, 1) & \cdots & \min(0.33, 0.33) \\
\min(0.67, 0) & \cdots & \min(0.67, 1) & \cdots & \min(0.67, 0.33) \\
\vdots & \vdots & \vdots & \vdots & \vdots \\
\min(0, 0) & \cdots & \min(0, 1) & \cdots & \min(0, 0.33)
\end{bmatrix}
\tag{10.17}
$$

which gives

$$
L \times M =
\begin{bmatrix}
0 & 0 & 0.33 & 0.33 & 0.33 & 0.33 & 0.33 \\
0 & 0 & 0.33 & 0.67 & 0.67 & 0.67 & 0.33 \\
0 & 0 & 0.33 & 0.67 & 1 & 0.67 & 0.33 \\
0 & 0 & 0.33 & 0.67 & 0.67 & 0.67 & 0.33 \\
0 & 0 & 0.33 & 0.33 & 0.33 & 0.33 & 0.33 \\
0 & 0 & 0 & 0 & 0 & 0 & 0 \\
0 & 0 & 0 & 0 & 0 & 0 & 0
\end{bmatrix}
\tag{10.18}
$$

Several such statements would form a control strategy and would be linked by their union

$$A = A_1 + A_2 + A_3 + \cdots + A_n \tag{10.19}$$

## 10.2.4 Fuzzy logic control

The basic structure of a Fuzzy Logic Control (FLC) system is shown in Figure 10.7.

### The fuzzification process
Fuzzification is the process of mapping inputs to the FLC into fuzzy set membership values in the various input universes of discourse. Decisions need to be made regarding

(a) number of inputs
(b) size of universes of discourse
(c) number and shape of fuzzy sets.

A FLC that emulates a PD controller will be required to minimize the error $e(t)$ and the rate of change of error $de/dt$, or $ce$.

The size of the universes of discourse will depend upon the expected range (usually up to the saturation level) of the input variables. Assume for the system about to be considered that $e$ has a range of $\pm 6$ and $ce$ a range of $\pm 1$.

The number and shape of fuzzy sets in a particular universe of discourse is a trade-off between precision of control action and real-time computational complexity. In this example, seven triangular sets will be used.

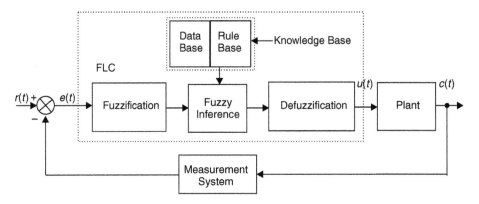

**Fig. 10.7** Fuzzy Logic Control System.

Each set is given a linguistic label to identify it, such as Positive Big (PB), Positive Medium (PM), Positive Small (PS), About Zero (Z), Negative Small (NS), Negative Medium (NM) and Negative Big (NB). The seven set fuzzy input windows for $e$ and $ce$ are shown in Figure 10.8. If at a particular instant, $e(t) = 2.5$ and $de/dt = -0.2$, then, from Figure 10.8, the input fuzzy set membership values are

$$\mu_{PS}(e) = 0.7 \quad \mu_{PM}(e) = 0.4$$
$$\mu_{NS}(ce) = 0.6 \quad \mu_{Z}(ce) = 0.3 \tag{10.20}$$

### The fuzzy rulebase
The fuzzy rulebase consists of a set of antecedent–consequent linguistic rules of the form

$$\text{IF } e \text{ is PS AND } ce \text{ is NS THEN } u \text{ is PS} \tag{10.21}$$

This style of fuzzy conditional statement is often called a 'Mamdani'-type rule, after Mamdani (1976) who first used it in a fuzzy rulebase to control steam plant.

The rulebase is constructed using *a priori* knowledge from either one or all of the following sources:

(a) Physical laws that govern the plant dynamics
(b) Data from existing controllers
(c) Imprecise heuristic knowledge obtained from experienced experts.

If (c) above is used, then knowledge of the plant mathematical model is not required.

The two seven set fuzzy input windows shown in Figure 10.8 gives a possible $7 \times 7$ set of control rules of the form given in equation (10.21). It is convenient to tabulate the two-dimensional rulebase as shown in Figure 10.9.

### Fuzzy inference
Figure 10.9 assumes that the output window contains seven fuzzy sets with the same linguistic labels as the input fuzzy sets. If the universe of discourse for the control signal $u(t)$ is $\pm 9$, then the output window is as shown in Figure 10.10.

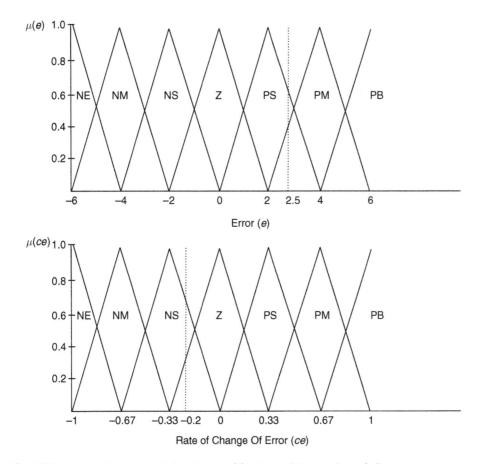

**Fig. 10.8** Seven set fuzzy input windows for error (e) and rate of change of error (ce).

Assume that a certain rule in the rulebase is given by equation (10.22)

$$\text{OR IF } e \text{ is A AND } ce \text{ is B THEN } u = C \qquad (10.22)$$

From equation (10.5) the Boolean OR function becomes the fuzzy max operation, and from equation (10.6) the Boolean AND function becomes the fuzzy min operation. Hence equation (10.22) can be written as

$$\mu_C(u) = \max[\min(\mu_A(e), \mu_B(ce))] \qquad (10.23)$$

Equation (10.23) is referred to as the max–min inference process or max–min fuzzy reasoning.

In Figure 10.8 and equation (10.20) the fuzzy sets that were 'hit' in the error input window when $e(t) = 2.5$ were PS and PM. In the rate of change input window when $ce = -0.2$, the fuzzy sets to be 'hit' were NS and Z. From Figure 10.9, the relevant rules that correspond to these 'hits' are

| ce \ e | NB | NM | NS | Z | PS | PM | PB |
|--------|----|----|----|----|----|----|----|
| NB | NB | NB | NB | NM | Z | PM | PB |
| NM | NB | NB | NB | NM | PS | PM | PB |
| NS | NB | NB | NM | NS | PS | PM | PB |
| Z | NB | NM | NS | Z | PS | PM | PB |
| PS | NB | NM | NS | PS | PM | PB | PB |
| PM | NB | NM | NS | PM | PB | PB | PB |
| PB | NB | NM | Z | PM | PB | PB | PB |

**Fig. 10.9** Tabular structure of a linguistic fuzzy rulebase.

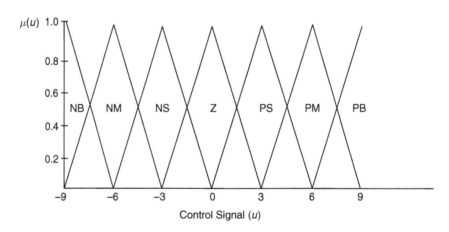

**Fig. 10.10** Seven set fuzzy output window for control signal ($u$).

$$\ldots \text{OR IF } e \text{ is PS AND } ce \text{ is NS}$$
$$\text{OR IF } e \text{ is PS AND } ce \text{ is Z}$$
$$\text{THEN } u = \text{PS} \tag{10.24}$$

$$\ldots \text{OR IF } e \text{ is PM AND } ce \text{ is NS}$$
$$\text{OR IF } e \text{ is PM and } ce \text{ is Z}$$
$$\text{THEN } u = \text{PM} \tag{10.25}$$

Applying the max–min inference process to equation (10.24)

$$\mu_{PS}(u) = \max[\min(\mu_{PS}(e), \mu_{NS}(ce)), \min(\mu_{PS}(e), \mu_Z(ce))] \qquad (10.26)$$

inserting values from equation (10.20)

$$\mu_{PS}(u) = \max[\min(0.7, 0.6), \min(0.7, 0.3)]$$
$$= \max[0.6, 0.3] = 0.6 \qquad (10.27)$$

Applying the max–min inference process to equation (10.25)

$$\mu_{PM}(u) = \max[\min(\mu_{PM}(e), \mu_{NS}(ce)), \min(\mu_{PM}(e), \mu_Z(ce))] \qquad (10.28)$$

inserting values from equation (10.20)

$$\mu_{PM}(u) = \max[\min(0.4, 0.6), \min(0.4, 0.3)]$$
$$= \max[0.4, 0.3] = 0.4 \qquad (10.29)$$

Fuzzy inference is therefore the process of mapping membership values from the input windows, through the rulebase, to the output window(s).

### The defuzzification process

Defuzzification is the procedure for mapping from a set of inferred fuzzy control signals contained within a fuzzy output window to a non-fuzzy (crisp) control signal. The *centre of area* method is the most well known defuzzification technique, which in linguistic terms can be expressed as

$$\text{Crisp control signal} = \frac{\text{Sum of first moments of area}}{\text{Sum of areas}} \qquad (10.30)$$

For a continuous system, equation (10.30) becomes

$$u(t) = \frac{\int u\mu(u)\mathrm{d}u}{\int \mu(u)\mathrm{d}u} \qquad (10.31)$$

or alternatively, for a discrete system, equation (10.30) can be expressed as

$$u(kT) = \frac{\sum_{i=1}^{n} u_i\mu(u_i)}{\sum_{i=1}^{n} \mu(u_i)} \qquad (10.32)$$

For the case when $e(t) = 2.5$ and $ce = -0.2$, as a result of the max–min inference process (equations (10.27) and (10.29)), the fuzzy output window in Figure 10.10 is 'clipped', and takes the form shown in Figure 10.11.

From Figure 10.11, using the equation for the area of a trapezoid

$$\text{Area}_{PS} = \frac{0.6(6 + 2.4)}{2} = 2.52$$
$$\text{Area}_{PM} = \frac{0.2(6 + 3.6)}{2} = 0.96 \qquad (10.33)$$

From equation (10.30)

$$u(t) = \frac{(2.52 \times 3) + (0.96 \times 6)}{2.52 + 0.96} = 3.83 \qquad (10.34)$$

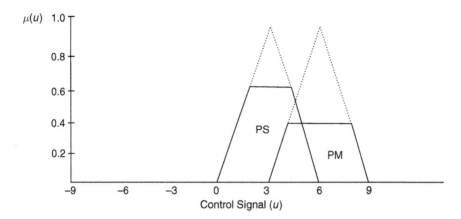

**Fig. 10.11** Clipped fuzzy output window due to fuzzy inference.

Hence, for given error of 2.5, and a rate of change of error of $-0.2$, the control signal from the fuzzy controller is 3.83.

*Example 10.2*
For the input and output fuzzy windows given in Figure 10.8 and 10.10, together with the fuzzy rulebase shown in Figure 10.9, determine

(a) the membership values of the input windows $e$ and $ce$.
(b) the max–min fuzzy inference equations
(c) the crisp control signal $u(t)$

when $e = -3$ and $ce = 0.3$

*Solution*
(a) When $e = -3$ and $ce = 0.3$ are mapped onto the input fuzzy windows, they are referred to as fuzzy singletons. From Figure 10.8

$$e = -3 \quad \mu_{NS}(e) = 0.5 \quad \mu_{NM}(e) = 0.5 \tag{10.35}$$

$ce = 0.3$, using similar triangles

$$\frac{1}{0.33} = \frac{\mu_Z(ce)}{(0.33 - 0.3)}$$
$$\mu_Z(ce) = 0.09 \tag{10.36}$$

and

$$\frac{1}{0.33} = \frac{\mu_{PS}(ce)}{0.3}$$
$$\mu_{PS}(ce) = 0.91 \tag{10.37}$$

(b) The rules that are 'hit' in the rulebase in Figure 10.9 are

$$\ldots \text{ OR IF } e \text{ is NS and } ce \text{ is Z}$$
$$\text{OR IF } e \text{ is NS and } ce \text{ is PS}$$
$$\text{THEN } u = \text{NS} \tag{10.38}$$

$$\ldots \text{ OR IF } e \text{ is NM and } ce \text{ is Z}$$
$$\text{OR IF } e \text{ is NM and } ce \text{ is PS}$$
$$\text{THEN } u = \text{NM} \tag{10.39}$$

Applying max–min inference to equation (10.38)

$$\mu_{NS}(u) = \max[\min(\mu_{NS}(e), \mu_Z(ce)), \min(\mu_{NS}(e), \mu_{PS}(ce))] \tag{10.40}$$

Inserting values into (10.40)

$$\mu_{NS}(u) = \max[\min(0.5, 0.09), \min(0.5, 0.91)]$$
$$= \max[0.09, 0.5] = 0.5 \tag{10.41}$$

and similarly with equation (10.39)

$$\mu_{NM}(u) = \max[\min(\mu_{NM}(e), \mu_Z(ce)), \min(\mu_{NM}(e), \mu_{PS}(ce))]$$
$$= \max[\min(0.5, 0.09), \min(0.5, 0.91)] \tag{10.42}$$
$$= \max[0.09, 0.5] = 0.5$$

Using equations (10.41) and (10.42) to 'clip' the output window in Figure 10.10, the output window is now as illustrated in Figure 10.12.

(c) Due to the symmetry of the output window in Figure 10.12, from observation, the crisp control signal is

$$u(t) = -4.5$$

*Example 10.3* (See also Appendix 1, *examp103.m*)
Design a fuzzy logic controller for the inverted pendulum system shown in Figure 10.13 so that the pendulum remains in the vertical position.

The inverted pendulum problem is a classic example of producing a stable closed-loop control system from an unstable plant.

Since the system can be modelled, it is possible to design a controller using the pole placement techniques discussed in Chapter 8. Neglecting friction at the pivot and the wheels, the equations of motion from Johnson and Picton (1995) are

$$\ddot{x} = \frac{F + m\ell(\dot{\theta}^2 \sin\theta - \ddot{\theta}\cos\theta)}{M + m} \tag{10.43}$$

$$\ddot{\theta} = \frac{g\sin\theta + \cos\theta\left(\frac{-F - m\ell\dot{\theta}^2 \sin\theta}{M+m}\right)}{\ell\left(\frac{4}{3} - \frac{m\cos^2\theta}{M+m}\right)} \tag{10.44}$$

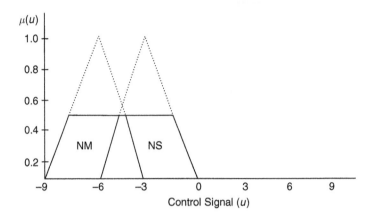

**Fig. 10.12** Fuzzy output window for Example 10.2.

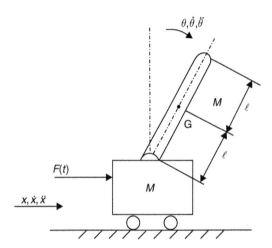

**Fig. 10.13** An inverted pendulum.

In equations (10.43) and (10.44), $m$ is the mass and $\ell$ is the half-length of the pendulum and $M$ is the mass of the trolley. $F(t)$ is the applied force to the trolley in the $x$-direction. If it is assumed that $\theta$ is small and second-order terms ($\dot{\theta}^2$) can be neglected, then

$$\ddot{x} = \frac{F - m\ell\ddot{\theta}}{M + m} \qquad (10.45)$$

$$\ddot{\theta} = \frac{g\theta + \left(\frac{-F}{M+m}\right)}{\ell\left(\frac{4}{3} - \frac{m}{M+m}\right)} \qquad (10.46)$$

If the state variables are

$$x_1 = \theta, \quad x_2 = \dot{\theta}, \quad x_3 = x \quad \text{and} \quad x_4 = \dot{x}$$

and the control variable is

$$u = F(t)$$

then from equations (10.45) and (10.46), the state equations become

$$
\begin{bmatrix} \dot{x}_1 \\ \dot{x}_2 \\ \dot{x}_3 \\ \dot{x}_4 \end{bmatrix} = \begin{bmatrix} 0 & 1 & 0 & 0 \\ a_{21} & 0 & 0 & 0 \\ 0 & 0 & 0 & 1 \\ a_{41} & 0 & 0 & 0 \end{bmatrix} \begin{bmatrix} x_1 \\ x_2 \\ x_3 \\ x_4 \end{bmatrix} + \begin{bmatrix} 0 \\ b_2 \\ 0 \\ b_4 \end{bmatrix} u \tag{10.47}
$$

where

$$a_{21} = \frac{3g(M+m)}{\ell\{4(M+m) - 3m\}}$$

$$a_{41} = \frac{-3gm}{4(M+m) - 3m}$$

$$b_2 = \frac{-3}{\ell\{4(M+m) - 3m\}} \tag{10.48}$$

$$b_4 = \left(\frac{1}{M+m}\right)\left\{1 + \frac{3m}{4(M+m) - 3m}\right\}$$

and the output equation is

$$y = \mathbf{Cx} \tag{10.49}$$

where $\mathbf{C}$ is the identity matrix. For a regulator, with a scalar control variable

$$u = -\mathbf{Kx}$$

The elements of $\mathbf{K}$ can be obtained by selecting a set of desired closed-loop poles as described in section 8.4.2, and applying one of the three techniques discussed.

Data for simulation

$$\ell = 1\,m \quad M = 1\,kg \quad m = 0.5\,kg$$

$$a_{21} = \frac{3 \times 9.81(1.5)}{1\{(4 \times 1.5) - 1.5\}} = 9.81$$

$$a_{41} = \frac{-3 \times 9.81 \times 0.5}{(4 \times 1.5) - 1.5} = -3.27$$

$$b_2 = \frac{-3}{1\{(4 \times 1.5) - 1.5\}} = -0.667$$

$$b_4 = \left(\frac{1}{1.5}\right)\left\{1 + \frac{1.5}{(4 \times 1.5) - 1.5}\right\} = 0.889$$

If the required closed-loop poles are
$s = -2 \pm j2$ for the pendulum, and
$s = -4 \pm j4$ for the trolley, then the closed-loop characteristic equation is

$$s^4 + 12s^3 + 72s^2 + 192s + 256 = 0 \qquad (10.50)$$

Using Ackermann's Formula in equations (8.103) and (8.104), the state feedback matrix becomes

$$\mathbf{K} = [-174.83 \quad -57.12 \quad -39.14 \quad -29.36] \qquad (10.51)$$

Using the fuzzy logic approach suggested by Johnson and Picton (1995), four, three set input windows (one for each state variable) and one, three set output window has been selected as shown in Figure 10.14. Using heuristic knowledge from broom-balancing experiments, the following Mamdani-type rulebase was constructed:

1.  IF $\theta$ is PB and $\dot{\theta}$ is PB then $F$ is PB

2.  IF $\theta$ is PB and $\dot{\theta}$ is Z then $F$ is PB

3.  IF $\theta$ is PB and $\dot{\theta}$ is NB then $F$ is Z

4.  IF $\theta$ is Z and $\dot{\theta}$ is PB then $F$ is PB

5.  IF $\theta$ is Z and $\dot{\theta}$ is Z then $F$ is Z

6.  IF $\theta$ is Z and $\dot{\theta}$ is NB then $F$ is NB $\qquad (10.52)$

7.  IF $\theta$ is NB and $\dot{\theta}$ is PB then $F$ is Z

8.  IF $\theta$ is NB and $\dot{\theta}$ is Z then $F$ is NB

9.  IF $\theta$ is NB and $\dot{\theta}$ is NB then $F$ is NB

10.  IF $\dot{\theta}$ is PB then $F$ is PB

11.  IF $\dot{\theta}$ is NB then $F$ is NB

The rulebase can be extended up to 22 rules by a further set of 11 rules replacing $\theta$ with $x$ and $\dot{\theta}$ with $\dot{x}$.
For the rulebase given in equation (10.52), the fuzzy max–min inference process is

$$\mu_{PB}(u) = \max[\mu_{PB}(\dot{\theta}), \min(\mu_{PB}(\theta), \mu_{PB}(\dot{\theta})), \min(\mu_{PB}(\theta), \mu_{Z}(\dot{\theta})), \min(\mu_{Z}(\theta), \mu_{PB}(\dot{\theta}))]$$

$$\mu_{NB}(u) = \max[\mu_{NB}(\dot{\theta}), \min(\mu_{Z}(\theta), \mu_{NB}(\dot{\theta})), \min(\mu_{NB}(\theta), \mu_{Z}(\dot{\theta})), \min(\mu_{NB}(\theta), \mu_{NB}(\dot{\theta}))]$$

$$\mu_{Z}(u) = \max[\min(\mu_{PB}(\theta), \mu_{NB}(\dot{\theta})), \min(\mu_{Z}(\theta), \mu_{Z}(\dot{\theta})), \min(\mu_{NB}(\theta), \mu_{PB}(\dot{\theta}))]$$

Again, a similar inference process occurs with $x$ and $\dot{x}$. Following defuzzification, a crisp control force $F(t)$ is obtained.

Figure 10.15 shows the time response of the inverted pendulum state variables from an initial condition of $\theta = 0.1$ radians. On each graph, three control strategies are shown, the 11 set rulebase of equation (10.52), the 22 set rulebase that includes $x$ and $\dot{x}$, and the state feedback method given by equation (10.51).

For the pendulum angle, shown in Figure 10.15(a), the 11 set rulebase gives the best results, the state feedback being oscillatory and the 22 set rulebase diverging

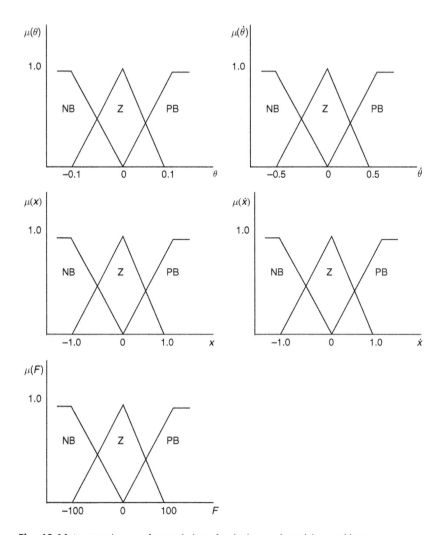

**Fig. 10.14** Input and output fuzzy windows for the inverted pendulum problem.

after a while. The same comments apply to the pendulum angular rate, given in Figure 10.15(b).

With the trolley displacement and velocity shown in Figures 10.15(c) and (d), the state feedback, although oscillatory, give the best results since there is no steady-state error. The positional error for both rulebases increases with time, and there is a constant velocity steady-state error for the 11 set rulebase, and increasing error for the 22 set rulebase. Figure 10.15(e) shows the control force for each of the three strategies.

The 11 and 22 set rulebase simulations were undertaken using SIMULINK, together with the fuzzy logic toolbox for use with MATLAB. More details on the

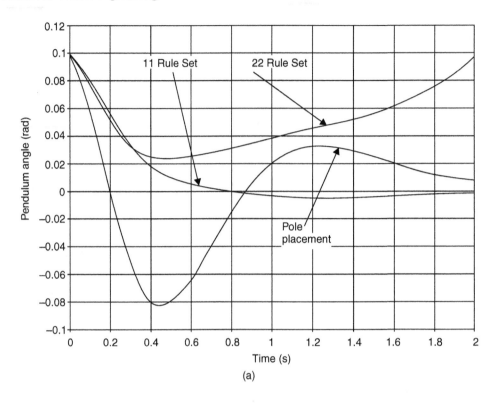

(a)

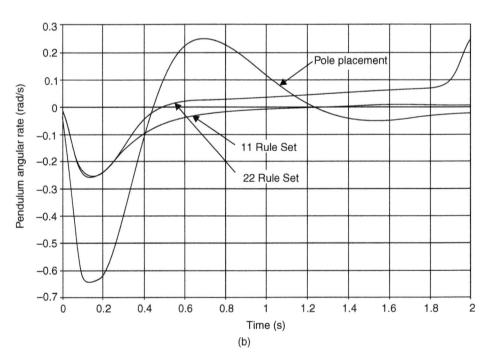

(b)

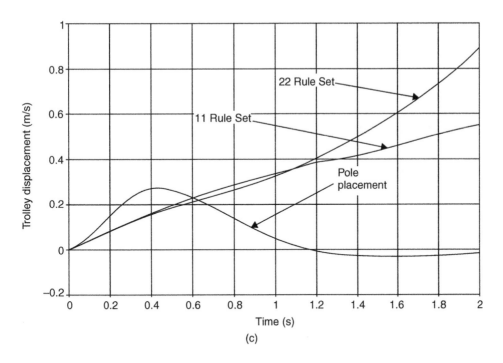

(c)

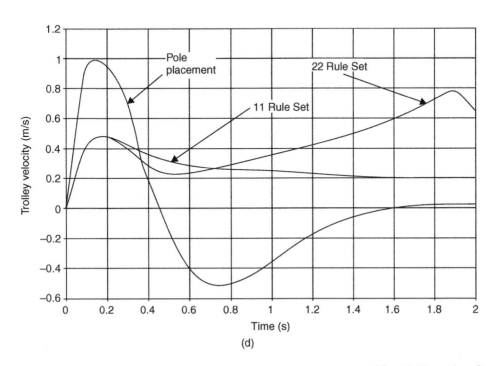

(d)

(**Fig. 10.15** *continued*)

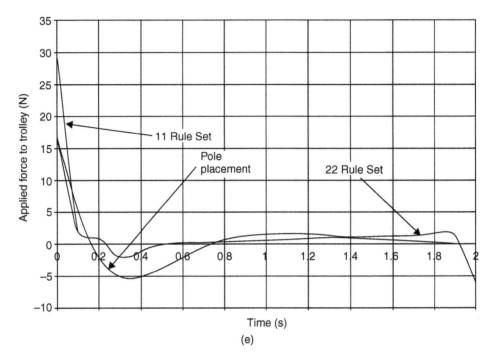

**Fig. 10.15** Inverted pendulum state variable time response for three control strategies.

MATLAB Fuzzy Inference System (FIS) editor can be found in Appendix 1. Figure 10.16 shows the control surface for the 11 set rulebase fuzzy logic controller.

### 10.2.5 Self-organizing fuzzy logic control

Self-Organizing Fuzzy Logic Control (SOFLC) is an optimization strategy to create and modify the control rulebase for a FLC as a result of observed system performance. The SOFLC is particularly useful when the plant is subject to time-varying parameter changes and unknown disturbances.

**Structure**

A SOFLC is a two-level hierarchical control system that is comprised of:

(a) a learning element at the top level
(b) a FLC at the bottom level.

The learning element consists of a Performance Index (PI) table combined with a rule generation and modification algorithm, which creates new rules, or modifies existing ones. The structure of a SOFLC is shown in Figure 10.17. With SOFLC it is usual to express the PI table and rulebase in numerical, rather than linguistic format. So, for

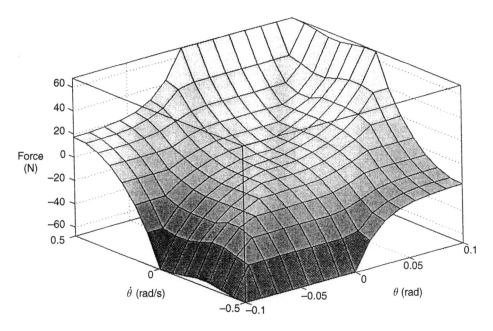

**Fig. 10.16** Control surface for 11 set rulebase fuzzy logic controller.

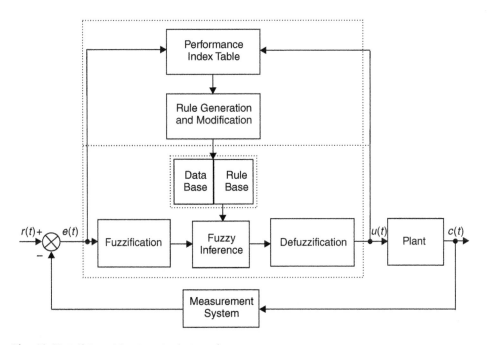

**Fig. 10.17** Self-Organizing Fuzzy Logic Control system.

| ce \ e | NB | NM | NS | Z | PS | PM | PB |
|---|---|---|---|---|---|---|---|
| NB | −50 | −40 | −30 | −20 | −10 | 0 | 10 |
| NM | −42 | −32 | −22 | −12 | −2 | 8 | 18 |
| NS | −36 | −26 | −16 | −6 | 4 | 14 | 24 |
| Z | −30 | −20 | −10 | 0 | 10 | 20 | 30 |
| PS | −24 | −14 | −4 | 6 | 16 | 26 | 36 |
| PM | −18 | −8 | 2 | 12 | 22 | 32 | 42 |
| PB | −10 | 0 | 10 | 20 | 30 | 40 | 50 |

**Fig. 10.18** Tabular structure of a numerical fuzzy rulebase.

example, the fuzzy rulebase in Figure 10.9, might take the form as shown in Figure 10.18.

### Rulebase modification

If the numerical structure of the fuzzy rulebase does not give an acceptable response, then the values in certain cells will need to be adjusted.

Let the error, rate of change of error and control signal at time $t$ be $e(t)$, $ce(t)$ and $u(t)$ respectively, and assume that a given PI is a function of $e(t)$ and $ce(t)$.

If there are unacceptable errors at time $t$, because of the dynamics of the plant, these will be as a result of control action taken $d$ seconds previously, or at time $(t - d)$. The parameter $d$ is a 'delay in reward' parameter and is related to the settling time of the plant, having a typical value of between $3T$ and $5T$, where $T$ is the dominant time constant of the plant.

The value of the PI is therefore determined using $e(t - d)$ and $ce(t - d)$ and applied to $u(t)$ as a correction factor to the rulebase in the form

$$\text{IF } e(t) \text{ is } \ldots \text{ and } ce(t) \text{ is } \ldots \text{ THEN } u(t) = \cdots + \text{PI} \tag{10.53}$$

where the PI is read from a Performance Index table of the form shown in Figure 10.19. When the values of $e(t - d)$ and $ce(t - d)$ are within an acceptable range, the PI tends to zero and the fuzzy rulebase settles down and convergence for the self-organizing process has been achieved. The PI table is usually designed heuristically, based upon an intuitive understanding of the learning process, and the trade-off between speed of learning and stability of the rulebase.

| ce \ e | NB | NM | NS | Z | PS | PM | PB |
|--------|----|----|----|----|----|----|----|
| NB | −5 | −4 | −3 | −3 | −2 | −1 | 0 |
| NM | −4 | −3 | −3 | −2 | −1 | 0 | 1 |
| NS | −3 | −3 | −2 | −1 | 0 | 1 | 2 |
| Z | −3 | −2 | −1 | 0 | 1 | 2 | 3 |
| PS | −2 | −1 | 0 | 1 | 2 | 3 | 3 |
| PM | −1 | 0 | 1 | 2 | 3 | 3 | 4 |
| PB | 0 | 1 | 2 | 3 | 3 | 4 | 5 |

**Fig. 10.19** Performance Index table.

## 10.3  Neural network control systems

### 10.3.1  Artificial neural networks

The human brain is comprised of many millions of interconnected units, known individually as biological neurons. Each neuron consists of a cell to which is attached several dendrites (inputs) and a single axon (output). The axon connects to many other neurons via connection points called synapses. A synapse produces a chemical reaction in response to an input. The biological neuron 'fires' if the sum of the synaptic reactions is sufficiently large. The brain is a complex network of sensory and motor neurons that provide a human being with the capacity to remember, think, learn and reason.

Artificial Neural Networks (ANNs) attempt to emulate their biological counterparts. McCulloch and Pitts (1943) proposed a simple model of a neuron, and Hebb (1949) described a technique which became known as 'Hebbian' learning. Rosenblatt (1961), devised a single layer of neurons, called a Perceptron, that was used for optical pattern recognition.

One of the first applications of this technology for control purposes was by Widrow and Smith (1964). They developed an ADaptive LINear Element (ADLINE) that was taught to stabilize and control an inverted pendulum. Kohonen (1988) and Anderson (1972) investigated similar areas, looking into 'associative' and 'interactive' memory, and also 'competitive learning'. The back propagation training algorithm was investigated by Werbos (1974) and further developed by Rumelhart (1986) and others, leading to the concept of the Multi-Layer Perceptron (MLP).

Artificial Neural Networks have the following potential advantages for intelligent control:

- They learn from experience rather than by programming.
- They have the ability to generalize from given training data to unseen data.
- They are fast, and can be implemented in real-time.
- They fail 'gracefully' rather than 'catastrophically'.

### 10.3.2   Operation of a single artificial neuron

The basic model of a single artificial neuron consists of a weighted summer and an activation (or transfer) function as shown in Figure 10.20. Figure 10.20 shows a neuron in the $j$th layer, where

$x_1 \dots x_i$ are inputs
$w_{j1} \dots w_{ji}$ are weights
$b_j$ is a bias
$f_j$ is the activation function
$y_j$ is the output

The weighted sum $s_j$ is therefore

$$s_j(t) = \sum_{i=1}^{N} w_{ji} x_i(t) + b_j \qquad (10.54)$$

Equation (10.54) can be written in matrix form

$$s_j(t) = \mathbf{W}_j \, \mathbf{x} + b_j \qquad (10.55)$$

The activation function $f(s)$ (where $s$ is the weighted sum) can take many forms, some of which are shown in Figure 10.21. From Figure 10.21 it can be seen that the bias $b_j$ in equations (10.54) and (10.55) will move the curve along the $s$ axis, i.e. effectively

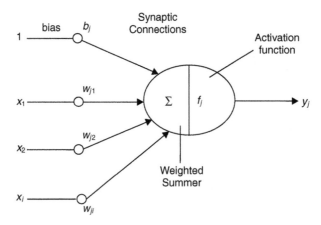

**Fig. 10.20** Basic model of a single artificial neuron.

setting the threshold at which the neuron 'fires'. So in the case of the hard-limiting function, if $b_j = 0$, the neuron will 'fire' when $s_j(t)$ changes from negative to positive.

The sigmoid activation function is popular for neural network applications since it is differentiable and monotonic, both of which are a requirement for the back-propagation algorithm. The equation for a sigmoid function is

$$f(s) = \frac{1}{1 + e^{-s_j}} \qquad (10.56)$$

### 10.3.3  Network architecture

#### *Feedforward networks*

An ANN is a network of single neurons jointed together by synaptic connections. Figure 10.22 shows a three-layer feedforward neural network.

The feedforward network shown in Figure 10.22 consists of a three neuron input layer, a two neuron output layer and a four neuron intermediate layer, called a hidden layer. Note that all neurons in a particular layer are fully connected to all neurons in the subsequent layer. This is generally called a fully connected multilayer network, and there is no restriction on the number of neurons in each layer, and no restriction on the number of hidden layers.

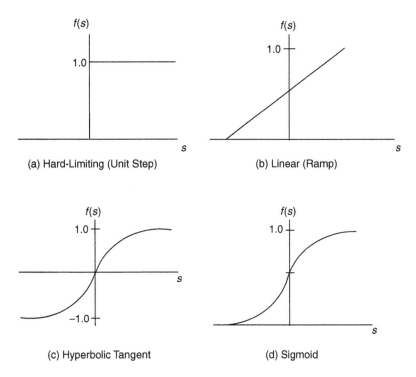

**Fig. 10.21** Activation functions.

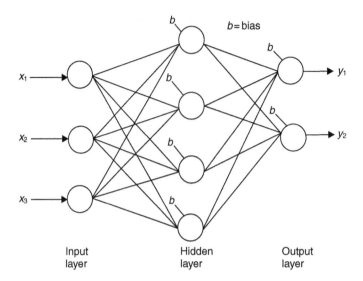

**Fig. 10.22** Three-layer feedforward neural network.

### Feedback (recurrent) networks

Recurrent networks are based on the work of Hopfield and contain feedback paths. Figure 10.23 shows a single-layer fully-connected recurrent network with a delay $(z^{-1})$ in the feedback path.

If, in Figure 10.23, the inputs occur at time $(kT)$ and the outputs are predicted at time $(k+1)T$, then the network can be represented in matrix form by

$$\mathbf{y}(k+1)T = \mathbf{W}_1\mathbf{y}(kT) + \mathbf{W}_2\mathbf{x}(kT) \tag{10.57}$$

Equation (10.57) is in the same form as the discrete-time solution of the state equation (8.76).

## 10.3.4  Learning in neural networks

Learning in the context of a neural network is the process of adjusting the weights and biases in such a manner that for given inputs, the correct responses, or outputs are achieved. Learning algorithms include:

(a) *Supervised learning*: The network is presented with training data that represents the range of input possibilities, together with the associated desired outputs. The weights are adjusted until the error between the actual and desired outputs meets some given minimum value.

(b) *Unsupervised learning*: Also called open-loop adaption because the technique does not use feedback information to update the network's parameters. Applications for unsupervised learning include speech recognition and image compression. Important unsupervised networks include the Kohonen Self-Organizing Map (KSOM) which is a competitive network, and the Grossberg Adaptive Resonance Theory (ART), which can be used for on-line learning.

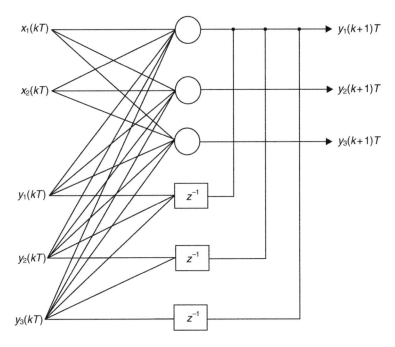

**Fig. 10.23** Recurrent neural network.

## 10.3.5   Back-Propagation

The Back-Propagation Algorithm (BPA) is a supervised learning method for training ANNs, and is one of the most common forms of training techniques. It uses a gradient-descent optimization method, also referred to as the delta rule when applied to feedforward networks. A feedforward network that has employed the delta rule for training, is called a Multi-Layer Perceptron (MLP).

If the performance index or cost function $J$ takes the form of a summed squared error function, then

$$J = \frac{1}{2} \sum_{j=1}^{M} (d_j - y_j)^2 \tag{10.58}$$

where $d_j$ and $y_i$ are the desired and actual network outputs respectively. Using gradient-descent, the weight increment $\Delta w_{ji}$ is proportional to the (negative) slope

$$\Delta w_{ji} = -\mu \frac{\partial J}{\partial w_{ji}} \tag{10.59}$$

where $\mu$ is a constant. From equations (10.58) and (10.59),

$$\frac{\partial J}{\partial w_{ji}} = \frac{1}{M} \sum_{j=1}^{M} \frac{\partial}{\partial w_{ji}} (d_j - y_j)^2$$

using the chain rule,

$$\frac{\partial J}{\partial w_{ji}} = \frac{1}{M} \sum_{j=1}^{M} \frac{\partial}{\partial y_j} (d_j - y_j)^2 \frac{\partial y_i}{\partial w_{ji}}$$

giving

$$\frac{\partial J}{\partial w_{ji}} = -\frac{2}{M} \sum_{j=1}^{M} (d_j - y_j) \frac{\partial y_j}{\partial w_{ji}} \qquad (10.60)$$

If the activation function is the sigmoid function given in equation (10.56), then its derivative is

$$\frac{\partial f}{\partial s_j} = -\frac{e^{-s_j}}{(1 + e^{-s_j})^2} = \frac{1}{1 + e^{-s_j}} - \left(\frac{1}{1 + e^{-s_j}}\right)^2$$

or

$$\frac{\partial f}{\partial s} = f(s) - f(s)^2 \qquad (10.61)$$

Since $f(s)$ is the neuron output $y_j$, then equation (10.61) can be written as

$$\frac{\partial y_j}{\partial s_j} = y_j(1 - y_j) \qquad (10.62)$$

From equation (10.60), again using the chain rule,

$$\frac{\partial y_j}{\partial w_{ji}} = \frac{\partial y_j}{\partial s_j} \frac{\partial s_j}{\partial w_{ji}} \qquad (10.63)$$

If, in equation (10.54), the bias $b_j$ is called $w_{j0}$, then equation (10.54) may be written as

$$s_j = \sum_{i=0}^{N} w_{ji} x_i \qquad (10.64)$$

thus

$$\frac{\partial s_j}{\partial w_{ji}} = \frac{\partial}{\partial w_{ji}} \sum_{i=0}^{N} w_{ji} x_i = \sum_{i=0}^{N} \frac{\partial w_{ji}}{\partial w_{ji}} x_i = x_i \qquad (10.65)$$

Substituting equations (10.62) and (10.65) into (10.63) gives

$$\frac{\partial y_i}{\partial w_{ji}} = y_j(1 - y_j)x_i \qquad (10.66)$$

Putting equation (10.66) into (10.60) gives

$$\frac{\partial J}{\partial w_{ji}} = -\frac{2}{M} \sum_{j=1}^{M} (d_j - y_j) y_j (1 - y_j) x_i \qquad (10.67)$$

or

$$\frac{\partial J}{\partial w_{ji}} = -\frac{2}{M} \sum_{j=1}^{M} \delta_j x_i \qquad (10.68)$$

where

$$\delta_j = (d_j - y_j) y_j (1 - y_j) \qquad (10.69)$$

Substituting equation (10.68) into (10.59) gives

$$\Delta w_{ji} = \eta \sum_{j=1}^{M} \delta_j x_i \qquad (10.70)$$

where

$$\eta = \frac{2\mu}{M}$$

This leads to a weight increment, called the delta rule, for a particular neuron

$$\Delta w_{ji}(kT) = \eta \delta_j x_i \qquad (10.71)$$

where $\eta$ is the learning rate and has a value of between 0 and 1. Hence the new weight becomes

$$w_{ji}(kT) = w_{ji}(k-1)T + \Delta w_{ji}(kT)$$

or

$$w_{ji}(kT) = w_{ji}(k-1)T + \eta \delta_j x_i \qquad (10.72)$$

Consider a three layer network. Let the input layer be layer one ($\ell = 1$), the hidden layer be layer two ($\ell = 2$) and the output layer be layer three ($\ell = 3$). The back-propagation commences with layer three where $d_j$ is known and hence $\delta_j$ can be calculated using equation (10.69), and the weights adjusted using equation (10.71). To adjust the weights on the hidden layer ($\ell = 2$) equation (10.69) is replaced by

$$[\delta_j]_\ell = [y_j(1 - y_j)]_\ell \left[ \sum_{j=1}^{N} w_{ji} \delta_j \right]_{\ell+1} \qquad (10.73)$$

Hence the $\delta$ values for layer $\ell$ are calculated using the neuron outputs from layer $\ell$ (hidden layer) together with the summation of $w$ and $\delta$ products from layer $\ell + 1$ (output layer). The back-propagation process continues until all weights have been adjusted. Then, using a new set of inputs, information is fed forward through the network (using the new weights) and errors at the output layer computed. The process continues until

(i) the performance index $J$ reaches an acceptable low value
(ii) a maximum iteration count (number of epochs) has been exceeded
(iii) a training-time period has been exceeded.

For either (ii) or (iii), it may well be that a local minima has been located. Under these conditions the BPA may be re-started, and if again unsuccessful, a new training set may be required.

The equations that govern the BPA can be summarized as

Single neuron summation

$$s_j = \sum_{i=1}^{N} w_{ji} x_i + b_j \tag{10.74}$$

Sigmoid activation function

$$y_j = \frac{1}{1 + e^{-s_j}} \tag{10.75}$$

Delta rule

$$\Delta w_{ji}(kT) = \eta \delta_j x_i \tag{10.76}$$

New weight

$$w_{ji}(kT) = w_{ji}(k-1)T + \Delta w_{ji}(kT) \tag{10.77}$$

Output layer

$$\delta_j = y_j(1 - y_j)(d_j - y_j) \tag{10.78}$$

$$J = \frac{1}{2} \sum_{j=1}^{M} (d_j - y_j)^2 \tag{10.79}$$

Other layers

$$[\delta_j]_\ell = [y_j(1 - y_j)]_\ell \left[ \sum_{j=1}^{N} w_{ji} \delta_j \right]_{\ell+1} \tag{10.80}$$

### *Learning with momentum*
When momentum is used in the BPA, the solution stands less chance of becoming trapped in local minima. It can be included by making the current change in weight equal to a proportion of the previous weight change summed with the weight change calculated using the delta rule.

The delta rule given in equation (10.76) can be modified to include momentum as indicated in equation (10.81).

$$\Delta w_{ji}(kT) = (1 - \alpha)\eta\delta_j x_i + \alpha\Delta w_{ji}(k-1)T \qquad (10.81)$$

where $\alpha$ is the momentum coefficient, and has a value of between 0 and 1.

*Example 10.4*
The neural network shown in Figure 10.24 is in the process of being trained using a BPA. The current inputs $x_1$ and $x_2$ have values of 0.2 and 0.6 respectively, and the desired output $d_j = 1$. The existing weights and biases are
Hidden layer

$$\mathbf{W}_j = \begin{bmatrix} 1.0 & 1.5 \\ 0.5 & 2.0 \\ 2.5 & 3.0 \end{bmatrix} \quad \mathbf{b}_j = \begin{bmatrix} 1.0 \\ -0.5 \\ 1.5 \end{bmatrix}$$

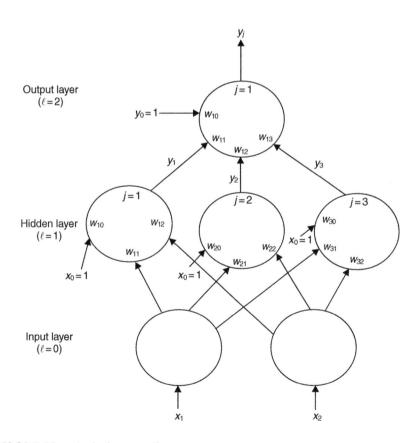

**Fig. 10.24** Training using back-propagation.

Output layer

$$\mathbf{W}_j = [3.0 \quad 2.0 \quad 1.0] \quad \mathbf{b}_j = [-4.0]$$

Calculate the output $y_j$ and hence the new values for the weights and biases. Assume a learning rate of 0.5.

*Solution*

*Forward propagation*
Hidden layer ($\ell = 1$): Single neuron summation

$$
\begin{aligned}
j = 1: & \quad s_1 = w_{10} + w_{11}x_1 + w_{12}x_2 \\
j = 2: & \quad s_2 = w_{20} + w_{21}x_1 + w_{22}x_2 \\
j = 3: & \quad s_3 = w_{30} + w_{31}x_1 + w_{32}x_2
\end{aligned}
$$

or

$$
\begin{bmatrix} s_1 \\ s_2 \\ s_3 \end{bmatrix} = \begin{bmatrix} w_{11} & w_{12} \\ w_{21} & w_{22} \\ w_{31} & w_{32} \end{bmatrix} \begin{bmatrix} x_1 \\ x_2 \end{bmatrix} + \begin{bmatrix} w_{10} \\ w_{20} \\ w_{30} \end{bmatrix}
\tag{10.82}
$$

Sigmoid activation functions ($j = 1$ to 3)

$$
y_1 = \frac{1}{1 + e^{-s_1}} \quad y_2 = \frac{1}{1 + e^{-s_2}} \quad y_3 = \frac{1}{1 + e^{-s_3}}
\tag{10.83}
$$

Output layer ($\ell = 2$)

$$
j = 1: \quad s_1 = w_{10} + w_{11}y_1 + w_{12}y_2 + w_{13}y_3
\tag{10.84}
$$

$$
y_j = y_1 = \frac{1}{1 + e^{-s_1}}
\tag{10.85}
$$

Inserting values into equations (10.82) and (10.83)

$$
\begin{bmatrix} s_1 \\ s_2 \\ s_3 \end{bmatrix} = \begin{bmatrix} 1.0 & 1.5 \\ 0.5 & 2.0 \\ 2.5 & 3.0 \end{bmatrix} \begin{bmatrix} 0.2 \\ 0.6 \end{bmatrix} + \begin{bmatrix} 1.0 \\ -0.5 \\ 1.5 \end{bmatrix}
$$

$$
\begin{aligned}
s_1 = 2.1 & \qquad s_2 = 0.8 & \qquad s_3 = 3.8 \\
y_1 = 0.891 & \qquad y_2 = 0.690 & \qquad y_3 = 0.978
\end{aligned}
$$

Inserting values into equations (10.84) and (10.85)

$$
s_1 = 1.031 \quad y_j = y_1 = 0.737
\tag{10.86}
$$

*Back propagation*
Output layer ($\ell = 2$): From equation (10.69)

$$
\delta_j = y_j(1 - y_j)(d_j - y_j)
$$

Since $j = 1$

$$
\begin{aligned}
\delta_1 & = 0.737(1 - 0.737)(1 - 0.737) \\
& = 0.051
\end{aligned}
\tag{10.87}
$$

Delta rule

$$\Delta w_{ji}(kT) = \eta \delta_j x_i$$
$$\Delta w_{10} = 0.5 \times 0.051 \times 1 = 0.0255$$
$$\Delta w_{11} = 0.5 \times 0.051 \times 0.891 = 0.0227$$
$$\Delta w_{12} = 0.5 \times 0.051 \times 0.69 = 0.0176$$
$$\Delta w_{13} = 0.5 \times 0.051 \times 0.978 = 0.0249$$

New weights and biases for output layer

$$\mathbf{w}_j = [3.0227 \quad 2.0176 \quad 1.0249] \quad \mathbf{b}_j = [-3.975] \tag{10.88}$$

Hidden layer ($\ell = 1$): From equation (10.80)

$$[\delta_j]_\ell = [y_j(1 - y_j)]_\ell \left[ \sum_{j=1}^{N} w_{ji} \delta_j \right]_{\ell+1}$$

To illustrate this equation, had there been two neurons in layer ($\ell + 1$), i.e. the output layer, then values for $\delta_1$ and $\delta_2$ for layer ($\ell + 1$) would have been calculated. Thus, for layer $\ell$ (the hidden layer), the $[\delta_j]_\ell$ values would be

$$j = 1: \quad [\delta_1]_\ell = [y_1(1 - y_1)]_\ell [w_{11}\delta_1 + w_{21}\delta_2]_{\ell+1}$$
$$j = 2: \quad [\delta_2]_\ell = [y_2(1 - y_2)]_\ell [w_{12}\delta_1 + w_{22}\delta_2]_{\ell+1}$$
$$j = 3: \quad [\delta_3]_\ell = [y_3(1 - y_3)]_\ell [w_{13}\delta_1 + w_{23}\delta_2]_{\ell+1}$$

However, since in this example there is only a single neuron in layer ($\ell + 1$), $\delta_2 = 0$. Thus the $\delta$ values for layer $\ell$ are

$$j = 1: \quad [\delta_1]_\ell = [0.891(1 - 0.891)][3.0227 \times 0.051] = 0.015$$
$$j = 2: \quad [\delta_2]_\ell = [0.690(1 - 0.690)][2.0176 \times 0.051] = 0.022 \tag{10.89}$$
$$j = 3: \quad [\delta_3]_\ell = [0.978(1 - 0.978)][1.0249 \times 0.051] = 0.001$$

Hence, using the delta rule, the weight increments for the hidden layer are

$$\Delta w_{10} = 0.5 \times 0.015 \times 1 = 0.0075$$
$$\Delta w_{11} = 0.5 \times 0.015 \times 0.2 = 0.0015$$
$$\Delta w_{12} = 0.5 \times 0.015 \times 0.6 = 0.0045$$
$$\Delta w_{20} = 0.5 \times 0.022 \times 1 = 0.0110$$
$$\Delta w_{21} = 0.5 \times 0.022 \times 0.2 = 0.0022$$
$$\Delta w_{22} = 0.5 \times 0.022 \times 0.6 = 0.0066$$
$$\Delta w_{30} = 0.5 \times 0.001 \times 1 = 0.0005$$
$$\Delta w_{31} = 0.5 \times 0.001 \times 0.2 = 0.0001$$
$$\Delta w_{32} = 0.5 \times 0.001 \times 0.6 = 0.0003$$

The new weights and biases for the hidden layer now become

$$\mathbf{W}_j = \begin{bmatrix} 1.0015 & 1.5045 \\ 0.5022 & 2.0066 \\ 2.5001 & 3.0003 \end{bmatrix} \quad \mathbf{b}_j = \begin{bmatrix} 1.0075 \\ -0.489 \\ 1.5005 \end{bmatrix} \quad (10.90)$$

## 10.3.6  Application of neural networks to modelling, estimation and control

An interesting and important feature of a neural network trained using back-propagation is that no knowledge of the process it is being trained to emulate is required. Also, since they learn from experience rather than programming, their use may be considered to be a 'black box' approach.

### Neural networks in modelling and estimation

Providing input/output data is available, a neural network may be used to model the dynamics of an unknown plant. There is no constraint as to whether the plant is linear or nonlinear, providing that the training data covers the whole envelope of plant operation.

Consider the neural network state observer shown in Figure 10.25. This is similar in operation to the Luenberge full-order state observer given in Figure 8.9. If the neural network in Figure 10.25 is trained using back-propagation, the algorithm will minimize the PI

$$J = \sum_{k=1}^{N} (\mathbf{y}(kT) - \hat{\mathbf{y}}(kT))^{\mathrm{T}} (\mathbf{y}(kT) - \hat{\mathbf{y}}(kT)) \quad (10.91)$$

Richter *et al.* (1997) used this technique to model the dynamic characteristics of a ship. The vessel was based on the Mariner Hull and had a length of 161 m and a displacement of 17 000 tonnes. The training data was provided by a three degree-of-freedom (forward velocity, or surge, lateral velocity, or sway and turn, or

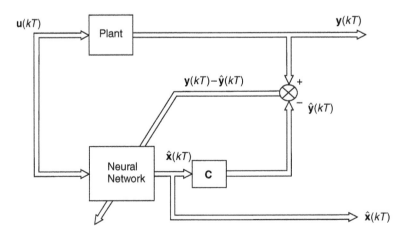

**Fig. 10.25** Neural network state observer.

yaw-rate) differential equation model produced by Burns and was based on previous work by Morse and Price.

The training file consisted of input data of the form: Time elapsed $t(kT)$, Rudder angle $\delta(kT)$, Engine speed $n(kT)$ with corresponding output data Forward velocity $u(kT)$, Lateral velocity $v(kT)$, Yaw-rate $r(kT)$.

With the engine speed held constant, the rudder was given step changes of $0°$, $\pm 10°$, $\pm 20°$ and $\pm 30°$. Figure 10.26 shows training and trained data for a rudder

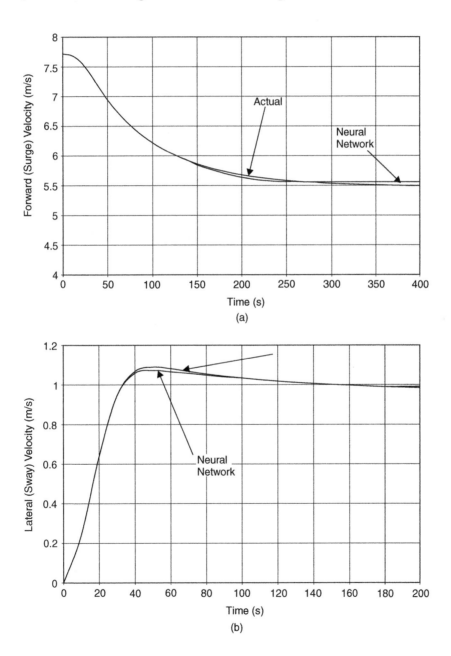

(a)

(b)

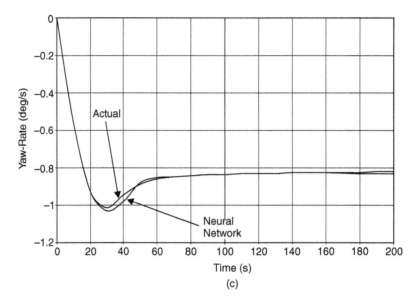

**Fig. 10.26** Training and trained data for a neural network model of a Ship's Hull.

angle of $+20°$ (where positive is to port, or left). The input and output data was sampled every 5 seconds, during the transient period of the turn (0–300 seconds).

The selected network had a 3-6-6-6-3 structure, i.e. input and output layers comprising 3 neurons in each, separated by three hidden layers of 6 neurons. During learning, 4 million epochs were trained. The learning rate and momentum were initially set at 0.3 and 0.8, but were reduced in three steps to final values of 0.05 and 0.4 respectively.

### Inverse models

The inverse model of a plant provides a control vector $\mathbf{u}(kT)$ for a given output vector $\mathbf{y}(kT)$ as shown in Figure 10.27.

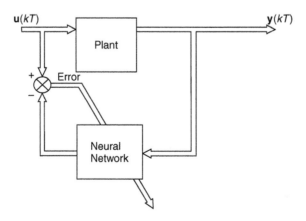

**Fig. 10.27** Neural network plant inverse model.

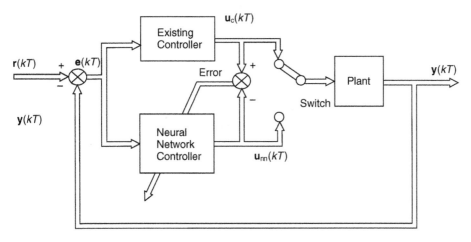

**Fig. 10.28** Training a neural network controller.

So, for example, with the ship model shown in Figure 10.26, the inverse model could be trained with time, forward velocity, lateral velocity and yaw-rate as input data and rudder angle and engine speed as output data.

### Neural networks in control

*Controller emulation*: A simple application in control is the use of neural networks to emulate the operation of existing controllers. It may be that a nonlinear plant requires several tuned PID controllers to operate over the full range of control actions. Or again, an LQ optimal controller has difficulty in running in real-time. Figure 10.28 shows how the control signal from an existing controller may be used to train, and to finally be replaced by, a neural network controller.

### Error back-propagation through plant model

All closed-loop control systems operate by measuring the error between desired inputs and actual outputs. This does not, in itself, generate control action errors that may be back-propagated to train a neural network controller. If, however, a neural network of the plant exists, back-propagation through this network of the system error $(\mathbf{r}(kT) - \mathbf{y}(kT))$ will provide the necessary control action errors to train the neural network controller as shown in Figure 10.29.

### Internal Model Control (IMC)

Internal Model Control was discussed in relation to robust control in section 9.6.3 and Figure 9.19. The IMC structure is also applicable to neural network control. The plant model $G_m(s)$ in Figure 9.19 is replaced by a neural network model and the controller $C(s)$ by an inverse neural network plant model as shown in Figure 10.30.

## 10.3.7 Neurofuzzy control

Neurofuzzy control combines the mapping and learning ability of an artificial neural network with the linguistic and fuzzy inference advantages of fuzzy logic. Thus

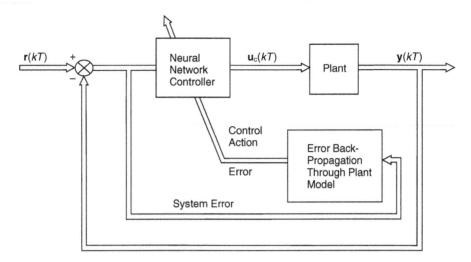

**Fig. 10.29** Control action error generated by system error back-propagation through plant model.

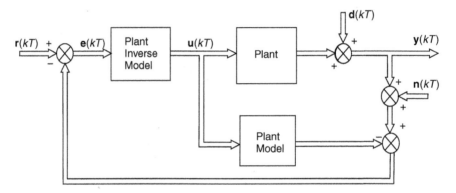

**Fig. 10.30** Application of neural networks to IMC.

a neurofuzzy controller has the potential to out-perform conventional ANN or fuzzy logic controllers. The general architecture of a neurofuzzy scheme is to employ neural network learning to upgrade either the membership functions or rulebase of the fuzzy logic element.

### The Adaptive Network based Fuzzy Inference System (ANFIS)

The ANFIS neurofuzzy controller was implemented by Jang (1993) and employs a Takagi–Sugeno–Kang (TSK) fuzzy inference system. The basic ANFIS architecture is shown in Figure 10.31.

Square nodes in the ANFIS structure denote parameter sets of the membership functions of the TSK fuzzy system. Circular nodes are static/non-modifiable and perform operations such as product or max/min calculations. A hybrid learning rule is used to accelerate parameter adaption. This uses sequential least squares in the forward pass to identify consequent parameters, and back-propagation in the backward pass to establish the premise parameters.

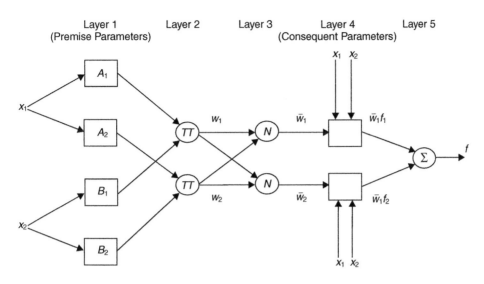

**Fig. 10.31** The Adaptive Network based Fuzzy Inference System (ANFIS) architecture (after Craven).

If the fuzzy inference system has inputs $x_1$ and $x_2$ and output $f$ as shown in Figure 10.31, then a first-order TSK rulebase might be

Rule 1:      If $x_1$ is $A_1$ and $x_2$ is $B_1$
then $f_1 = p_1 x_1 + q_1 x_2 + r_1$

Rule 2:      If $x_1$ is $A_2$ and $x_2$ is $B_2$
then $f_2 = p_2 x_1 + q_2 x_2 + r_2$

Rule $n$:      If $x_1$ is $A_n$ and $x_2$ is $B_n$
then $f_n = p_n x_1 + q_n x_2 + r_n$      (10.92)

Where $A_1 \ldots A_n$, $B_1 \ldots B_n$ are membership functions and $p_1 \ldots p_n$, $q_1 \ldots q_n$ and $r_1 \ldots r_n$ are constants within the consequent functions.

Layer 1 contains adaptive nodes that require suitable premise membership functions (triangular, trapezoidal, bell, etc). Hence

$$y_{1,i} = \mu_{8i}(x_i) \qquad (10.93)$$

Layer 2 undertakes a product or T-norm operation.

$$y_{2,i} = w_i = \mu_{Ai}(x_1)\mu_{Bi}(x_2) \ldots \mu_{pi}(x_n) \quad i = 1, 2 \ldots, n \qquad (10.94)$$

Layer 3 calculates the ratio of the firing strength of the rules

$$y_{3,i} = \bar{w}_i = \frac{w_i}{\sum_{i=1}^{n} w_i} \qquad (10.95)$$

Layer 4 generates the linear consequent functions as given in equation (10.92). Layer 5 sums all incoming signals

$$y_{5,i} = f = \sum_{i=1}^{n} \bar{w}_i f_i = \frac{\sum_{i=1}^{n} w_i f_i}{\sum_{i=1}^{n} w_i} \qquad (10.96)$$

A limitation of the ANFIS technique is that it cannot be employed on multivariable systems. The Co-active ANFIS (CANFIS) developed by Craven (1999) extends the ANFIS architecture to provide a flexible multivariable control environment. This was employed to control the yaw and roll channels of an Autonomous Underwater Vehicle (AUV) simultaneously.

### Predictive Self-Organizing Fuzzy Logic Control (PSOFLC)

This is an extension of the SOFLC strategy discussed in section 10.2.5 and illustrated in Figure 10.17. Predictive Self-Organizing Fuzzy Logic Control is particularly useful when the plant dynamics are time-varying, and the general architecture is shown in Figure 10.32.

In Figure 10.30 the predictive neural network model tracks the changing dynamics of the plant. Following a suitable time delay, $e_m(kT)$ is passed to the performance index table. If this indicates poor performance as a result of changed plant dynamics, the rulebase is adjusted accordingly. Richter (2000) demonstrated that this technique could improve and stabilize a SOFLC when applied to the autopilot of a small motorized surface vessel.

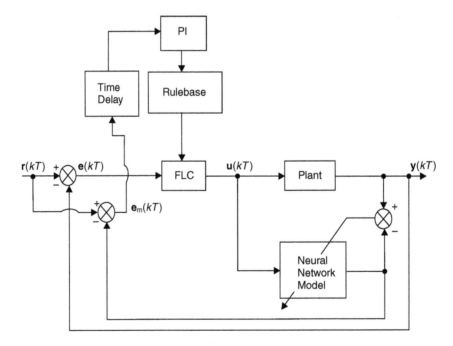

**Fig. 10.32** Predictive Self-Organizing Fuzzy Logic Control (PSOFLC).

## 10.4 Genetic algorithms and their application to control system design

### 10.4.1 Evolutionary design techniques

In any design problem there is a multi-dimensional space of possible solutions. Some of these solutions may be acceptable, but not the best (local optima) and there may exist a single best solution (global optimum).

It has been shown in Chapter 9 that it is possible to obtain an optimal mathematical solution for a control system with linear plant dynamics. An alternative approach is to use 'heuristics', or knowledge acquired through experience, to search for optimal solutions. One such technique is to employ a Genetic Algorithm (GA). This is a search algorithm based upon the evolutional process of natural selection of the fittest members of a given population to breed the next generation.

### 10.4.2 The genetic algorithm (GA)

In the early 1960s Rechenburg (1965) conducted studies at the Technical University of Berlin in the use of an evolutionary strategy to minimize drag on a steel plate. Genetic algorithms were used by Holland (1975) and his students at the University of Michigan in the late 1970s and early 1980s to analyse a range of engineering problems. In particular, Goldberg (1983) used GAs to optimize the design of gas pipeline systems.

The basic element of a GA is the chromosome. This contains the genetic information for a given solution and is typically coded as a binary string. For example, an eight digit binary number such as 11001001 represents a chromosome that contains eight genes. Initially, a population of chromosomes, created randomly, represent a number of solutions to a given problem.

A 'fitness function', which is in effect a performance index, is used to select the best solutions in the population to be parents to the offsprings that will comprise the next generation. The fitter the parent, the greater the probability of selection. This emulates the evolutionary process of 'survival of the fittest'. Parents are selected using a roulette wheel method as shown in Figure 10.33. Here there are four candidate parents P1, P2, P3 and P4, having selection probabilities (from the fitness function) of 0.5, 0.3, 0.15 and 0.05 respectively. For the example in Figure 10.33, if the roulette wheel is spun four times, P1 may be selected twice, P2 and P3 once, and P4 not at all.

Offsprings are produced by selecting parent chromosomes for breeding, and crossing over some of the genetic material. The amount of genetic material passed from parent to offspring is dictated by the random selection of a crossover point as indicated in Figure 10.34, where P1 and P2 are the parents, and 01 and 02 the offsprings.

Mutation is allowed to occur in some of the offsprings, the amount being controlled by the mutation rate, typically a very small number. This results in the random change in a gene in an offspring, i.e. from 0 to 1.

The breeding of successive generations continues until all offsprings are acceptably fit. In some cases, all offsprings will eventually have the same genetic structure,

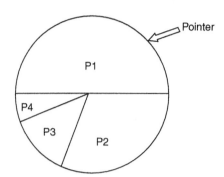

**Fig. 10.33** Roulette wheel selection.

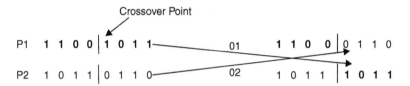

**Fig. 10.34** Genetic material transfer during breeding.

representing a global optimum, or in other cases, several solutions, called clustering may evolve. In this latter case, the system designer makes the decision as to which is the best solution.

*Example 10.5*
A system has a fitness function

$$J = 1 + \sin ax \qquad (10.97)$$

as shown in Figure 10.35. Assume that the solution space has 31 values and that each solution can be represented by a five digit binary string ranging from 00000 to 11111. The value of $a$ in equation (10.97) is therefore 11.6° (0.203 rad). If the population has four members, spinning a coin (heads $= 1$, tails $= 0$) produced the following initial population

$$00101 \quad 11110 \quad 00001 \quad 00011$$

Determine the offsprings from the initial generation and the subsequent generation.

*Solution*
From Figure 10.35 it can be seen that the optimal solution occurs when $x_{10} = 8$, or $x_2 = 01000$.

Table 10.1 shows the selection of parents for mating from the initial population. If a random number generator is used to generate numbers between 0.0 and 1.0, then the cumulative probability values in Table 10.1 is used as follows:

Values between 0 and 0.342, Parent 1 selected
Values between 0.343 and 0.488, Parent 2 selected

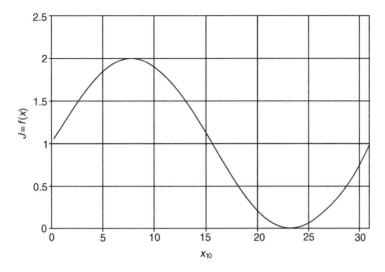

**Fig. 10.35** Fitness function for Example 10.5.

**Table 10.1** Selection of parents for mating from initial population

| Parent | $x_2$ | $x_{10}$ | $J = f(x)$ | $p = J/\Sigma J$ | Cumulative probability | Roulette wheel hits |
|--------|-------|----------|------------|------------------|------------------------|---------------------|
| 1 | 00101 | 5 | 1.848 | 0.342 | 0.342 | 2 |
| 2 | 11110 | 30 | 0.792 | 0.146 | 0.488 | 1 |
| 3 | 00001 | 1 | 1.201 | 0.222 | 0.710 | 0 |
| 4 | 00011 | 3 | 1.571 | 0.290 | 1.000 | 1 |
| Total | | | 5.412 | 1.000 | | 4 |
| Mean | | | 1.353 | 0.250 | | 1 |
| Maximum | | | 1.848 | 0.342 | | 2 |

Values between 0.489 and 0.710, Parent 3 selected
Values between 0.711 and 1.000, Parent 4 selected

The random number generator produced the following values: 0.326, 0.412, 0.862 and 0.067. Hence Parent 1 was selected twice, Parents 2 and 4 once and Parent 3 not at all. The selected parents were randomly mated with random choice of crossover points. The fitness of the first generation of offsprings is shown in Table 10.2.

From Tables 10.1 and 10.2 the total fitness of the initial population was 5.412, whereas the total fitness of their offsprings was 5.956, an increase of 10%. Note that if each offspring was perfect, they would have a value of $8_{10}$, or $01000_2$, thus giving the maximum fitness that any population could have of $2 \times 4 = 8.0$.

The next spin of the random number generator produced values: 0.814, 0.236, 0.481 and 0.712, giving the roulette wheel hits shown in Table 10.2.

The second generation of offsprings is shown in Table 10.3. From Table 10.3 the total fitness of the second generation of offsprings is 7.204, or an increase of 33% above the initial population. As things stand, since the two most significant binary digits in the second generation offsprings are 00, subsequent breeding will not

**Table 10.2** Fitness of first generation of offsprings

| Parent | Breeding | Offspring | $x_2$ | $x_{10}$ | $J = f(x)$ | $p = J/\sum J$ | Cumulative probability | Roulette wheel hits |
|---|---|---|---|---|---|---|---|---|
| 1 | 001⦚01 ╳ | 1 | 00110 | 6 | 1.937 | 0.325 | 0.325 | 1 |
| 2 | 111⦚10 | 2 | 11101 | 29 | 0.600 | 0.101 | 0.426 | 0 |
| 1 | 001⦚01 ╳ | 3 | 00011 | 3 | 1.571 | 0.264 | 0.690 | 1 |
| 4 | 000⦚11 | 4 | 00101 | 5 | 1.848 | 0.310 | 1.000 | 2 |
| Total | | | | | 5.956 | 1.000 | | |
| Mean | | | | | 1.489 | 0.250 | | |
| Maximum | | | | | 1.937 | 0.325 | | |

**Table 10.3** Fitness of second generation of offsprings

| Parent | Breeding | Offspring | $x_2$ | $x_{10}$ | $J = f(x)$ |
|---|---|---|---|---|---|
| 1 | 001⦚10 ╳ | 1 | 00101 | 5 | 1.848 |
| 4 | 001⦚01 | 2 | 00110 | 6 | 1.937 |
| 3 | 000⦚11 ╳ | 3 | 00101 | 5 | 1.848 |
| 4 | 001⦚01 | 4 | 00011 | 3 | 1.571 |
| Total | | | | | 7.204 |
| Mean | | | | | 1.801 |
| Maximum | | | | | 1.937 |

produce strings greater than 00111, or $7_{10}$. If all four offsprings had this value, then a total fitness for the population would be 7.953, which is close to, but not the ideal fitness of 8.0 as explained earlier. However, if mutation in a particular offspring changed one of the two most significant digits from 0 to 1, a perfect optimal value could still be achieved. Also to bear in mind, is the very limited size of the population used in this example.

*Example 10.6*
The block diagram for an angular positional control system is shown in Figure 10.36. The system parameters are:

Amplifier gain $K_2 = 3.5$
Servomotor constant $K_3 = 15\,\text{Nm/A}$
Field resistance $R_f = 20\,\Omega$
Gear ratio $n = 5$
Equivalent moment of inertia of motor and load $I_e = 1.3\,\text{kgm}^2$
Sampling time $T = 0.05$ seconds.

Using a GA with a population of 10 members, find the values of the controller gain $K_1$ and the tachogenerator constant $K_4$ that maximizes the fitness function

$$J = 100 \left/ \sum_{k=0}^{N-1} \{(\theta_i(kT) - \theta_0(kT))^2 T \right. \tag{10.98}$$

when the system is subjected to a unit step at time $kT = 0$. Perform the summation over a time period of 2 seconds ($N = 40$). Allow a search space of 0–15 for $K_1$ and

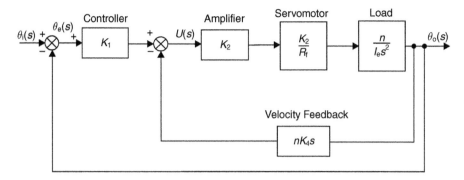

**Fig. 10.36** Angular positional control system.

0–1 for $K_4$. Assume that the solution space has 255 values and that each solution can be represented by an eight digit binary string, the first four digits representing $K_1$ and the second four representing $K_4$.

*Solution*
The plant transfer function is

$$\frac{\theta_0}{U}(s) = \frac{K_2 K_3 n}{R_f I_e s^2} \qquad (10.99)$$

If the state variables are $x_1 = \theta_0(t)$ and $x_2 = \dot\theta_0(t) = \dot x_1$, then the state equations become

$$\begin{bmatrix} \dot x_1 \\ \dot x_2 \end{bmatrix} = \begin{bmatrix} 0 & 1 \\ 0 & 0 \end{bmatrix} \begin{bmatrix} x_1 \\ x_2 \end{bmatrix} + \begin{bmatrix} 0 \\ \dfrac{K_2 K_3 n}{R_f I_e} \end{bmatrix} u$$

and the output equations

$$\begin{bmatrix} \theta_0 \\ \dot\theta_0 \end{bmatrix} = \begin{bmatrix} 1 & 0 \\ 0 & 1 \end{bmatrix} \begin{bmatrix} x_1 \\ x_2 \end{bmatrix}$$

Inserting values into the state equations

$$\begin{bmatrix} \dot x_1 \\ \dot x_2 \end{bmatrix} = \begin{bmatrix} 0 & 1 \\ 0 & 0 \end{bmatrix} \begin{bmatrix} x_1 \\ x_2 \end{bmatrix} + \begin{bmatrix} 0 \\ 10.096 \end{bmatrix} u \qquad (10.100)$$

For a sampling time of 0.05 seconds, the discrete form of equation (10.100) is

$$\begin{bmatrix} x_1(k+1)T \\ x_2(k+1)T \end{bmatrix} = \begin{bmatrix} 1 & 0.05 \\ 0 & 1 \end{bmatrix} \begin{bmatrix} x_1(kT) \\ x_2(kT) \end{bmatrix} + \begin{bmatrix} 0.0126 \\ 0.5048 \end{bmatrix} u(kT) \qquad (10.101)$$

From Figure 10.36 the control law is

$$u(kT) = K_1(\theta_i(kT) - x_1(kT)) - nK_4 x_2(kT) \qquad (10.102)$$

where $\theta_i(kT) = 1.0$ for all $kT > 0$.

Hence values of $K_1$ and $K_4$ generated by the GA are inserted into equation (10.102) and the control $u(kT)$ used to drive the discrete plant equation (10.101). The fitness function $J$ is updated at each sampling instant to give an overall value at the end of each simulation. For a population of 10 members, 10 simulations are required per generation.

Since the required search space is $K_1$ 0–15, $K_4$ 0–1, the following are examples of population membership

$$
\begin{array}{llll}
K_1 & K_4 & & \\
1111 & 1111 & K_1 = 15 & K_4 = 1 \\
0011 & 0111 & K_1 = 3 & K_4 = 7/15 = 0.467 \\
1001 & 1100 & K_1 = 9 & K_4 = 12/15 = 0.8
\end{array}
$$

Table 10.4 shows the parent selection for mating from a randomly seeded initial population. The random numbers (0–1) from the roulette wheel spins were: 0.145, 0.422, 0.977, 0.339, 0.607, 0.419, 0.075, 0.027, 0.846, 0.047.

The fitness of the first generation of offsprings is given in Table 10.5. Further breeding produced the sixth generation of offsprings given in Table 10.6. Inspection of Table 10.6 reveals that values of $K_1 = 15$, $K_4 = 0.333$ produces a global maximum of $J = 1065$. Figure 10.37 shows the unit step response of the system in Example 10.6 for

(a) the maximum of the first generation of offsprings ($J = 841$)
(b) the global maximum of the sixth generation of offsprings ($J = 1065$)

*Use of Schemata* (Similarity templates): As the progression through generations of solutions takes place, there evolves certain similarities between genes within chromosomes. These similarities can be exploited using a similarity template or schema, that sits within a schemata framework.

A schema employs a don't care symbol '*', so, for example, the sixth generation of offsprings in Table 10.6 could have employed the template

$$111^{**}1^{**}$$

The use of schemata will aid the speed of convergence.

**Table 10.4** Parent selection from initial population for Example 10.6

| Parent | $K_1$ | $K_4$ | $J$ | $p = J/\sum J$ | Cumulative probability | Roulette wheel hits |
|--------|-------|-------|-----|----------------|------------------------|---------------------|
| 1011 1001 | 11 | 0.600 | 647 | 0.157 | 0.157 | 4 |
| 0100 1010 | 4 | 0.667 | 233 | 0.056 | 0.213 | 0 |
| 0001 0111 | 1 | 0.400 | 112 | 0.027 | 0.240 | 0 |
| 1001 1010 | 9 | 0.667 | 497 | 0.122 | 0.362 | 1 |
| 1110 0111 | 14 | 0.467 | 923 | 0.224 | 0.586 | 2 |
| 0101 0001 | 5 | 0.067 | 375 | 0.092 | 0.678 | 1 |
| 1001 1101 | 9 | 0.867 | 18 | 0.004 | 0.682 | 0 |
| 0110 1101 | 6 | 0.867 | 46 | 0.011 | 0.693 | 0 |
| 1101 1010 | 13 | 0.667 | 691 | 0.168 | 0.861 | 1 |
| 0101 0100 | 5 | 0.267 | 573 | 0.139 | 1.000 | 1 |
| Total | | | 4115 | 1.000 | | 10 |
| Mean | | | 411.5 | 0.100 | | 1 |
| Maximum | | | 923 | 0.224 | | 4 |

**Table 10.5** Fitness of first generation of offsprings for Example 10.6

| Parent | Offspring | $K_1$ | $K_4$ | $J$ |
|---|---|---|---|---|
| 101\|1 1001 | 101\|0 0111 | 10 | 0.467 | 712 |
| 111\|0 0111 | 111\|1 1001 | 15 | 0.600 | 841 |
| 0101 0\|001 | 0101 0\|100 | 5 | 0.267 | 573 |
| 0101 0\|100 | 0101 0\|001 | 5 | 0.067 | 375 |
| 1011\|1001 | 1011\|1010 | 11 | 0.667 | 596 |
| 1001\|1010 | 1001\|1001 | 9 | 0.600 | 541 |
| 1101 10\|10 | 1101 10\|01 | 13 | 0.600 | 747 |
| 1011 10\|01 | 1011 10\|10 | 11 | 0.667 | 596 |
| 101\|1 1001 | 101\|0 0111 | 10 | 0.467 | 713 |
| 111\|0 0111 | 111\|1 1001 | 15 | 0.600 | 841 |
| Total | | | | 6535 |
| Mean | | | | 653.5 |
| Maximum | | | | 841 |

**Table 10.6** Fitness of sixth generation of offsprings for Example 10.6

| Offspring | $K_1$ | $K_4$ | $J$ |
|---|---|---|---|
| 1110 0101 | 14 | 0.333 | 1030 |
| 1111 0110 | 15 | 0.400 | 1029 |
| 1110 0100 | 14 | 0.267 | 1021 |
| 1111 0101 | 15 | 0.333 | 1065 |
| 1111 0111 | 15 | 0.467 | 970 |
| 1111 0100 | 15 | 0.267 | 1041 |
| 1110 1000 | 14 | 0.533 | 858 |
| 1110 0111 | 14 | 0.467 | 923 |
| 1111 1000 | 15 | 0.533 | 905 |
| 1111 0101 | 15 | 0.333 | 1065 |
| Total | | | 9907 |
| Mean | | | 990.7 |
| Maximum | | | 1065 |

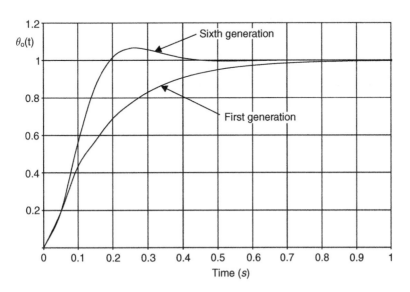

**Fig. 10.37** Comparison between best first generation and best sixth generations solutions for Example 10.6.

### *Other applications of Genetic Algorithms*

(a) *Optimal control*:  Optimal control problems such as the linear quadratic regulator discussed in section 9.2 can also be solved using GAs. The discrete quadratic performance index given in equation (9.28) can be employed as a (minimum) fitness function directly. Alternatively, as shown in Example 10.6 the reciprocal of equation (9.28) provides a (maximum) fitness function.

(b) *Self-Organizing Fuzzy Logic Control*:  Genetic Algorithms may be used to adapt both membership functions and rulebase structures in a SOFLC system.

Figure 10.38 shows an input window with three triangular fuzzy sets NB, Z and PB. Each set is positioned in its regime of operation by the centre parameter c so that, for example, NB can only operate on the negative side of the universe of discourse. The width of each set is controlled by parameter $w$.

The chromosome string could take the form

$$[c_1 \ w_1 \ c_2 \ w_2 \ c_3 \ w_3] \tag{10.103}$$

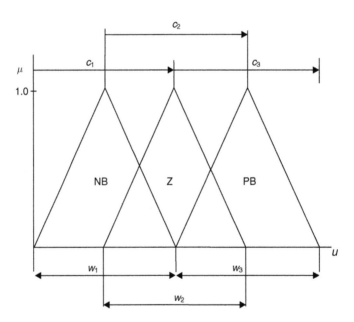

**Fig. 10.38** Adaption of membership function features using genetic algorithms.

If each parameter is a 4-bit string, the configuration in Figure 10.38 would be represented by

$$[100011111000111110001111] \tag{10.104}$$

## 10.4.3.   Alternative search strategies

In addition to evolutionary search strategies such as GAs, there are a number of other search techniques that are employed for design optimization.

### Simulated annealing

The physical annealing process consists of heating a metal up to a prescribed temperature and then allowing it to cool slowly. During cooling, the molecules form crystals which are in a minimum energy condition. The metal, therefore, settles to a global minimum energy state.

With simulated annealing, an energy term $E$ is defined, which then becomes the performance index to be minimized. For a given energy change $\Delta E$ at temperature $T$, the probability P of accepting a solution is

$$P = e^{-\Delta E/kT} \tag{10.105}$$

where k is the Boltzmann constant. If the energy of the system has decreased, then the system may move to the new state, based on the probability given by equation (10.105). As cooling proceeds along a prescribed schedule, the system avoids local minima, and settles down to a global minimum energy state.

### Tabu search

Like simulated annealing, tabu search is a technique designed to avoid the problem of becoming trapped in local optima. The procedure is basically hill-climbing, which commences at an initial solution and searches the neighbourhood for a better solution. However, the process will recognize, and avoid areas of the solution space that it has already encountered, thus making these areas 'tabu'. The tabu moves are kept in a finite list, which is updated as the search proceeds.

## 10.5  Further problems

*Example 10.7*
A fuzzy logic controller has input and output fuzzy windows as shown in Figure 10.39. The fuzzy rulebase is given in Figure 10.40. If defuzzification is by the centre of area method, calculate crisp control signals $u(t)$ when the error $e(t)$ and the rate of change of error $ce(t)$ have the following values:

|      | $e(t)$ | $ce(t)$ | [Answer $u(t)$] |
|------|--------|---------|-----------------|
| (i)  | 0      | −2      | −6.67           |
| (ii) | −4     | −1      | −2.66           |
| (iii)| 1      | 1.5     | 3.01            |
| (iv) | 4      | −1.5    | 0.106           |

*Note*: For the centre of area method, use only those values inside the dotted lines in the output window.

*Example 10.8*
The angular positional control system shown by the block diagram in Figure 10.36 is to have the velocity feedback loop removed and controller $K_1$ replaced by a fuzzy logic controller (FLC) as demonstrated by Barrett (1992). The inputs to the FLC

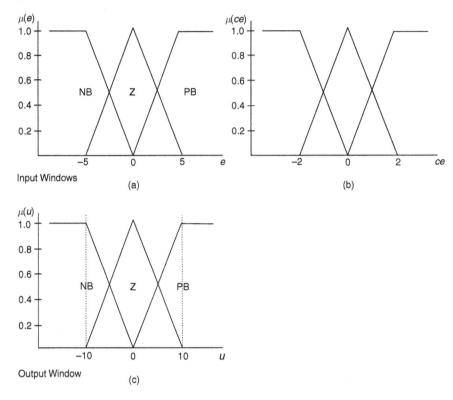

**Fig. 10.39** Input and output fuzzy windows for Example 10.7.

|  e ce | NB | Z | PB |
|---|---|---|---|
| NB | NB | NB | Z |
| Z | NB | Z | PB |
| PB | Z | PB | PB |

**Fig. 10.40** Fuzzy rulebase for Example 10.7.

are angular error $e(kT)$ and rate-of-change of angular error $ce(kT)$. The output from the FLC is a control signal $u(kT)$. The input and output fuzzy windows have the same number of fuzzy sets, but different scales for the universes of discourse as shown in Figure 10.41. The fuzzy rulebase is given in Figure 10.42. If the state variables are

$$x_1 = \theta_0(t)$$
$$x_2 = \dot{\theta}_0(t) = \dot{x}_1$$

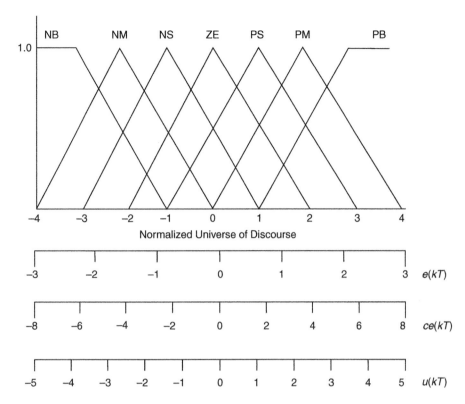

**Fig. 10.41** Input and output fuzzy windows for Example 10.8.

| ce \ e | NB | NM | NS | ZE | PS | PM | PB |
|--------|----|----|----|----|----|----|----|
| NB     | NB | NB | NM | NS | ZE | PM | PB |
| NM     | NB | NB | NM | NM | PS | PM | PB |
| NS     | NM | NS | NS | NS | PS | PM | PM |
| ZE     | NM | NS | NS | ZE | PS | PS | PM |
| PS     | NM | NM | NS | PS | PS | PS | PM |
| PM     | NB | NM | NS | PM | PM | PB | PB |
| PB     | NB | NM | ZE | PS | PM | PB | PB |

**Fig. 10.42** Fuzzy rulebase for Example 10.8.

then the state and output equations for the plant are as given in equation (10.100)

$$\begin{bmatrix} \dot{x}_1 \\ \dot{x}_2 \end{bmatrix} = \begin{bmatrix} 0 & 1 \\ 0 & 0 \end{bmatrix} \begin{bmatrix} x_1 \\ x_2 \end{bmatrix} + \begin{bmatrix} 0 \\ 10.096 \end{bmatrix} u$$

$$\begin{bmatrix} \theta_1 \\ \dot{\theta}_2 \end{bmatrix} = \begin{bmatrix} 1 & 0 \\ 0 & 1 \end{bmatrix} \begin{bmatrix} x_1 \\ x_2 \end{bmatrix} \tag{10.106}$$

and, for a sampling time of 0.2 seconds, the discrete form of equation (10.106) is

$$\begin{bmatrix} x_1(k+1)T \\ x_2(k+1)T \end{bmatrix} = \begin{bmatrix} 1 & 0.2 \\ 0 & 1 \end{bmatrix} \begin{bmatrix} x_1(kT) \\ x_2(kT) \end{bmatrix} + \begin{bmatrix} 0.2019 \\ 2.0192 \end{bmatrix} u(kT) \tag{10.107}$$

If the system, which is initially at rest when $kT = 0$, is given a unit step input at this time, determine $e(kT)$, $ce(kT)$, $u(kT)$, $x_1(kT)$ and $x_2(kT)$ for values of $kT = 0$, 0.2, 0.4 and 0.6 seconds.

Assume that

$$e(kT) = (\theta_i - \theta_0) = 1 - x_1(kT)$$

$$ce(kT) = d/dt(\theta_i - \theta_0)$$

$$= \frac{d\theta_i}{dt} - \frac{d\theta_0}{dt}$$

$$= 0 - \frac{d\theta_0}{dt} = -x_2(kT)$$

*Solution*

**Table 10.7** Solution to Example 10.8

| $kT$ | $x_1(kT)$ | $x_2(kT)$ | $e(kT)$ | $ce(kT)$ | $u(kT)$ | $x_1(k+1)T$ | $x_2(k+1)T$ |
|------|-----------|-----------|---------|----------|---------|-------------|-------------|
| 0    | 0         | 0         | 1       | 0        | 1.083   | 0.2187      | 2.187       |
| 0.2  | 0.2187    | 2.187     | 0.7813  | -2.187   | 0.666   | 0.7906      | 3.532       |
| 0.4  | 0.7906    | 3.532     | 0.2094  | -3.532   | -1.32   | 1.2305      | 0.867       |
| 0.6  | 1.2305    | 0.867     | -0.2305 | -0.867   | -0.511  | 1.3007      | -0.165      |

*Example 10.9*
A neural network has a structure as shown in Figure 10.43. Assuming that all the activation functions are sigmoids, calculate the values of $y_{12}$ and $y_{22}$ when the inputs are

(a) $x_1 = 0.3$, $x_2 = 0.5$
(b) $x_1 = 0.9$, $x_2 = 0.1$

when the weights and biases are

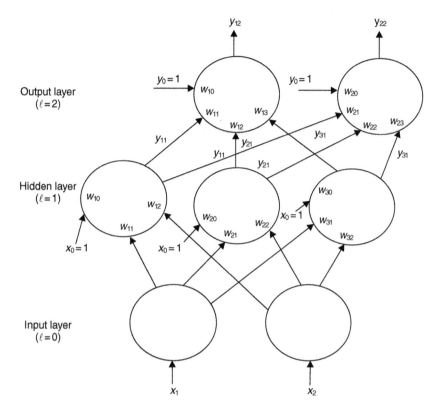

**Fig. 10.43** Neural network structure for Example 10.9.

Hidden layer

$$\mathbf{W}_{j1} = \begin{bmatrix} 1 & 2 \\ 3 & 3 \\ 2 & 1 \end{bmatrix} \quad \mathbf{b}_{j1} = \begin{bmatrix} 2 \\ -2 \\ 1 \end{bmatrix}$$

Output layer

$$\mathbf{W}_{j2} = \begin{bmatrix} -3 & -2 & 1 \\ 1 & 2 & 3 \end{bmatrix} \quad \mathbf{b}_{j2} = \begin{bmatrix} 3 \\ -3 \end{bmatrix}$$

*Solution*
(a) $y_{12} = 0.450$, $y_{22} = 0.862$
(b) $y_{12} = 0.405$, $y_{22} = 0.906$

*Example 10.10*
In Example 10.9(b), if the target values for the outputs are $d_{12} = 0$ and $d_{22} = 1$, calculate new values for the weights and biases using the back-propagation algorithm. Assume a learning rate of 0.5 with no momentum term.

*Solution*
Output layer: $\delta_1 = -0.098$, $\delta_2 = 0.008$ (Equation 10.78)

$$\mathbf{W}_{j2} = \begin{bmatrix} -3.047 & -2.036 & 0.954 \\ 1.004 & 2.003 & 3.004 \end{bmatrix} \quad \mathbf{b}_{j2} = \begin{bmatrix} 2.951 \\ -2.996 \end{bmatrix}$$

Hidden layer: $\delta_1 = 0.0016$, $\delta_2 = 0.042$, $\delta_3 = -0.003$ (Equation 10.80)

$$\mathbf{W}_{j1} = \begin{bmatrix} 1.0007 & 2.00008 \\ 3.019 & 3.0021 \\ 1.999 & 0.9999 \end{bmatrix} \quad \mathbf{b}_{j1} = \begin{bmatrix} 2.0008 \\ -1.979 \\ 0.9985 \end{bmatrix}$$

*Example 10.11*
A system has a parabolic fitness function

$$J = -x^2 + 2x \tag{10.108}$$

Using a genetic algorithm, find the value of $x$ that maximizes $J$ in equation (10.108). Assume that the solution space has 31 values in the range $x = 0$ to 2 and that each solution can be represented by a five digit binary string ranging from 00000 to 11111. Let the population have four members and it is initially seeded by spinning a coin (heads $= 1$, tails $= 0$).

*Solution*

| $x(2)$ | 01110 | 01111 | 10000 | 10001 |
|---|---|---|---|---|
| $x(10)$ | 14 | 15 | 16 | 17 |
| $x$ | 0.903 | 0.968 | 1.032 | 1.097 |
| $J$ | 0.990591 | 0.998976 | 0.998976 | 0.990591 |

Hence $x = 0.968$ and 1.032 both give optimum solutions.

*Example 10.12*
The closed-loop control system analysed by the root-locus method in Example 5.8 can be represented by the block diagram shown in Figure 10.44. Using root-locus, the best setting for $K_1$ was found to be 11.35, representing a damping ratio of 0.5.

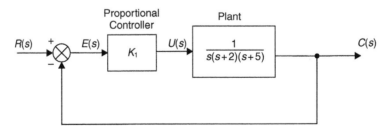

**Fig. 10.44** Block diagram for Example 5.8.

Validate, or otherwise, using a genetic algorithm that this value of $K_1$ maximizes the fitness function

$$J = 100 \Big/ \sum_{k=0}^{N-1}\{(r(kT) - c(kT))^2 T\} \qquad (10.109)$$

when the system is subjected to a unit step at time $kT = 0$. Using a sampling time $T = 0.05$ seconds, perform the summation over a time period of 10 seconds ($N = 200$). Allow a search space of 0–100 for $K_1$. Assume that the solution space has 255 values and that each solution can be represented by an eight digit binary string chromosome. Use a randomly seeded initial population of 10 members.

If the state variables are

$$x_1 = c(t)$$
$$x_2 = \dot{c}(t) = \dot{x}_1$$
$$x_3 = \ddot{c}(t) = \dot{x}_2$$

From Figure 10.44, the plant differential equation may be written as

$$\dot{x}_3 = 0x_1 - 10x_2 - 7x_3 + u$$

The plant state and output equations are then

$$\begin{bmatrix} \dot{x}_1 \\ \dot{x}_2 \\ \dot{x}_3 \end{bmatrix} = \begin{bmatrix} 0 & 1 & 0 \\ 0 & 0 & 1 \\ 0 & -10 & -7 \end{bmatrix} \begin{bmatrix} x_1 \\ x_2 \\ x_3 \end{bmatrix} + \begin{bmatrix} 0 \\ 0 \\ 1 \end{bmatrix} u$$

$$C = [1 \quad 0 \quad 0]\mathbf{x} \qquad (10.110)$$

For a sampling time of 0.05 seconds, the discrete form of equation (10.110) is

$$\begin{bmatrix} x_1(k+1)T \\ x_2(k+1)T \\ x_3(k+1)T \end{bmatrix} = \begin{bmatrix} 1.0 & 0.0498 & 0.0011 \\ 0.0 & 0.9889 & 0.0420 \\ 0.0 & -0.4201 & 0.6948 \end{bmatrix} \begin{bmatrix} x_1(kT) \\ x_2(kT) \\ x_3(kT) \end{bmatrix} + \begin{bmatrix} 0.00002 \\ 0.00111 \\ 0.04201 \end{bmatrix} u(kT)$$

$$(10.111)$$

*Solution*

| $K_1(2)$ | 00011101 | 00101001 | 00110011 | 00111000 | 10000001 | 10110011 |
|---|---|---|---|---|---|---|
| $K_1(10)$ | 29 | 41 | 51 | 56 | 129 | 179 |
| $K_1$ | 11.37 | 16.08 | 20.0 | 21.96 | 50.59 | 70.20 |
| $J$ | 112.99 | 124.51 | 126.63 | 125.82 | 48.18 | 7.77 |

Thus the value of $K_1$ that maximizes the fitness function in equation (10.109) is $K_1 = 20.0$. This produces an overshoot of 38% when $kT = 2.0$ seconds, and represents an $\omega_n$ of 1.847 rad/s and a $\zeta$ of 0.303.

# Appendix 1
# Control system design using MATLAB

## A1.1 Introduction

MATLAB, its Toolboxes and SIMULINK have become, over a number of years, the industry standard software package for control system design. The purpose of this Appendix is to introduce the reader to some of the more useful aspects of MATLAB, and to illustrate how the software may be used to solve examples given in the main text of the book.

### A1.1.1 Getting started

The examples given in this Appendix were generated using MATLAB Version 5.3. Once the software has been installed, MATLAB is most easily entered by clicking the MATLAB icon. Alternatively, in a Windows environment, MATLAB can be entered by clicking the following sequence

$$\text{Start} \longrightarrow \text{Programs} \longrightarrow \text{MATLAB for Windows} \longrightarrow \text{MATLAB 5.3}$$

The user should now be in the MATLAB command window, which contains some helpful comments together with the MATLAB prompt ». MATLAB commands are typed after the prompt, and entered using 'Return' (or 'Enter'). Terminating the command with ';' will suppress the result of the command being printed in the command window. Comments are preceded by the '%' symbol.

## A1.2 Tutorial 1: Matrix operations

This tutorial introduces the reader to matrix operations using MATLAB. All text in courier font is either typed into, or printed into the command window.

```
»% Matrix Operations
»% To enter a matrix
»A=[1 3;5 9];
»B=[4 -7;10 0];
»A
```

```
A=

      1        3
      5        9
»B
B=

      4       -7
     10        0
»% Note that MATLAB is case sensitive
»% Matrix Addition
»A+B % Terminating with ';' will suppress 'ans'
ans=

      5       -4
     15        9
»% Matrix Multiplication
»A*B
ans=

     34       -7
    110      -35
»% Determinant of a Matrix
»det(A)
ans=
     -6
»% Inverse of a Matrix
»inv(A)
ans=
  -1.5000   0.5000
   0.8333  -0.1667
»% Check
»C=inv(A);
»A*C
ans=
   1.0000  0.0000
   0.0000  1.0000
»% Solve A^-1*B
»A\B
ans=
  -1.0000  10.5000
   1.6667  -5.8333
»% Solve A*B^-1
»A/B
ans=
  -0.4286  0.2714
  -1.2857  1.0143
»% Eigenvalues of a Matrix
»eig(A)
ans=
  -0.5678
  10.5678
»% Coefficients of Characteristic Equation (as^2+bs+c)
»poly(A)
ans=
   1.0000  -10.0000  -6.0000
»% Roots of Characteristic Equation (as^2+bs+c=0)
»ce=[1 -10 -6];
»roots(ce)
```

```
ans=
   10.5678
   -0.5678
»% Note that roots(ce) and eig(A) give the same result
»% Transpose of a Matrix
»A'
ans=
   1   5
   3   9
»% Rank of a Matrix
»rank(A)
ans=
   2
»% Create an Identity Matrix
»I=eye(3);
»I
I=
   1 0 0
   0 1 0
   0 0 1
»% Condition of a Matrix
»% The higher the Condition Number, the more ill-conditioned the
»% matrix is
»% Log10 of Condition Number gives approx. number of decimal
»% places
»% lost due to round-off errors
»cond(A)
ans=
   19.2815
»% Tutorial End
```

The above session may be printed by clicking

   File ➞ Print

MATLAB may be closed by clicking

   File ➞ Exit MATLAB

## A1.3  Tutorial 2: Time domain analysis

This tutorial introduces the reader to time domain analysis using MATLAB. It uses commands from the Control System Toolbox. A list of the commands can be found using

```
»help control
```

More information on individual commands can be obtained, for example

```
»help step
```

will provide more detail on how to use the step command.

*Script files*: A script file is an ASCII text file of MATLAB commands, that can be created using

(a) a text editor
(b) the MATLAB editor/debugger
(c) a word processor that can save as pure ASCII text files.

A script file should have a name that ends in '.m', and is run by typing the name of the file (without '.m') after the MATLAB prompt, or by typing the sequence

File ⟶ Run Script ⟶ enter file name

The advantage of a script file is that it only needs to be created once and saves the labour of continually typing lists of commands at the MATLAB prompt.

The examples given in this tutorial relate to those solved in Chapter 3. Consider a first-order transfer function

$$G(s) = 1/1 + s$$

The impulse response function (Example 3.4, Figure 3.11) can be created by the following script file

*File name: examp34.m*

```
%impulse response of transfer function G(s)=num(s)/den(s)
%num and den contain polynomial coefficients
%in descending powers of s
clf
num=[1];
den=[1 1];
impulse (num,den);
grid;
printsys(num,den,'s');
```

This shows how a transfer function is entered into MATLAB, where num = [1] is the numerator ($K = 1$) and den=[1 1] represents the '$s$' coefficient and the '$s^0$' coefficient respectively. 'Impulse (num, den) computes and plots the impulse response and grid produces a rectangular grid on the plot. Printsys (num, den, 's') prints the transfer function at the MATLAB prompt. A hard copy can be obtained by selecting, from the screen plot

File ⟶ Print

The step response of a first-order system (Example 3.5, Figure 3.13) is obtained using the step command

*File name: examp35.m*

```
% step response of transfer function G(s)=num(s)/den(s)
%num and den contain polynomial coefficients
%in descending powers of s
clf
num=[1];
den=[1 1];
step(num,den);
grid;
printsys(num,den,'s');
```

*SIMULINK*: The Control System Toolbox does not possess a 'ramp' command, but the ramp response of a first-order system (Example 3.6, Figure 3.15) can be obtained using SIMULINK, which is an easy to use Graphical User Interface (GUI). SIMULINK allows a block diagram representation of a control system to be constructed and real-time simulations performed.

With MATLAB Version 5.3, typing `simulink` at the MATLAB prompt brings up the SIMULINK Library Browser. Clicking on the 'Create new model' icon in the top left-hand corner creates a new window called 'untitled'.

Clicking on the ⊞ icon attached to SIMULINK lists the SIMULINK options, and ⊞ Continuous lists the continuous systems options. To obtain a transfer function block, click and hold left mouse button on the 'Transfer Fcn' icon under 'Continuous', and drag to 'untitled' window.

Click on ⊟ to close down 'Continuous' and click on ⊞ 'Sources' to drag 'Ramp' from Browser to 'untitled' window. Close down 'Sources' and click on ⊞ 'Sinks' to drag 'Scope' to 'untitled'. Holding down left mouse button connect 'Ramp', 'Transfer Fcn' and 'Scope' together as shown in Figure A1.1.

Double click on 'Scope' to bring up scope screen, and, in 'untitled', click 'Simulation' and 'Start'. The ramp response should appear on the scope screen. Click on the scope screen and choose autoscale (binoculars). Click print icon to obtain a hard copy of the ramp response. In the 'untitled' window, click on 'File' and 'Save As' and save as a '.mdl' (model) file in a directory of your choice (i.e. *examp36.mdl* in 'work').

The transfer function for a second-order system can easily be obtained in terms of $\omega_n$ and $\zeta$ using the `ord2` command. This has $\omega_n$ and $\zeta$ as input arguments and generates the numerator and denominator of the equivalent transfer function. The script file *sec_ord.m* shows how Figure 3.19 can be generated using the `ord2` and step MATLAB commands

*File name: sec_ord.m*

```
%Second-order system
t=[0:0.1:15];
wn=1;
zeta=0.2;
[num,den]=ord2(wn,zeta);
[y,x,t]=step(num,den,t);
zeta = 0.4;
[num1,den1]=ord2(wn,zeta);
[y1,x,t]=step(num1,den1,t);
zeta=0.6;
[num2,den2]=ord2(wn,zeta);
[y2,x,t]=step(num2,den2,t);
plot(t,y,t,y1,t,y2);
grid;
```

In *sec_ord.m* a user supplied common time base ($t = 0$–15 seconds in 0.1 second intervals) has been set up. The `plot` command superimposes the step responses for zeta = 0.2, 0.4 and 0.6.

The step response for Example 3.8, the resistance thermometer and valve, shown in Figure 3.23 can be generated with script file *examp38.m*.

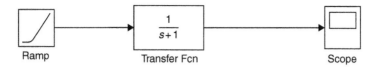

**Fig. A1.1** SIMULINK model for the ramp response of a first-order system.

*File name: examp38.m*

```
%Step response of a third-order system
%G(s)=12.5/(s+0.5)(s^2+s+25)
clf
num=[12.5];
den=conv([1 0.5],[1 1 25]);
step(num,den);
grid;
```

The transfer function is based on equation (3.75), with the input term $X_i(s) = 1/s$ removed. The denominator is input using the `conv` command (short for convolution), which multiplies two polynomial expressions together.

## A1.4   Tutorial 3: Closed-loop control systems

This tutorial shows how MATLAB can be used to build up and test closed-loop control systems. The examples given relate to those solved in Chapter 4. The commands used in MATLAB to create a single model from the elements in the control loop are

Command     Operation

`series`     Combines blocks in cascade, see Transformation 1, Table 4.1
`parallel`   Combines blocks in parallel, see Transformation 2, Table 4.1
`feedback`   Eliminates a feedback loop, see Transformation 4, Table 4.1
`cloop`      Closes the loop with unity feedback.

Example 4.5 is a PI controlled liquid-level system shown in Figure 4.26. In the block diagram representation, Figure 4.27, the system parameters are

$$K_1 = 1; \quad T_i = 5 \text{ seconds}; \quad K_v = 0.1 \text{m}^3/\text{sV}$$
$$R_f = 15 \text{ s/m}^2; \quad A = 2 \text{ m}^2; \quad H_1 = 1$$

The PI controller and control valve transfer function is therefore

$$G_c(s) = \frac{0.5s + 0.1}{5s}$$

and the plant transfer function is

$$G_p(s) = \frac{15}{30s + 1}$$

Since $H_1 = 1$, the system has unity feedback, and the closed-loop transfer function and step response is given by

*Filename: examp45.m*

```
%Example 4.5 (Liquid-Level Process Control System)
%Use of series and cloop
numc=[0.5  0.1];
denc=[5  0];
nump=[15];
denp=[30  1];
[numol,denol]=series(numc,denc,nump,denp);
[numcl,dencl]=cloop(numol,denol);
printsys(numcl,dencl,'s');
step (numcl,dencl);
grid;
```

Case study Example 4.6.1 is a CNC machine-tool positional control system, whose block diagram representation is shown in Figure 4.31. When system parameter values are inserted, the block diagram is as shown in Figure A1.2.

The closed-loop transfer function and step response is given by

*Filename: examp461.m*

```
%Case Study Example 4.6.1 (CNC Machine-Tool Positional Control)
%Use of series and feedback
nump=[80];
denp=[0.45  0];
nump1=[0.365];
denp1=[1  0];
numvf=[0.697];
denvf=[1];
numpf=[60];
denpf=[1];
[num,den]=feedback(nump,denp,numvf,denvf);
[numfp,denfp]=series(num,den,nump1,denp1);
[numcl,dencl]=feedback(numfp,denfp,numpf,denpf);
printsys(numcl,dencl,'s');
step(numcl,dencl);
grid;
```

In this example, the inner loop is solved first using feedback. The controller and integrator are cascaded together (nump1,denp1) and then series is used to find the forward-path transfer function (numfp,denfp). Feedback is then used again to obtain the closed-loop transfer function.

The time response of the CNC control system is also obtained using SIMULINK as shown in Figure A1.3. Note that the variables $t$ and $x_0$ have been sent to workspace, an area of memory that holds and saves variables. The commands who and whos lists the variables in the workspace. The system time response can be obtained by the command

```
plot(t,x₀)
```

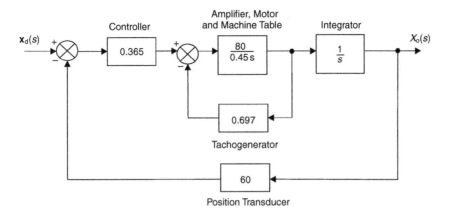

**Fig. A1.2** CNC machine-tool positional control system.

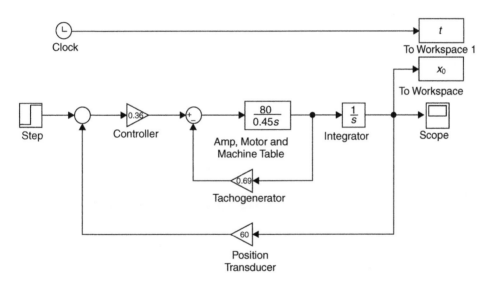

**Fig. A1.3** SIMULINK simulation of CNC machine-tool positional control system.

Case study Example 4.6.2 is a **PID** temperature control system, and is represented by the block diagram in Figure 4.33. Here, the PID control transfer function is not 'proper', and the `series` command does not work. The forward path transfer function, with values inserted, is

$$G_{fp}(s) = \frac{29.412s^2 + 39.216s + 13.072}{3s(4s + 1)(8s + 1)}$$

The closed-loop transfer function and step response is given by

*Filename: examp462.m*

```
%Case study example 4.6.2 (Temperature Control)
%Use of feedback
```

```
numfp=[29.412  39.216  13.072];
denfp=conv([3  0],conv([4  1],[8  1]));
numtf=[1];
dentf=[1];
[numcl,dencl]=feedback(numfp,denfp,numtf,dentf);
printsys(numcl,dencl,'s');
step(numcl,dencl);
grid;
```

Case study Example 4.6.3 is a ship autopilot control system. The block diagram representation is given by Figure 4.37. Inserting system parameters gives a forward-path transfer function

$$G_{fp}(s) = \frac{0.2}{20s^2 + 2s}$$

The closed-loop transfer function and step response is given by

*Filename: examp463.m*

```
%Case study example 4.6.3 (Ship Autopilot Control System)
%Use of feedback
numfp=[0.2];
denfp=[20  2  0];
numhf=[1];
denhf=[1];
[numcl,dencl]=feedback(numfp,denfp,numhf,denhf);
printsys(numcl,dencl,'s');
step(numcl,dencl);
grid;
```

## A1.5  Tutorial 4: Classical design in the s-plane

The tutorial demonstrates how MATLAB is used to generate root locus diagrams, and hence how to design control systems in the *s*-plane. Examples given in Chapter 5 are used to illustrate the MATLAB commands. The roots of the characteristic equation (or any polynomial) can be found using the `roots` command.

*Example 5.1*
Check the stability of the system that has the characteristic equation

$$s^4 + 2s^3 + s^2 + 4s + 2 = 0$$

At the MATLAB prompt type

```
»ce=[1  2  1  4  2];
»roots(ce)
ans=
-2.1877
 0.3516+1.2843i
 0.3516-1.2843i
-0.5156
```

Two roots have positive real parts, hence the system is unstable.

*Example 5.2*

What values of $K_1$ make the following system unstable

$$G(s)H(s) = \frac{8K_1}{s(s^2 + s + 2)}$$

*Filename: examp52.m*

```
%Roots of denominator of closed-loop transfer function
k1=0.25
num=[8*k1];
den=conv([1  0],[1  1  2]);
[numcl,dencl]=cloop(num,den);
roots(dencl)
```

Running *examp52.m* with $K_1$ set to 0.25, 0.15 and 0.35 gives

```
»examp52
k1=
    0.2500

ans=

   0.0000+1.4142i
   0.0000-1.4142i
  -1.0000

»examp52
k1=
    0.1500

ans=

  -0.1630+1.3243i
  -0.1630-1.3243i
  -0.6740

»examp52
k1=
    0.3500

ans=

   0.1140+1.5057i
   0.1140-1.5057i
  -1.2280
```

Hence, with $K_1 = 0.15$, the system is stable, $K_1 = 0.25$, the system has marginal stability and $K_1 = 0.35$, the system is unstable.

*Example 5.8*

The root locus for

$$G(s)H(s) = \frac{K}{s(s + 2)(s + 5)}$$

can be drawn using the r locus command as shown in *examp58.m*

*Filename: examp58.m*

```
%Example 5.8 Simple root locus
%G(s)=K/s(s+2)(s+5)
clf
num=[1];
den=conv(conv([1 0],[1 2]),[1 5]);
rlocus(num,den);
axis([-9 2 -4 4]);
%axis('square');
```

With no axis command, the axes are generated automatically. Alternatively, user defined axes can be generated with $axis$ ($[x_{min} \, x_{max} \, y_{min} \, y_{max}]$). Using the command 'square' generates the same scales on both $x$ and $y$ axes. With *examp58.m* it is easy to experiment with all three methods.

Using MATLAB to design a system, it is possible to superimpose lines of constant $\zeta$ and $\omega_n$ on the root locus diagram. It is also possible, using a cursor in the graphics window, to select a point on the locus, and return values for open-loop gain $K$ and closed-loop poles using the command

```
[k,poles]=relocfind(num,den)
```

The script file *examp58a.m* shows how this is achieved

*Filename: examp58a.m*

```
%Example 5.8 Root locus
%G(s)=K/s(s+2)(s+5)
clf
num=[1];
den=conv(conv)[1 0],[1 2]),[1 5]);
rlocus(num,den);
axis([-9 2 -4 4]);
zeta=0.5;
wn=2:2:8;
sgrid(zeta,wn);
[k,poles]=rlocfind(num,den)
```

Here a line of $\zeta = 0.5 \, (\beta = 60°)$ is drawn together with four $\omega_n$ circles ($\omega_n = 2, 4, 6$ and $8 \, \text{rad/s}$.) At the MATLAB prompt, the user is asked to select a point in the graphics window. If the intersection of the complex locus with the $\zeta = 0.5$ line is selected (see Figure 5.14), the following response is obtained

```
»examp58a
Select a point in the graphics window

selected point=
  -0.7143+1.2541i
k=
  11.5754
poles=
  -5.5796
  -0.7102+1.2531i
  -0.7102-1.2531i
```

Example 5.9 is solved using *examp59.m* to create Figure 5.16.

*Filename : examp59.m*

```
%Example 5.9 Root Locus
%G(s)=K/s(s^2+4s+13)
clf
num=[1];
den=conv([1 0],[1 4 13]);
rlocus(num,den);
axis([-9 2 -4 4]);
zeta=0.25;
wn=3;
sgrid(zeta,wn);
k=rlocfind(num,den)        %Select k from graphic window
```

When run, the program invites the user to select a point in the graphics window, which may be used to find the value of $K$ when $\zeta = 0.25$. If the last line of *examp59.m* is typed at the MATLAB prompt, the cursor re-appears, and a further selection can be made, in this case to select the value of $K$ for marginal stability. This is demonstrated below

```
»examp59
Select a point in the graphics window
selected_point=
  -0.7429+2.9175i                    .
k=
  22.9452
»k=rlocfind(num,den)
Select a point in the graphics window
selected_point=
  0+3.6304i
k=
  52.7222
»
```

Example 5.10 uses the root locus method to design a PD controller that will allow the system

$$G(s)H(s) = \frac{1}{s(s+2)(s+5)}$$

To meet a given time-domain specification. In *examp510.m*, the PD controller takes the form

$$G_c(s) = K_1(s+2)$$

*Filename: examp510.m*

```
%Example 5.10 Root Locus
%G(s)=K(s+2)/s(s+2)(s+5)
clf
num=[1 2];
den=conv(conv([1 0],[1 2]),[1 5]);
rlocus(num,den);
axis([-9 2 -4 4]);
zeta=0.7;
wn=3.5;
sgrid(zeta,wn);
k=rlocfind(num,den) %Select k from screen
```

```
num=[k k*2];
[numcl,dencl]=cloop(num,den);
step(numcl,dencl);
grid;
```

This produces the (pole-zero cancellation) root locus plot shown in Figure 5.18. When run, *examp510.m* allows the user to select the value of $K_1$ that corresponds to $\zeta = 0.7$, and then uses this selected value to plot the step response. The text that appears in the command window is

```
»examp510
Select a point in the graphics window
selected_point=
   -2.4857+2.5215i
k=
   12.6077
»
```

Case study Example 5.11 uses root locus to design a ship roll stabilization system. The script file *examp5ll.m* considers a combined PD and PID (PIDD) controller of the form

$$G_c(s) = \frac{K_1(s+2)(s^2+4s+8)}{s}$$

*Filename: examp511.m*

```
%Example 5.11 Root Locus
%G(s)=K(s+2)(s^2+4s+8)/s(s+1)(s^2+0.7s+2)
clf
num=conv([1 2],[1 4 8]);
den=conv(conv([1 0],[1 1]),[1 0.7 2]);
rlocus(num,den);
axis([-9 2 -4 4]);
zeta=0.7;
wn=4.2;
sgrid(zeta,wn);
k=rlocfind(num,den) % Select k from screen
num=k*conv([1 2],[1 4 8]);
[numcl,dencl]=cloop(num,den);
step(numcl,dencl);
grid;
```

This script file produces the root locus shown in Figure 5.24 and allows the user to select the value of $K$ furthest from the imaginary axis that corresponds to $\zeta = 0.7$. The command window text is

```
»examp511
Select a point in the graphics window
selected_point=
   -3.2000+3.2607i
k=
   10.2416
»
```

When $K$ has been selected, the roll angle step response, as shown in Figure 5.25 is plotted.

## A1.6   Tutorial 5: Classical design in the frequency domain

This tutorial shows how MATLAB can be used to construct all the classical frequency domain plots, i.e. Bode gain and phase diagrams, Nyquist diagrams and Nichols charts. Control system design problems from Chapter 6 are used as examples.

*Example 6.1*
Construct the Bode diagram for

$$G(s) = \frac{2}{1 + 0.5s}$$

*Filename: examp61.m*

```
%Example 6.1 Bode Diagram
%First-order system
num=[2];
den=[0.5  1];
bode(num,den);
```

The Nyquist diagram for the same system is

*Filename: examp61a.m*

```
%Example 6.1(a) Nyquist Diagram
%First-order system
num=[2];
den=[0.5  1];
nyquist(num,den);
```

The script file *examp61b.m* shows how it is possible to customize a Bode diagram

*Filename: examp61b.m*

```
%Example 6.1(b)
%Customizing a Bode Diagram
clf
num=[2];
den=[0.5  1];
w=logspace(-1,2,200);%w from 10^-1 to 10^2,200 points
[mag,phase,w]=bode(num,den,w) % cal mag and phase
semilogx(w,20*log10(mag)),grid;
xlabel('Frequency (rad/s)'),ylabel('Gain dB');
```

The `logspace` command allows a vector of frequencies to be specified by the user, in this case, 200 points between $10^{-1}$ and $10^2$ (rad/s). The `bode` command has a left-hand argument [`mag`,`phase`,`w`] which allows the magnitude and phase to be calculated, but not displayed. The `semilogx` command plots the magnitude (converted to deciBels) against log frequency, and `grid` provides a log-linear grid. Script file *examp62.m* produces a second-order Bode gain diagram for Example 6.2.

*Filename: examp62.m*

```
%Example 6.2 Bode Diagram
%Second-order system
clf
num=[4];
den=[0.25 0.2 1];
w=logspace(-1,2,200);
[mag,phase,w]=bode(num,den,w);
semilogx(w,20*log10(mag)),grid;
xlabel('Frequency (rad/s)'),ylabel ('Gain dB')
```

The Nyquist diagram ($\omega$ varying from $-\infty$ to $+\infty$) is produced by *examp64.m* where

$$G(s)H(s) = \frac{K}{s(s^2 + 2s + 4)}$$

*Filename: examp64.m*

```
%Example 6.4 Nyquist Diagram
%Third-order type one system
num=[1];
den=conv([1 0],[1 2 4]);
nyquist(num,den);
axis([-0.6 0.1 -0.6 0.1]);
```

Note that to obtain a reasonable diagram, it is usually necessary for the user to define the scales of the *x* and *y* axes using the axis command. The script file *examp64a.m* produces the Bode gain diagrams for the same system when $K = 4$ and 8, see Figure 6.23(a).

*Filename: examp64a.m*

```
%Example 6.4(a)
G(s)=K/s(s^2+2s+4)
%Creates Bode Gain Diagrams for K=4 and 8
clf
num=[4];
den=conv([1 0],[1 2 4]);
w=logspace(-1,1,200);
mag,phase,w]=bode(num,den,w);
semilogx(w,20*log10(mag));
num=[8];
[mag1,phase1,w]=bode(num,den,w);
semilogx(w,20*log10(mag),w,20*log10(mag1)),grid;
xlabel('Frequency (rad/s)'),ylabel ('Gain (dB)')
```

Using the command margin (mag,phase,w) gives

(a) Bode gain and phase diagrams showing, as a vertical line, the gain and phase margins.
(b) a print-out above the plots of the gain and phase margins, and their respective frequencies.

This is illustrated by running *examp64b.m*, which confirms the values given in Figure 6.23, when $K = 4$.

*Filename: examp64b.m*

```
%Example 6.4(b)
%G(s)=K/s(s^2+2s+4)
%Creates a Bode Gain and Phase Diagrams for K=4
%Determines Gain and Phase Margins
clf
num=[4];
den=conv([1 0],[1 2 4]);
w=logspace(-1,1,200);
[mag,phase,w]=bode(num,den,w);
margin(mag,phase,w);
xlabel('Frequency(rad/s)');
```

Script file *fig627.m* produces the Nichols chart for Example 6.4 when $K = 4$, as illustrated in Figure 6.27. The command `ngrid` produces the closed-loop magnitude and phase contours and axis provides user-defined axes. Some versions of MATLAB appear to have problems with the `nichols` command.

*Filename: fig627.m*

```
%Example 6.4 as displayed in Figure 6.27
%G(s)=K/s(s^2+2s+4) where K=4
%Nichols Chart
num=[4];
den=conv([1 0],[1 2 4]);
nichols(num,den);
ngrid;
axis([-210 0 -30 40]);
```

Running script file *fig629.m* will produce the closed-loop frequency response gain diagrams shown in Figure 6.29 for Example 6.4 when $K = 3.8$ and 3.2 (value of $K$ for best flatband response).

*Filename: fig629.m*

```
%Example 6.4 as displayed in Figure 6.29
%G(s)=K/s(s^2+2s+4)
%Creates closed-loop Bode Gain Diagrams for K=3.8 and 3.2
%Prints in Command Window Mp,k,wp and bandwidth
clf
num=[3.8];
den=conv([1 0],[1 2 4]);
w=logspace(-1,1,200);
[numcl,dencl]=cloop(num,den);
[mag,phase,w]=bode(numcl,dencl,w);
[Mp,k]=max(20*log10(mag))
wp=w(k)
n=1;
while 20*log10(mag(n))>=-3,n=n+1; end
bandwidth=w(n)
num=[3.2];
[num2cl,den2cl]=cloop(num,den);
[mag1,phase1,w]=bode(num2cl,den2cl,w);
semilogx(w,20*log10(mag),w,20*log10(mag1)),grid;
xlabel('Frequency (rad/s)'),ylabel ('Gain)dB)')
```

The command `cloop` is used to find the closed-loop transfer function. The command `max` is used to find the maximum value of 20 $\log_{10}$ (mag), i.e. $M_p$ and the frequency at which it occurs i.e. $\omega_p = \omega(k)$. A `while` loop is used to find the $-3$ dB point and hence `bandwidth=w(n)`. Thus, in addition to plotting the closed-loop frequency response gain diagrams, *fig629.m* will print in the command window:

```
»
Mp=
   3.1124
k=
   121
wp=
   1.6071
bandwidth=
   2.1711
```

### Case study

*Example 6.6*
This is a laser guided missile with dynamics

$$G(s)H(s) = \frac{20}{s^2(s+5)}$$

The missile is to have a series compensator (design one) of the form

$$G(s) = \frac{K(1+s)}{(1+0.25s)}$$

Figure 6.34 is generated using *fig634.m* and shows the Nichols Chart for the uncompensated system. Curve (a) is when the compensator gain $K = 1$, and curve (b) is when $K = 0.537$ (a gain reduction of 5.4 dB).

*Filename: fig634.m*

```
%Nichols Chart for Case Study Example 6.6
%Lead Compensator Design One, Figure 6.34
clf
%Uncompensated System
num=[20];
den=[1 5 0 0];
w=logspace(-1,1,200);
[mag,phase,w]=nichols(num,den,w);
%Compensator Gain set to unity
num1=[20 20];
den1=conv([0.25 1],[1 5 0 0]);
[mag1,phase1,w]=nichols(num1,den1,w);
%Compensator Gain reduced by 5.4dB
num2=[10.74 10.74];
den2=conv([0.25 1],[1 5 0 0]);
[mag2,phase2,w]=nichols(num2,den2,w);
plot(phase,20*log10(mag),phase1,20*log10(mag1),
phase2,20*log10(mag2))
ngrid;
axis([-240 -60 -20 30]);
```

## A1.7   Tutorial 6: Digital control system design

This tutorial looks at the application of MATLAB to digital control system design, using the problems in Chapter 7 as design examples.

*Example 7.3*
To obtain the *z*-transform of a first-order sampled data system in cascade with a zero-order hold (zoh), as shown in Figure 7.10.

*Filename: examp73.m*

```
%Example 7.3 Transfer Function to z-Transform
%Continuous and Discrete Step Response
num=[1];
den=[1 1];
Ts=0.5;
[numd,dend]=c2dm(num,den,Ts,'zoh');
printsys(num,den,'s')
printsys(numd,dend,'z')
subplot(121), step(num,den);
subplot(122), dstep(numd,dend);
```

The continuous to discrete command c2dm employs a number of conversion methods in the last term of the right-hand argument. These include

| | |
|---|---|
| 'zoh' | zero-order hold |
| 'foh' | first-order hold |
| 'tustin' | tustin's rule (see equation 7.102) |

The command subplot (mnp) creates an m-by-n array of equal sized graphs, the argument *p* indicates the current graph. Thus subplot (12*p*) produces a single row, two column array (i.e. side by side graphs). When *p* = 1, the left-hand graph is produced and when *p* = 2, the right-hand one is produced. Running *examp73.m* generates step response plots of both continuous and discrete systems, with the following text in the command window

```
»
num/den=
    1
   ___
   s+1
num/den=
    0.39347
   _____
   z-0.60653
```

*Example 7.4*
This is a closed-loop digital control system as shown in Figure 7.14, with a plant transfer function

$$G_{\mathrm{p}}(s) = \frac{1}{s(s+2)}$$

*Filename: examp74.m*

```
%Example 7.4 Open and Closed-Loop Pulse Transfer Functions
%Discrete Step Response
```

```
num=[1];
den=conv[1  0],[1  2];
Ts=0.5;
[numd,dend]=c2dm(num,den,Ts,'zoh');
[numc1d,denc1d]=cloop(numd,dend);
printsys(num,den,'s')
printsys(numd,dend,'z')
printsys(numc1d,denc1d,'z')
dstep(numc1d,denc1d);
```

Note that the command `cloop` works with both continuous and discrete systems. Running *examp74.m* will produce a step response plot of the closed-loop discrete system, and the open and closed-loop pulse transfer functions will be written in the command window

```
»
num/den=
   1
 ─────
 s^2+2s
num/den=
```

$$\frac{0.09197z+0.06606}{z^2-1.3679z+0.36788}$$  %see equation (7.53)

```
num/den=
```

$$\frac{0.09197z+0.06606}{z^2-1.2759z+0.43394}$$  %see equation (7.55)

The script file *examp75.m* simulates the Jury stability test undertaken in Example 7.5. With the controller gain $K$ in Example 7.5 (Figure 7.14) set to 9.58 for marginal stability see equation (7.75), the roots of the denominator of the closed-loop pulse transfer function are calculated, and found to lie on the unit circle in the $z$-plane.

*Filename: examp75.m*

```
%Example 7.5 Stability in the z-plane
K=9.58
num=[K];
den=conv([1  0],[1  2]);
Ts=0.5;
[numd,dend]=c2dm(num,den,Ts,'zoh');
[numc1d,denc1d]=cloop(numd,dend);
printsys(numc1d,denc1d,'z')
roots(denc1d)
```

Running *examp75.m* produces command window text

```
»
K=
   9.5800
num/den=
```

$$\frac{0.88107z+0.63286}{z^2-0.48681z+1.0007}$$

```
ans=
   0.2434+0.9703i          %lies on unit circle i.e. √RP²+IP²=1
   0.2434-0.9703i          %see equation (7.87)
```

Example 7.6 constructs the root-locus in the *z*-plane for the digital control system in Example 7.4 (Figure 7.14).

*Filename: examp76.m*

```
%Example 7.6 Root Locus Analysis in the z-plane
num=[1];
den=conv([1 0],[1 2]);
Ts=0.5;
[numd,dend]=c2dm(num,den,Ts,'zoh');
rlocus(numd,dend);
axis('square')
zgrid;
k=rlocfind(numd,dend)
```

Note that `rlocus` and `rlocfind` works for both continuous and discrete systems. The statement `'square'` provides square axes and so provides a round unit circle. The command `zgrid` creates a unit circle together with contours of constant natural frequency and damping, within the unit circle. When *examp76.m* has been run, using `rlocfind` at the MATLAB prompt allows points on the loci to be selected and values of *K* identified (see Figure 7.20)

```
»Select a point in the graphics window
selected_point=
   0.2212+0.9591i
k=
   9.7078
»k=rlocfind(numd,dend)
Select a point in the graphics window
selected_point=
  -2.0876-0.0117i
k=
   60.2004
k = rlocfind(numd,dend)
Select a point in the graphics window
selected_point=
  -1.0092-0.0117i
k=
   103.3290
```

*Example 7.7*

A laser guided missile has dynamics

$$G(s)H(s) = \frac{20}{s^2(s+5)}$$

and a compensator

$$G_c(s) = \frac{0.8(1+s)}{(1+0.0625s)}$$

Script file *examp77.m* plots the closed-loop step responses of both the continuous system and discrete system (see Figure 7.21). In the latter case the plant pulse transfer function uses `zoh`, and the compensator is converted into discrete form using

Tustin's rule. **Subplot (211)** and **(212)** creates a 2 row, single column plot matrix (i.e. produces 2 plots, one beneath the other).

*Filename: examp77.m*

```
%Example 7.7 Digital Compensator using Tustin's Rule
%Laser Guided Missile
num=[20];
den=conv([1 0 0],[1 5]);
Ts=0.1;
ncomp=0.8*[1 1];
dcomp=[0.0625 1];
[nol,dol]=series(ncomp,dcomp,num,den);
[ncl,dcl]=cloop(nol,dol);
[numd,dend]=c2dm(num,den,Ts,'zoh');
[ncomd,dcomd]=c2dm(ncomp,dcomp,Ts,'tustin');
printsys(num,den,'s')
printsys(numd,dend,'z')
printsys(ncomp,dcomp,'s')
printsys(ncomd,dcomd,'z')
[nold,dold]=series(ncomd,dcomd,numd,dend);
[ncld,dcld]=cloop(nold,dold);
subplot(211), step(ncl,dcl);
subplot(212), dstep(ncld,dcld);
```

The open continuous transfer functions and pulse transfer functions for the plant and compensator are printed in the command window

```
»
num/den=
      20
   ---------
   s^3+5s^2
num/den=
   0.0029551z^2+0.010482z+0.002302            %see equation (7.108)
   ----------------------------------
   z^3-2.6065z^2+2.2131z-0.60653
num/den=
    0.8s+0.8
   ----------
   0.0625s+1
num/den=
   7.4667z-6.7556                             %see equation (7.110)
   --------------
     z-0.11111
```

*Example 7.8*
Here pole placement is used to design a digital compensator that produces exactly the step response of the continuous system.

*Filename: examp78.m*

```
%Example 7.8 Digital Compensator Design using Pole Placement
K=0.336
numk=K*[3];
num=[3];
```

```
den=conv([1 0],[1 1]);
Ts=0.5;
[ncl,dcl]=cloop(numk,den);
ncomd=0.327*[1-0.6065];
dcomd=[1-0.519];
[numd,dend]=c2dm(num,den,Ts,'zoh');
[nold,dold]=series(ncomd,dcomd,numd,dend);
[ncld,dcld]=cloop(nold,dold);
printsys(num,den,'s')
printsys(numd,dend,'z')
printsys(ncomd,dcomd,'z')
printsys(ncl,dcl,'s')
printsys(ncld,dcld,'z')
subplot(211), step(ncl,dcl);
subplot(212), dstep(ncld,dcld);
```

The continuous and discrete closed-loop systems are shown in Figures 7.22(a) and (b). The digital compensator is given in equation (7.128). Script file *examp78.m* produces the step response of both systems (Figure 7.25) and prints the open and closed-loop continuous and pulse transfer functions in the command window

```
»
K=
    0.3360
num/den=
    3
  ───────
  s^2+s
num/den=
    0.31959z+0.27061          %see equation (7.115)
  ──────────────────────
  z^2-1.6065z+0.60653
num/den=
  0.327z-0.19833              %see equation (7.128)
  ──────────────
    z-0.519
num/den=
      1.008
  ─────────────
  s^2+s+1.008
num/den=
  0.10451z^2+0.025107z-0.053669
  ─────────────────────────────
  z^3-2.021z^2+1.4654z-0.36846
```

## A1.8 Tutorial 7: State-space methods for control system design

This tutorial looks at how MATLAB commands are used to convert transfer functions into state-space vector matrix representation, and back again. The discrete-time response of a multivariable system is undertaken. Also the controllability and observability of multivariable systems is considered, together with pole placement design techniques for both controllers and observers. The problems in Chapter 8 are used as design examples.

*Example 8.4*

This converts a transfer function into its state-space representation using tf2ss(num, den) and back again using ss2tf(A,B,C,D,iu) when iu is the ith input u, normally 1.

*Filename: examp84.m*

```
%Example 8.4 transfer function to state space representation
%And back again
num=[4];
den=[1 3 6 2];
printsys(num,den,'s')
[A,B,C,D]=tf2ss(num,den)
[num1,den1]=ss2tf(A,B,C,D,1);
printsys(num1,den1,'s')
condition_number=cond(A)
```

The print-out in the command window is

```
»examp84
num/den=.

        4
 ─────────────────
 s^3+3s^2+6s+2
A=
  -3  -6  -2
   1   0   0
   0   1   0
B=
   1
   0
   0
C=
   0   0   4
D=
   0
num/den=

   4.441e-016s^2+5.329e-015s+4
  ─────────────────────────────
        s^3+3s^2+6s+2
condition_number=
   24.9599
```

Comparing the output with equation (8.35) it will be noticed that the top and bottom rows of **A** have been swapped, which means that the state variables $x_1$ and $x_3$ have also been exchanged. This means that if the user is expecting the state variables to represent particular parameters (say position, velocity and acceleration), then the rows of **A** and **B**, and the columns of **C** might have to be re-arranged.

The conversion from state-space to transfer function has produced some small erroneous numerator terms, which can be neglected. These errors relate to the condition of **A**, and will increase as the condition number increases.

Example 8.8 uses c2d to calculate the discrete-time state and control transition matrices $\mathbf{A}(T)$ and $\mathbf{B}(T)$ as given by equations (8.78), (8.80), (8.82) and (8.85). The matrix-vector difference equation (8.76) is then used to calculate the first five discrete values of the state variables when responding to a unit step input.

*Filename: examp88.m*

```
%Calculate state and control transition matrices
A=[0  1;-2  -3]
B=[0;1]
Ts=0.1
[AT,BT]=c2d(A,B,Ts)
%Compute discrete step response
kT=0
x=[0;0]
u=1
for i=1:5
xnew=AT*x+BT*u;      %Matrix-vector difference equation (8.76)
kT=kT+Ts
x=xnew
end
```

The command window text is

```
»examp88
A=
     0    1
    -2   -3
B=
     0
     1
Ts=
    0.1000
AT=

    0.9909   0.0861
   -0.1722   0.7326

BT=
    0.0045
    0.0861
kT=
     0
x=
     0
     0
u=
     1
kT=
    0.1000
x=
    0.0045
    0.0861
kT=
    0.2000
x=
    0.0164
    0.1484
```

```
kT=
   0.3000
x=
   0.0336
   0.1920
kT=
   0.4000
x=
   0.0543
   0.2210
kT=
   0.5000
x=
   0.0774
   0.2387
»
```

The for-end loop in *examp88.m* that employs equation (8.76), while appearing very simple, is in fact very powerful since it can be used to simulate the time response of any size of multivariable system to any number and manner of inputs. If **A** and **B** are time-varying, then $\mathbf{A}(T)$ and $\mathbf{B}(T)$ should be calculated each time around the loop. The author has used this technique to simulate the time response of a 14 state-variable, 6 input time-varying system. Example 8.10 shows the ease in which the controllability and observability matrices **M** and **N** can be calculated using ctrb and obsv and their rank checked.

*Filename: examp810.m*

```
%Example 8.10
%Controllability and observability
A=[-2  0;3  -5]
B=[1;0]
C=[1  -1]
D=[0]
M=ctrb(A,B)
rank_of_M=rank(M)
system_order=length(A)
N=(obsv(A,C))'
Rank_of_N=rank(N)
```

The command window will display

```
»examp810
A=
   -2   0
    3  -5
B=
    1
    0
C=
    1  -1
D=
    0
```

```
M=
   1  -2
   0   3
rank_of_M=
   2
system_order=
   2                    %rank of M = system order. System completely
                        %controllable.
N=
   1  -5
  -1   5
rank_of_N=
   1                    %rank of N< system order. System is unobservable.
```

Example 8.11 uses Ackermann's formula (`acker`) to calculate the elements of a regulator feedback matrix **K**, that places the closed-loop poles at some desired location in the *s*-plane.

*Filename: examp811.m*

```
%Example 8.11
%Regulator design using pole placement
%G(s)=1/s(s+4)
num=1;
den=conv([1 0],[1 4]);
printsys(num,den,'s')
%convert to state space
[A,B,C,D]=tf2ss(num,den);
%Check for controllability;
rank_of_M=rank(ctrb(A,B))
system_order=length(A)
%Enter desired characteristic equation
chareqn=[1 4 4]
%Calculate desired closed-loop poles
desiredpoles=roots(chareqn)
%Calculate feedback gain matrix using Ackermann's formula
K=acker(A,B,desiredpoles)
%Closed-loop state feedback system
Asf=A-B*K;Bsf=B;Csf=C;Dsf=0;
[numsf,densf]=ss2tf(Asf,Bsf,Csf,Dsf);
densf
roots(densf)
```

In *examp811.m*, the closed-loop transfer function is obtained and the roots of the denominator calculated to check that the closed-loop poles are at the desired locations. The output at the command window is

```
»examp811
num/den=
      1
   ------
   s^2+4s
rank_of_M=
   2
system_order{=}        % Rank of M = system order hence system
   2                    % is controllable
```

```
chareqn=
   1  4  4
desiredpoles=
  -2
  -2
K=
   0  4
densf=
   1  4  4
ans=
  -2
  -2        % Actual closed-loop poles = desired closed-loop poles
```

Example 8.12 shows how acker uses the transpose of the **A** and **C** matrices to design a full-order state observer.

*Filename: examp812.m*

```
%Example 8.12
%Full-order observer design using pole placement
A=[0  1;-2  -3]
B=[0;1]
C=[1  0]
D=[0]
%Check for observability;
N=obsv(A,C)
rank_or_N=rank(N)
system_order=length(A)
%Enter desired characteristic equation
chareqn=[1  10  100]
%Calculate desired observer eigenvalues
desired_eigenvalues=roots(chareqn)
%Calculate observer gain matrix using Ackermann's formula
Ke=acker(A',C',desired_eigenvalues)
```

The command window text is

```
» examp812
A=

   0   1
  -2  -2
B=
   0
   1
C=
   1  0
D=
   0
N=
   1  0
   0  1
rank_of_N=
   2
system_order=
   2            % Rank of N = system order hence system
                % is observable
```

```
chareqn=
   1   10   100
desired_eigenvalues=
 -5.0000+8.6603i
 -5.0000-8.6603i
Ke=
   7.0000   77.0000          % Agrees with equation (8.150)
```

Example 8.13 illustrates the design of a regulator combined with a reduced-order state observer.

*Filename: examp813.m*

```
%Example 8.13
%Regulator and reduced-order observer design
%Using pole placement
%G(s)=1/s(s+2)(s+5)
%Input in state-space format directly
A=[0  1  0;0  0  1;0  -10  -7]
B=[0;0;1]
C=[1  0  0]
D=[0]
%Regulator design
%Check for controllability;
rank_of_M=rank(ctrb(A,B))
%Enter desired characteristic equation
chareqn=[1  7  25  15]
%Calculate desired closed-loop poles
desiredpoles=roots(chareqn);
%Calculate feedback gain matrix using Ackermann's formula
K=acker(A,B,desiredpoles)
%Reduced-order observer design
A1E=[1  0]
AEE=[0  1;-10  -7] % See equation (8.163)
obchareqn=[1  63.2  2039.4]
observerpoles=roots(obchareqn);
%Calculate observer matrix using Ackermann's formula
Ke=acker(AEE', A1E', observerpoles)
```

Note that the transposes of the partitioned matrices $A_{1e}$ and $A_{ee}$ from equation (8.163) are used in acker to calculate $K_e$ for the reduced-order state observer. The command window output is

```
»examp813
A=
    0    1    0
    0    0    1
    0  -10   -7
B=
    0
    0
    1
C=
    1    0    0
D=
    0
```

```
chareqn=
   1    7   25   15
K=
   15.0000  15.0000  0           % Regulator feedback matrix
A1E=
   1    0
AEE=
   0    1
  -10   -7
obchareqn=
   1.0e+003*
   0.0010  0.0632  2.0394
Ke=
   1.0e+003*
   0.0562  1.6360              %Reduced-order observer matrix
```

## A1.9    Tutorial 8: Optimal and robust control system design

This tutorial uses the MATLAB Control System Toolbox for linear quadratic regulator, linear quadratic estimator (Kalman filter) and linear quadratic Gaussian control system design. The tutorial also employs the Robust Control Toolbox for multivariable robust control system design. Problems in Chapter 9 are used as design examples.

*Example 9.1*
This example uses the MATLAB command lqr to provide the continuous solution of the reduced matrix Riccati equation (9.25)

*Filename: examp91.m*

```
%Example 9.1
%Continuous Optimal Linear Quadratic Regulator (LQR) Design
A=[0  1;-1  -2]
B=[0;1]
Q=[2  0;0  1]
R=[1]
[K,P,E]=lqr(A,B,Q,R)
```

The output at the command window is

```
»examp91
A=
   0    1
  -1   -2
B=
   0
   1
```

```
Q=
  2 0                   %State weighting matrix, see equation (9.8)
  0 1
R=
  1                     %Control weighting matrix, see equation (9.8)
K=
  0.7321  0.5425        %State Feedback gain matrix, see equations
                        %(9.20) and (9.46)
P=
  2.4037  0.7321        %Riccati matrix, see equation (9.45)
  0.7321  0.5425
E=
  -1.2712+0.3406i       %Closed-loop eigenvalues
  -1.2712-0.3406i
```

*Example 9.2*

This example solves the discrete Riccati equation using a reverse-time recursive process, commencing with $\mathbf{P}(n) = \mathbf{0}$. Also tackled is the discrete state-tracking problem which solves an additional set of reverse-time state tracking equations (9.49) to generate a command vector $\mathbf{v}$.

*Filename: examp92.m*

```
%Example 9.2
%Discrete Solution of Riccati Equation
%Optimal Tracking Control Problem
A=[0  1;-1  -1];
B=[0;1];
Q=[10  0;0  1];
R=[1];
F=[0.9859 -0.27;0.08808  0.76677];
G=[-0.9952  0.01409;-0.04598  -0.08808];
S=[0;0];          %Initialize
T=0;
V=0;
Ts=0.1;
[AD,BD]=c2d(A,B,Ts);
P=[0  0;0  0];
H=BD'*P;
X=Ts*R+H*BD;
Y=H*AD;
K=X\Y;
%Discrete reverse-time solution of the Riccati equations (9.29)
%and (9.30)
%Discrete reverse-time solution of the state tracking equations
%(9.53) and (9.54)
for i=1:200
L=Ts*Q+K'*Ts*R*K;
M=AD-BD*K;
PP1=L+M'*P*M;           %Value of Riccati matrix at time (N-(k+1))T
RIN=[-sin(0.6284*T);0.6*cos(0.6284*T)];
SP1=F*S+G*RIN;
V=-B'*SP1;              %Value of command vector at time (N-(k+1))T
S=SP1;
```

```
T=T+Ts;
P=PP1;
H=BD'*P;
X=Ts*R+H*BD;
Y=H*AD;
K=X\Y;                          %Value of feedback gain matrix at time
                                %(N-(k+1))T
end
```

Checking values in the command window gives

```
»examp92
»A
A=
   0   1
  -1  -1
»B
B=
   0
   1
»Q
Q=
  10   0                        %State weighting matrix
   0   1
»R
R=
   1                            %Control weighting matrix
»Ts
Ts=
   0.1000
»AD
AD=
   0.9952  0.0950               %Discrete-time state transition matrix
  -0.0950  0.9002
»BD
BD=
   0.0048                       %Discrete-time control transition matrix
   0.0950
»K
K=
   2.0658  1.4880               %Discrete-time steady-state feedback gain
                                % matrix, after 200 reverse-time iterations
»P
P=
   8.0518  2.3145               %Steady-state value of Riccati matrix
   2.3145  1.6310
```

The reverse-time process is shown in Figure 9.3. The discrete-time steady-state feed-back matrix could also have been found using lqrd, but this would not have generated the command vector **v**. The forward-time tracking process is shown in Figure 9.4 using $\mathbf{K}(kT)$ and $\mathbf{v}(kT)$ to generate $\mathbf{u}_{opt}(kT)$ in equation (9.55). The script file *kalfilc.m* uses the MATLAB command lqe to solve the continuous linear quadratic estimator, or Kalman filter problem.

*Filename: kalfilc.m*

```
%Continuous Linear Quadratic Estimator (Kalman Filter)
A=[0  1;-1  -2]
B=[0;1]
C=[1  0;0  1]
D=[0]
R=[0.01  0;0  1]
Cd=[0.1  0;0  0.01]
Q=[0.1  0;0  0.1]
[K,P,E]=lqe(A,Cd,C,Q,R)
```

The command window text is

```
»kalfilc
A=
    0    1
   -1   -2
B=
    0
    1
C=
    1   0
    0   1
D=
    0
R=
    0.0100        0        %Measurement noise covariance matrix
         0   1.0000
Cd=
    0.1000        0        %Disturbance matrix
         0   0.0100
Q=
    0.1000        0        %Disturbance noise covariance matrix
         0   0.1000
K=
    0.1136  -0.0004        %Kalman gain matrix
   -0.0435   0.0002
P=
    0.0011  -0.0004        %Estimation error covariance matrix
   -0.0004   0.0002
E=
   -1.0569+0.2589i         %Closed-loop estimator eigenvalues
   -1.0569-0.2589i
```

The script file *kalfild.m* solves, in forward-time, the discrete solution of the Kalman filter equations, using equations (9.74), (9.75) and (9.76) in a recursive process. The MATLAB command `lqed` gives the same result.

*Filename: kalfild.m*

```
%Discrete Linear Quadratic Estimator (Kalman Filter)
%The algorithm uses discrete transition matrices A(T) and Cd(T)
A=[0  1;-2  -3]
```

```
Cd=[0.1 0;0 0.1]
C=[1 0;0 1]
ID=eye(2);
Ts=0.1
[AT,CD]=c2d(A,Cd,Ts)
R=[0.01 0;0 1.0]
Q=[0.1 0;0 0.1]
%Discrete solution of Kalman filter equations
%Initialize
P1=ID;          %Initial covariance matrix=identity matrix
P2=(AT*P1*AT')+(CD*Q*CD');
X=P2*C';
Y=(C*P2*C')+R;
K=X/Y;
P3=(ID-(K*C))*P2;
%Solve recursive equations
for i=1:20
P1=P3;
P2=(AT*P1*AT')+(CD*Q*CD');
X=P2*C';
Y=(C*P2*C')+R;
K=X/Y;
P3=(ID-(K*C))*P2;
end
```

Checking results in the command window

```
»kalfild
A=
   0   1
  -2  -3
Cd=
   0.1000        0      %Disturbance matrix
        0   0.1000
C=
   1   0
   0   1
Ts=
   0.1000
AT=
   0.9909  0.0861       %Discrete-time state transition matrix
  -0.1722  0.7326
CD=
   0.0100  0.0005       %Discrete-time disturbance transition
  -0.0009  0.0086       %matrix

R=
   0.0100        0      %Measurement noise covariance matrix
        0   1.0000
Q=
   0.1000        0      %Disturbance noise covariance matrix
        0   0.1000
»K
K=
   0.0260  -0.0002      %Kalman gain matrix after 21 iterations
  -0.0181   0.0002
```

## Case study

*Example 9.3*

This is a china clay band drying oven. The state and output variables are burner temperature, dryer temperature and clay moisture content. The control parameters are burner gas supply valve angle and clay feed-rate. A linear quadratic Gaussian control strategy is to be implemented. Script file *examp93.m* calculates the optimal feedback control matrix **K** and also the Kalman gain matrix $\mathbf{K}_e$ using the recursive equation (9.99).

*Filename: examp93.m*

```
%Linear Quadratic Gaussian (LQG) Design
%Case Study Example 9.3 Clay Drying Oven
%Optimal Controller
A=[-0.02128  0  0;0.0006  -0.005  0;0  -0.00038  -0.00227]
B=[8.93617;0;0]
Cd=[0.1  0  0;0  0.1  0;0  0  0.00132]
C=[1  0  0;0  1  0;0  0  1]
D=[0]
Ts=2;
[AT, BT] =c2d(A,B, Ts)
Q=[0  0  0;0  0.5  0;0  0  20]
R=[1]
[K,P,E]=lqr(A,B,Q,R)
%Kalman Filter
Re=[0.01  0  0;0  0.01  0;0  0  7.46]
Qe=[0.1  0  0;0  0.1  0;0  0  0.1]
[AT,CD]=c2d(A,Cd,Ts)
%Initialize
%Implement equations (9.99)
ID=eye(3);
P1=ID; %Initial covariance matrix=identity matrix
P2=(AT*P1*AT')+(CD*Qe*CD');
X=P2*C';
Y=(C*P2*C')+Re;
Ke=X/Y;
P3=(ID-(Ke*C))*P2;
%Solve recursive equations
for i=1:19
P1=P3;
P2=(AT*P1*AT')+(CD*Qe*CD');
X=P2*C';
Y=(C*P2*C')+Re;
Ke=X/Y;
P3=(ID-(Ke*C))*P2;
End
Ke
P3
```

The output at the command window is

```
»examp93
A=
```

$$
B =
\begin{array}{ccc}
-0.0213 & 0 & 0 \\
0.0006 & -0.0050 & 0 \\
0 & -0.0004 & -0.0023 \\
8.9362 & & \\
0 & & \\
0 & &
\end{array}
$$

Cd=

$$
\begin{array}{ccc}
0.1000 & 0 & 0 \\
0 & 0.1000 & 0 \\
0 & 0 & 0.0013
\end{array}
$$
%Disturbance matrix, equation

C=

$$
\begin{array}{ccc}
1 & 0 & 0 \\
0 & 1 & 0 \\
0 & 0 & 1
\end{array}
$$

D=

0

AT=

$$
\begin{array}{ccc}
0.9583 & 0 & 0 \\
0.0012 & 0.9900 & 0 \\
0.0000 & -0.0008 & 0.9955
\end{array}
$$
%A(T), Ts=2 seconds

BT=

$$
\begin{array}{c}
17.4974 \\
0.0105 \\
0.0000
\end{array}
$$
%B(T), equation (9.86)

Q=

$$
\begin{array}{ccc}
0 & 0 & 0 \\
0 & 0.5000 & 0 \\
0 & 0 & 20.0000
\end{array}
$$
%State weighting matrix

R=

1          %Control weighting matrix

K=

0.0072  0.6442  −1.8265  %Feedback gain matrix,
%equation (9.92)

P=

$$
1.0e+003*
$$
$$
\begin{array}{ccc}
0.0000 & 0.0001 & -0.0002 \\
0.0001 & 0.0108 & -0.0300 \\
-0.0002 & -0.0300 & 3.6704
\end{array}
$$
%Riccati matrix, equation
%(9.91)

E=

−0.0449+0.0422i          %Closed-loop eigenvalues,
%equation (9.93)

−0.0449−0.0422i
−0.0033

Re=

$$
\begin{array}{ccc}
0.0100 & 0 & 0 \\
0 & 0.0100 & 0 \\
0 & 0 & 7.4600
\end{array}
$$
%Measurement noise covariance
%matrix, equation (9.100)

Qe=

$$
\begin{array}{ccc}
0.1000 & 0 & 0 \\
0 & 0.1000 & 0 \\
0 & 0 & 0.1000
\end{array}
$$
%Disturbance noise covariance
%matrix, equation(9.101)

CD=

$$
\begin{array}{ccc}
0.1958 & 0 & 0 \\
0.0001 & 0.1990 & 0
\end{array}
$$
%Discrete-time disturbance
%transition matrix, equation
%(9.95)

```
        0.0000  −0.0001  0.0026
Ke=
        0.4408   0.0003  0.0000    %Kalman gain matrix after 20
                                   %iterations,
        0.0003   0.4579  0.0000    %equation (9.102)
        0.0000  −0.0006  0.0325    % and Figure 9.16
P3=
        0.0044   0.0000  0.0000    %Estimation error covariance
                                   %matrix,
        0.0000   0.0046  0.0000    %after 20 iterations,
        0.0000   0.0000  0.2426    %equation (9.102)
```

The implementation of the LQG design is shown in Figures 9.12 through to 9.17.

*Example 9.6*
This is a multivariable robust control problem that calculates the optimal $H_\infty$ controller. The MATLAB command hinfopt undertakes a number of iterations by varying a parameter $\gamma$ until a best solution, within a given tolerance, is achieved.

*Filename: examp96.m*

```
%Example 9.6
%Multivariable robust control using H infinity
%Singular value loop shaping using the weighted mixed
%sensitivity approach
nug=200;
dng=[1  3  102  200];
[ag,bg,cg,dg]=tf2ss(nug,dng);
ss_g=mksys(ag,bg,cg,dg);
w1=[1  100;100  1];
nw1neg=[100  1];
dw1neg=[1  100];
w2=[1;1];
w3=[100  1;1  100];
nw3neg=[1  100];
dw3neg=[100  1];
[TSS_1]=augtf(ss_g,w1,w2,w3);
[ap,bp,cp,dp]=branch(TSS_);
[gamopt,ss_f,ss_cl]=hinfopt(TSS_,1);
[acp,bcp,ccp,dcp]=branch(ss_f);
[acl,bcl,ccl,dcl]=branch(ss_cl);
[numcp,dencp]=ss2tf(acp,bcp,ccp,dcp,1);
printsys(nug,dng,'s')
printsys(nw1neg,dw1neg,'s')
printsys(nw3neg,dw3neg,'s')
w=logspace(-3,3,200);
%[mag,phase,w]=bode(nw1neg,dw1neg,w);
%[mag1,phase1,w]=bode(nw3neg,dw3neg,w);
%semilogx(w,20*log10(mag),w,20*log10(mag1)),grid;
[sv,w]=sigma(ss_g);
%[sv,w]=sigma(ss_cl);
%[sv,w]=sigma(ss_f);
semilogx(w,20*log10(sv)),grid;
ag
bg
```

```
cg
dg
acp
bcp
ccp
dcp
printsys(numcp,dencp,'s')
```

The command mksys packs the plant state-space matrices **ag, bg, cg** and **dg** into a tree structure ss_g.

The command augtf augments the plant with the weighting functions as shown in Figure 9.31. The branch command recovers the matrices **ap, bp, cp** and **dp** packed in TSS_. The hinfopt command produces the following output in the command window

| No | Gamma | D11<=1 | P-Exist | P>=0 | S-Exist | s>0 | lam(PS)<1 | C.L. |
|----|-------|--------|---------|------|---------|-----|-----------|------|
| 1 | 1.0000e+000 | OK | OK | FAIL | OK | OK | OK | UNST |
| 2 | 5.0000e−001 | OK | OK | FAIL | OK | OK | OK | UNST |
| 3 | 2.5000e−001 | OK | OK | FAIL | OK | OK | OK | UNST |
| 4 | 1.2500e−001 | OK | OK | OK | OK | OK | OK | STAB |
| 5 | 1.8750e−001 | OK | OK | FAIL | OK | OK | OK | UNST |
| 6 | 1.5625e−001 | OK | OK | FAIL | OK | OK | OK | UNST |
| 7 | 1.4063e−001 | OK | OK | FAIL | OK | OK | OK | UNST |
| 8 | 1.3281e−001 | OK | OK | FAIL | OK | OK | OK | UNST |
| 9 | 1.2891e−001 | OK | OK | OK | OK | OK | OK | STAB |
| 10 | 1.3086e−001 | OK | OK | FAIL | OK | OK | OK | UNST |
| 11 | 1.2988e−001 | OK | OK | OK | OK | OK | OK | STAB |

Iteration no. 11 is your best answer under the tolerance: 0.0100.

After eleven iterations, hinfopt identifies that $\gamma$ in equation (9.176) has a best value of 0.13. The command sigma calculates the data for a singular value Bode diagram as shown in Figures 9.32, 9.34 and 9.35. Other information printed in the command window is given below

num/den=

$$\frac{200}{s^3+3s^2+102s+200}$$   %Plant transfer function

num/den=

$$\frac{100s+1}{s+100}$$   %$W_s^{-1}(s)$

num/den=

$$\frac{s+100}{100s+1}$$   %$W_T^{-1}(s)$

ag=
```
   −3  −102  −200
    1     0     0
    0     1     0
```

```
bg=
     1
     0
     0
cg=
     0   0   200
dg=
     0
acp=
    −0.0100  −0.0022   0.0039      0.0148      0.1367      %Controller
                                                           %matrices,
    −0.0086  −7.7634   21.6533     37.1636     621.8932    %See equation
                                                           %(9.178)
    −0.0461  −2.0317  −3.1477     −2.2338     −81.8751
     0.4353   0.1067   8.8071    −102.2531    −37.7767
     0.6756   0.1970   13.5575    −3.7282    −162.5129
bcp=
    −7.9756
     0.0906
     0.5820
    −8.8009
   −13.6595
ccp=
    −0.0861   0.1698   0.1213     −0.6544    −11.1708
dcp=
     0
num/den=
```

$$\frac{159.1184s^4+16389.1905s^3+63965.5785s^2+1654830.8855s+3182367.0874}{s^5+275.687s^4+20439.8141s^3+324835.0198s^2+3782679.1393s+37794.3283}$$

```
                    %Controller transfer function,
                    %See equation (9.179)
```

## A1.10   Tutorial 9: Intelligent control system design

This tutorial uses the MATLAB Control System Toolbox, the Fuzzy Logic Toolbox and the Neural Network Toolbox. Problems in Chapter 10 are used as design examples.

*Example 10.3*
This is the inverted pendulum control problem and, as a benchmark, is initially solved as a multivariable control problem, using pole placement (Ackermann's formula) to calculate the feedback gain matrix in *examp103.m*.

*Filename: examp103m*

```
%Regulator design using pole placement
%Inverted pendulum problem
A=[0  1  0  0;9.81  0  0  0;0  0  0  1;−3.27  0  0  0]
B=[0;−0.667;0;0.889]
C=[1  0  0  0]
D=0
%Check for controllability;
rank_of_M=rank(ctrb(A,B))
```

```
%Enter desired characteristic equation
chareqn=[1 12 72 192 256]
%Calculate desired closed-loop poles
desiredpoles=roots(chareqn)
%Calculate feedback gain matrix using Ackermann's formula
K=acker(A,B,desiredpoles)
```

The output at the command window is

```
» examp103
A=
          0  1.0000        0        0
     9.8100        0        0        0
          0        0        0  1.0000
    -3.2700        0        0        0
B=
          0
    -0.6670
          0
     0.8890
C=
          1        0        0        0
D=
          0
rank_of_M=
          4
chareqn=
     1  12  72  192  256
desiredpoles=
     -4.0000+4.0000i
     -4.0000-4.0000i
     -2.0000+2.0000i
     -2.0000-2.0000i
K=
    -174.8258  -57.1201  -39.1437  -29.3578
```

The state-variable feedback solution was implemented in SIMULINK as shown in Figure A1.4.

*Inverted Pendulum, Fuzzy Logic Controller Design*:  The fuzzy logic controller used for the inverted pendulum problem has four input windows and one output window as shown in Figure 10.14. One version of the controller has a Mamdani-type rulebase consisting of 11 rules as given in equation (10.52). The MATLAB Fuzzy Inference System (FIS) Editor can be entered by typing at the MATLAB prompt

```
» fuzzy
```

This brings up the main menu screen as shown in Figure A1.5. Double-clicking on the 'theta' icon brings up the membership function editor. Figure A1.6 shows the NB fuzzy set being given the trapizoidal (trapmf) membership function $[-1 \ -1 \ -0.10]$. Other membership functions include gaussmf (Gaussian), trimf (triangular) and gbellmf (generalized bell). For further information type

```
» help fuzzy
```

Returning to the main menu and double-clicking on the centre box entitled 'Pendulum 11' (mamdani) will bring up the FIS rule editor, as shown in Figure A1.7. Highlighted are the antecedents and the consequent of Rule 1.

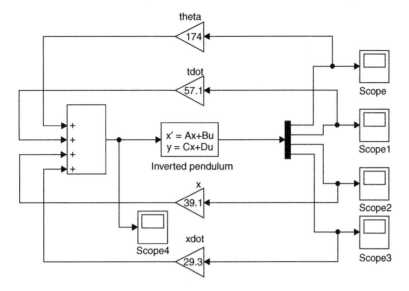

**Fig. A1.4** Inverted pendulum control system design using pole placement

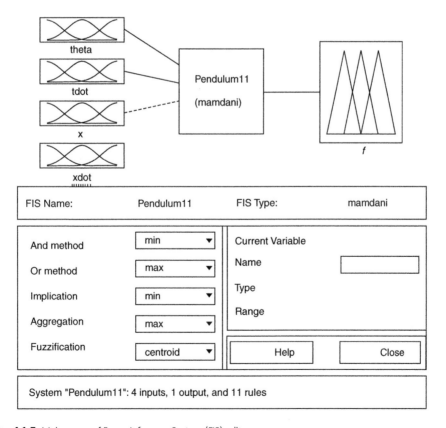

**Fig. A1.5** Main menu of Fuzzy Inference System (FIS) editor.

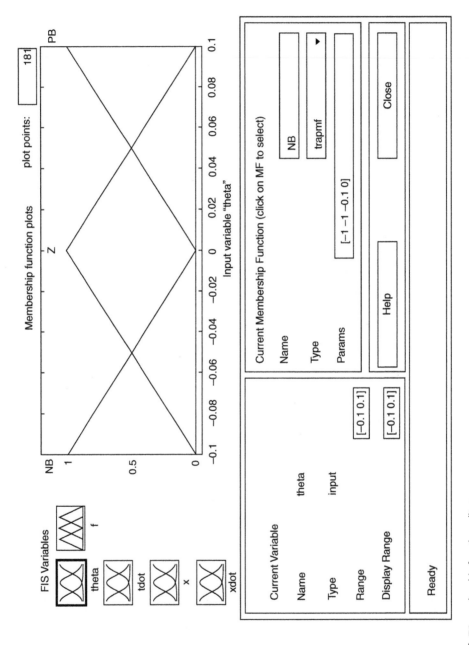

**Fig. A1.6** FIS membership function editor.

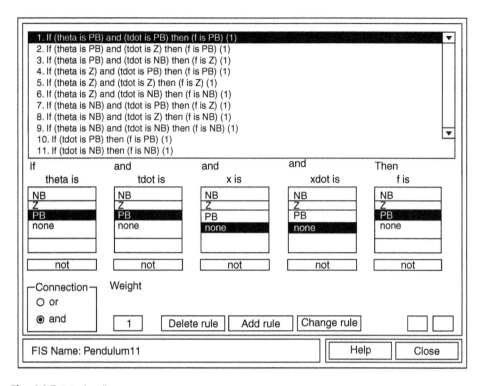

**Fig. A1.7** FIS rule editor.

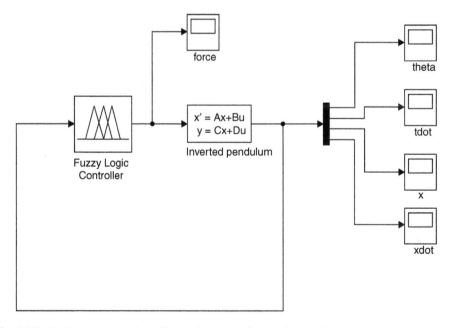

**Fig. A1.8** Simulink implementation of inverted pendulum fuzzy logic control problem.

When editing is complete, the controller is stored as a *.fis* file, in this case *pendulum11.fis*. The inverted pendulum fuzzy logic control problem can now be implemented in SIMULINK as shown in Figure A1.8. The fuzzy logic icon is obtained by opening the fuzzy logic toolbox within the Simulink Library Browser, and dragging it across. Running the simulation as it stands in Figure A1.8 will bring up a MATLAB error because the properties of the fuzzy logic controller have not been defined. At the MATLAB prompt, type

```
» fismat=readfis
```

This will allow you to select from a directory a suitable stored filename, i.e. *pendulum11.fis*. The system will respond with

```
fismat=
name:         'Pendulum11'
type:         'mamdani'
andMethod:    'min'
orMethod:     'max'
defuzzMethod: 'centroid'
impMethod:    'min'
aggMethod:    'max'
input:        [1×4 struct]
output:       [1×1 struct]
rule:         [1×11 struct]
```

This means that the fuzzy logic controller parameters have been placed in the workspace under 'fismat', and that the simulation can now proceed. More details on the properties of the fuzzy logic controller can be found by typing

```
» out=getfis(fismat)
```

The system will respond with

```
Name  = Pendulum11
Type  = mamdani
NumInputs=4
InLabels=
    theta
    tdot
    x
    xdot
NumOutputs=1
OutLabels=
    f
NumRules      =11
AndMethod     =min
OrMethod      =max
ImpMethod     =min
AggMethod     =max
DefuzzMethod =centroid
Out=
Pendulum 11
```

The results of the pole placement and the 11 rule and 22 rule fuzzy logic controllers are compared in Figure 10.15. Figure 10.16 shows the $\theta$–$\dot{\theta}$ control surface, also generated using the FIS Editor.

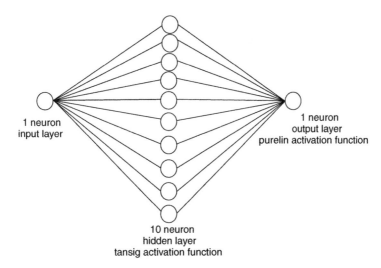

**Fig. A1.9** Demonstration neural network.

*The MATLAB Neural Network Toolbox*: This Toolbox was not used in the Examples given in Chapter 10. For details on the Toolbox, type

» help nnet

To demonstrate how the Toolbox is used, consider a neural network with a structure shown in Figure A1.9.

The script file *neural1.m* trains and runs the network.

*Filename: neural1.m*

```
%Feedforward Back-propagation Neural Network
%Network structure: 1:10(tansig):1(purelin)
%P=Input values
P=[0  1  2  3  4  5  6  7  8];
%T=Target values
T=[0  0.84  0.91  0.14  -0.77  -0.96  -0.28  0.66  0.99];
plot(P,T,'o');
%Use Levenberg-Marquardt back-propagation training
net=newff([0  8],[10  1],{'tansig''purelin'},'trainlm');
Y1=sim(net,P);
plot(P,T,P,Y1,'o');
net.trainParam.epochs=50;
net=train(net,P,T);
Y2=sim(net,P);
plot(P,T,P,Y2,'o');
```

**P** is a vector of inputs and **T** a vector of target (desired) values. The command newff creates the feed-forward network, defines the activation functions and the training method. The default is Levenberg–Marquardt back-propagation training since it is fast, but it does require a lot of memory. The train command trains the network, and in this case, the network is trained for 50 epochs. The results before and after training are plotted.

# Appendix 2
# Matrix algebra

*Matrix*: An $m \times n$ matrix $\mathbf{A}$ is an array of elements with $m$ rows and $n$ columns, and is written as

$$\mathbf{A} = \begin{bmatrix} a_{11} & a_{12} & \cdots & a_{1n} \\ a_{21} & a_{22} & \cdots & a_{2n} \\ \vdots & \vdots & & \vdots \\ a_{m1} & a_{m2} & \cdots & a_{mn} \end{bmatrix} \tag{A2.1}$$

If $m = n$, $\mathbf{A}$ is a square matrix.

*Vector*: A column vector $\mathbf{x}$ is

$$\mathbf{x} = \begin{bmatrix} x_1 \\ x_2 \\ \vdots \\ x_m \end{bmatrix} \tag{A2.2}$$

A row vector $\mathbf{y}$ is

$$\mathbf{y} = [y_1 \; y_2 \; \cdots \; y_n] \tag{A2.3}$$

*Null matrix*: This has all elements equal to zero.

$$\mathbf{B} = \begin{bmatrix} 0 & 0 & \cdots & 0 \\ 0 & 0 & \cdots & 0 \\ \vdots & \vdots & & \\ 0 & 0 & \cdots & 0 \end{bmatrix} \tag{A2.4}$$

*Diagonal matrix*: This is a square matrix with all elements off the diagonal equal to zero.

$$\mathbf{C} = \begin{bmatrix} c_{11} & 0 & 0 \\ 0 & c_{22} & 0 \\ 0 & 0 & c_{33} \end{bmatrix} \tag{A2.5}$$

*Identity matrix*: This is a diagonal matrix with all diagonal elements equal to unity, and is normally denoted by $\mathbf{I}$.

$$\mathbf{I} = \begin{bmatrix} 1 & 0 & 0 \\ 0 & 1 & 0 \\ 0 & 0 & 1 \end{bmatrix} \tag{A2.6}$$

*Symmetric matrix*: This is a square matrix where elements $a_{ij} = a_{ji}$.

$$\mathbf{S} = \begin{bmatrix} 5 & 8 & 9 \\ 8 & 4 & 2 \\ 9 & 2 & 3 \end{bmatrix} \tag{A2.7}$$

## A2.2   Matrix operations

*Transpose of a matrix*: The transpose of a matrix $\mathbf{A}$, denoted by $\mathbf{A}^{\mathrm{T}}$, is formed by interchanging its rows and columns.

$$\mathbf{A} = \begin{bmatrix} 2 & 4 & 5 \\ 3 & 1 & 9 \\ 6 & 8 & 4 \end{bmatrix} \tag{A2.8}$$

$$\mathbf{A}^{\mathrm{T}} = \begin{bmatrix} 2 & 3 & 6 \\ 4 & 1 & 8 \\ 5 & 9 & 4 \end{bmatrix} \tag{A2.9}$$

*Determinant of a matrix*: Given a $2 \times 2$ matrix

$$\mathbf{A} = \begin{bmatrix} a_{11} & a_{12} \\ a_{21} & a_{22} \end{bmatrix} \tag{A2.10}$$

Its determinant is

$$\det \mathbf{A} = \begin{vmatrix} a_{11} & a_{12} \\ a_{21} & a_{22} \end{vmatrix} = a_{11}a_{22} - a_{21}a_{12} \tag{A2.11}$$

*Minors of an Element*: The minor $M_{ij}$ of element $a_{ij}$ of $\det \mathbf{A}$ is the determinant formed by removing the $i$th row and the $j$th column from $\det \mathbf{A}$.

$$\det \mathbf{A} = \begin{vmatrix} 8 & 6 & 2 \\ 4 & 3 & 1 \\ 9 & 5 & 7 \end{vmatrix}$$

$$M_{23} = \begin{vmatrix} 8 & 6 \\ 9 & 5 \end{vmatrix} = -14 \tag{A2.12}$$

*Cofactor of an Element*: The cofactor $C_{ij}$ of element $a_{ij}$ of $\det \mathbf{A}$ is defined as

$$\mathbf{C}_{ij} = (-1)^{(i+j)} M_{ij}$$

In general, the determinant of a square matrix is given by

$$\det \mathbf{A} = \sum_{k=1}^{n} a_{ik} C_{ik} \text{ (expanding along the } i\text{th row)}$$

or

$$\det \mathbf{A} = \sum_{k=1}^{m} a_{kj} C_{kj} \text{ (expanding along the } j\text{th column)} \tag{A2.13}$$

Given

$$\mathbf{A} = \begin{bmatrix} 2 & 4 & 5 \\ 1 & 3 & 0 \\ 6 & 8 & 9 \end{bmatrix}$$

Expanding along the first column gives

$$\det \mathbf{A} = 2 \begin{vmatrix} 3 & 0 \\ 8 & 9 \end{vmatrix} - 1 \begin{vmatrix} 4 & 5 \\ 8 & 9 \end{vmatrix} + 6 \begin{vmatrix} 4 & 5 \\ 3 & 0 \end{vmatrix}$$

$$= 2(27) - 1(-4) + 6(-15)$$

$$= -32 \tag{A2.14}$$

*Singular matrix*: A matrix is singular if its determinant is zero.

*Nonsingular matrix*: A matrix is nonsingular if its determinant is not zero.

*Adjoint of a matrix*: The adjoint of an $n \times n$ matrix is the transpose of the matrix when all elements have been replaced by their cofactors.

$$\text{adj}\,\mathbf{A} = \begin{bmatrix} C_{11} & C_{12} & \cdots & C_{1n} \\ C_{21} & C_{22} & \cdots & C_{2n} \\ \vdots & \vdots & & \vdots \\ C_{n1} & C_{n2} & \cdots & C_{nn} \end{bmatrix}^{\mathrm{T}} \tag{A2.15}$$

*Inverse of a matrix*: An $n \times n$ matrix $\mathbf{A}$ has an inverse $\mathbf{A}^{-1}$, which satisfies the relationship

$$\mathbf{A}^{-1}\mathbf{A} = \mathbf{A}\mathbf{A}^{-1} = \mathbf{I} \tag{A2.16}$$

The inverse of $\mathbf{A}$ is given by

$$\mathbf{A}^{-1} = \frac{\text{adj}\,\mathbf{A}}{\det \mathbf{A}} \tag{A2.17}$$

*Addition and subtraction of matrices*: The sum of two matrices

$$\mathbf{A} + \mathbf{B} = \mathbf{C}$$

is given by

$$a_{ij} + b_{ij} = c_{ij} \tag{A2.18}$$

the difference between two matrices

$$\mathbf{A} - \mathbf{B} = \mathbf{C}$$

is given by

$$a_{ij} - b_{ij} = c_{ij} \tag{A2.19}$$

*Multiplication of matrices*: The product of two matrices

$$\mathbf{AB} = \mathbf{C}$$

is given by

$$c_{ij} = \sum_{k=1}^{n} a_{ik} b_{ki} \tag{A2.20}$$

Multiplication by a constant

$$\mathbf{A} = \begin{bmatrix} a_{11} & a_{12} \\ a_{21} & a_{22} \end{bmatrix}$$

$$k\mathbf{A} = \begin{bmatrix} ka_{11} & ka_{12} \\ ka_{21} & ka_{22} \end{bmatrix} \tag{A2.21}$$

Note that in general, multiplication is not commutative, i.e. $\mathbf{AB} \neq \mathbf{BA}$.

*Trace of a matrix*: The trace of an $n \times n$ matrix is the sum of the diagonal elements of the matrix.

$$\operatorname{tr} \mathbf{A} = a_{11} + a_{22} + \cdots + a_{nn} \tag{A2.22}$$

*Rank of a matrix*: The rank of a matrix is equal to the number of linearly independent rows or columns. The rank can be found by determining the largest square submatrix that is nonsingular.

If

$$\mathbf{A} = \begin{bmatrix} 1 & 2 & 3 \\ 0 & 4 & 1 \\ 1 & 2 & 3 \end{bmatrix} \tag{A2.23}$$

since det $\mathbf{A}$ is zero, the $3 \times 3$ matrix is singular. Consider a submatrix

$$\mathbf{A}_{\text{sub}} = \begin{bmatrix} 1 & 2 \\ 0 & 4 \end{bmatrix} \tag{A2.24}$$

The determinant of $\mathbf{A}_{\text{sub}}$ is 4, hence the rank of $\mathbf{A}$ is the size of $\mathbf{A}_{\text{sub}}$, i.e. 2.

# References and further reading

Ackermann, J. (1972) Der Entwurf Linearer Regelungssysteme im Zustandsraum, *Regelungstechnik und Prozessdatenverarbeitung*, **7**, pp. 297–300.

Anderson, J.A. (1972) A Simple Neural Network Generating an Interactive Memory, *Mathematical Biosciences*, **14**, pp. 197–220.

Anderson, B.D.O. and Moore, J.B. (1979) *Optimal Filtering*, Prentice-Hall, Englewood Cliffs, NJ.

Anderson, B.D.O. and Moore, J.B. (1990) *Optimal Control*, Prentice-Hall, Englewood Cliffs, NJ.

Åström, K.J. and Wittenmark, B. (1984) *Computer Controlled Systems*, Prentice-Hall, Inc., Englewood Cliffs, NJ.

Åström, K.J. and Wittenmark, B. (1989) *Adaptive Control*, Addison-Wesley, Reading, Mass.

Athans, M. (1971) The Role and Use of the Stochastic Linear-Quadratic-Gaussian Problem in Control System Design, *IEEE Trans. on Automatic Control* AC-16, **6**, pp. 529–551.

Atherton, D.P. (1982) *Nonlinear Control Engineering*, Van Nostrand Reinhold, London.

Bellman, R. (1957) *Dynamic Programming*, Princeton University Press, Princeton, NJ.

Bennett, S. (1984) Nicholas Minorsky and the automatic steering of ships, *IEEE Control Systems Magazine*, **4**, pp. 10–15.

Bode, H.W. (1945) *Network Analysis and Feedback Amplifier Design*, Van Nostrand, Princeton, NJ.

Brown, M. and Harris, C. (1994) *Neurofuzzy Adaptive Modelling and Control*, Prentice-Hall International (UK) Hemel Hempstead, UK.

Burns, R.S. (1984) The Automatic Control of Large Ships in Confined Waters, PhD Thesis, University of Plymouth (then Plymouth Polytechnic).

Burns, R.S. (1989) Application of the Riccati Equation in the Control and Guidance of Marine Vehicles. In: *The Riccati Equation in Control, Systems, and Signals*, Pitagora Editrice, Bologna, Italy, pp. 18–23.

Burns, R.S. (1990) The Design, Development and Implementation of an Optimal Guidance System for Ships in Confined Waters, *Proc: Ninth Ship Control Systems Symposium*, Naval Sea Systems Command, Department of the Navy, Bethesda, USA, 9–14 September, **3**, pp. 386–401.

Burns, R.S. (1991) A Multivariable Mathematical Model for Simulating the Total Motion of Surface Ships. In: *Proc. European Simulation Multiconference*, The Society for Computer Simulation International, Copenhagen, Denmark, 17–19 June.

Burns, R.S. (1995) The Use of Artificial Networks for the Intelligent Optimal Control of Surface Ships, *IEEE Journal of Oceanic Engineering*, **20**(1); Special Issue: *Advanced Control & Signal Processing for Oceanic Applications*, **20**(1), pp. 66–72.

Burns, R.S. (1997) The Application of Artificial Intelligence Techniques to Modelling and Control of Surface Ships. In: *Proc. 11th Ship Control Systems Symposium*, Southampton UK, April, **1**, 77–83.

Burns, R.S. (1997) Intelligent Manufacturing, *Journal of Aircraft Engineering and Aerospace Technology*, MCB University Press, **69**(5), pp. 440–446.

Burns, R.S. and Richter, R. (1995) A Neural Network Approach to the Control of Surface Ships, *Journal of Control Engineering Practice*, International Federation of Automatic Control, Elsevier Science, **4**(3), pp. 411–416.

Burns, R.S., Richter, R. and Polkinghorne, M.N. (1995) A Multivariable Neural Network Ship Mathematical Model. In: *Marine Technology and Transportation*, Graczyk, T., Jastrzebski, T., Brebbia, C.A. and Burns R.S. (eds.), Southampton U.K., Computational Mechanics.

Burns, R.S., Sutton, R. and Craven, P.J. (2000) A Multivariable Online Intelligent Autopilot Design Study. In: *The Ocean Engineering Handbook*, El-Hawary, F. (ed.), CRC Press, Boca Raton, Florida, pp. 252–259.

Cadzow, J.A. and Martens, H.R. (1970) *Discrete-Time and Computer Control Systems*, Prentice-Hall, Inc., Englewood Cliffs, N.J.

Chiang, R.Y. (1988) Modern Robust Control Theory, PhD Dissertation, USC.

Chiang, R.Y. and Safonov, M.G. (1992) *Robust Control Toolbox for Use with MATLAB. Users Guide*, MathWorks.

Chu, C.C. (1985) H∞-Optimization and Robust Multivariable Control, PhD Thesis, University of Minnesota, Minneapolis, MN.

Craven, P.J. (1999) Intelligent Control Strategies for an Autonomous Underwater Vehicle, PhD Thesis, Department of Mechanical and Marine Engineering, University of Plymouth, UK.

Craven, P.J., Sutton, R. and Burns, R.S. (1997) Intelligent Course Changing Control of an Autonomous Underwater Vehicle. In: *Twelfth International Conference on Systems Engineering*, Coventry, UK, September, **1**, pp. 159–164.

Demuth, H. and Beale, M. (1993) *Neural Network Toolbox for Use with MATLAB – User's Guide*, The MathWorks, Inc., Natick, Mass.

Dorato, P. (ed.) (1987) *Robust Control*, IEEE Press, New York.

Dorf, R.C. (1992) *Modern Control Systems*, 6th ed., Addison-Wesley, Reading, Mass.

Dove, M.J., Burns, R.S. and Evison, J.L. (1986) The use of Kalman Filters in Navigation Systems – Current Status and Future Possibilities. In: *Proc. of the Int. Conf. on Computer Aided Design, Manf. and Operation in the Marine and Offshore Industries*, Keramidas, G.A. and Murthy, T.K.S. (eds.), Springer-Verlag, Washington, DC, pp. 361–374.

Drew, B.L. and Burns, R.S. (1992) Simulation of Optimal Control and Filtering of an Industrial China Clay Dryer. In: *European Simulation Symposium on Simulation and AI in Computer Aided Techniques*, The Society for Computer Simulation, Dresden, 5–8 November, pp. 531–535.

Dreyfus, S.E. (1962) Variational Problems with Inequality Constraints, *J. Math. Anal. Appl*, **4**, p. 297.

Dutton, K., Thompson, S. and Barraclough, B. (1997) *The Art of Control Engineering*, Addison-Wesley-Longman, Harlow, Essex.

Evans, W.R. (1948) Graphical Analysis of Control Systems, *Transactions of the AIEE*, **67**, pp. 547–551.

Francis, B.A. (1987) *A Course in H∞ Control Theory*, Springer-Verlag, New York.

Franklin, G.F., Powell, J.D. and Workman, M.L. (1990) *Digital Control of Dynamic Systems*, 2nd ed., Addison-Wesley, Menlow Park, CA.

Goldberg, D.E. (1983) Computer-aided gas pipeline operation using genetic algorithms and rule learning (Doctoral dissertation, University of Michigan). *Dissertation Abstracts International*, **44**(10), p. 3174B (University Microfilms No. 8402282).

Goldberg, D.E. (1989) *Genetic Algorithms in Search, Optimization & Machine Learning*, Addison-Wesley Publishing Company, Inc., Reading, Mass.

Grace, A., Lamb, A.J., Little, J.N. and Thompson, C.M. (1992) *Control System Toolbox for Use with MATLAB – User's Guide*, The MathWorks, Inc., Natick, Mass.

Grimble, M.J., Patton, R.J. and Wise, D.A. (1979) The design of dynamic ship positioning control systems using extended Kalman filtering techniques, *IEEE Conference, Oceans '79*, CA, San Diego.

Grimble, M.J. and Johnson, M.A. (1988) *Optimal Control and Stochastic Estimation: Theory and Application, Vols 1 and 2*, John Wiley & Sons, Chichester, UK.

Healey, M. (1967) *Principles of Automatic Control*, The English Universities Press, London.

Hebb, D.O. (1949) *The Organization of Behaviour*, Wiley, New York.

Holland, J.H. (1975) *Adaption in natural and artificial systems*, The University of Michigan Press, Ann Arbor.

Hughs, T.P. (1971) *Elmer Sperry: Inventor and Engineer*, Johns Hopkins Press, Baltimore. pp. 232–233.

James, H.M., Nichols, N.B. and Phillips, R.S. (1947) *Theory of Servomechanisms*, McGraw-Hill, New York.

Jang, J.S.R. (1993) ANFIS: Adaptive Network-based Fuzzy Inference System, *IEEE Transactions on Systems, Man and Cybernetics*, **23**, pp. 665–685.

Johnson, J. and Picton, P. (1995) *Designing Intelligent Machines, Vol. 2, Concepts in Artificial Intelligence*, Butterworth-Heineman (in association with The Open University), Oxford, UK.

Jury, E.I. (1958) *Sampled-Data Control Systems*, John Wiley & Sons, New York.

Kalman, R.E. (1961) On the General Theory of Control Systems. In: *Proceedings of the First International Congress IFAC*, Moscow 1960: *Automatic and Remote Control*, Butterworth & Co., London, pp. 481–492.

Kalman, R.E. and Bucy, R.S. (1961) New Results in Linear Filtering and Prediction Theory, *J. of Basic Eng., Trans. of the Am. Soc. of Mech. Eng.*, pp. 95–108.

Kohonen, T. (1988) *Self-Organization and Associative Memory*, Springer-Verlag, Berlin.

Kuo, B.C. (1980) *Digital Control Systems*, Holt, Rinehart and Winston, Inc., New York.

Kurzweil, R. (1990) *The Age of Intelligent Machines*, M.I.T. Press, Cambridge, Mass.

Laub, A.J. (1979) A Schur Method for Solving Riccati Equations, *IEEE Trans. on Automat. Contr.*, **AC-24**, pp. 913–921.

Lehtomaki, N.A., Sandell, Jr., N.R. and Athans, M. (1981) Robustness Results in Linear-Quadratic Gaussian Based Multivariable Control Designs, *IEEE Trans. on Automat. Contr.*, **AC-26**(1), pp. 75–92.

Leigh, J.R. (1985) *Applied Digital Control*, Prentice-Hall International, Englewood Cliffs, NJ.

Luenberger, D.G. (1964) Observing the State of a Linear System, *IEEE Trans. Military Electronics*, **MIL-8**, pp. 74–80.

Lyapunov, A.M. (1907) Problème général de la stabilité du mouvement. In: *Ann. Fac. Sci.*, Toulouse, **9**, pp. 203–474 (Translation of the original paper published in 1893 in *Comm. Soc. Math. Kharkow*).

MacFarlane, A.G.J. and Kouvaritakis, B. (1977) A design technique for linear multivariable feedback systems, *International Journal of Control*, **25**, pp. 837–874.

Mamdani, E.H. (1976) Advances in linguistic synthesis of fuzzy controllers, *Int. J. Man & Mach. Studies*, **8**(6), pp. 669–678.

Maxwell, J.C. (1868) On Governors. In: *Proceedings of the Royal Society of London*, **16**.

Mayr, O. (1970) *The Origins of Feedback Control*, M.I.T. Press, Cambridge, Mass.

McCulloch, W.S. and Pitts, W.H. (1943) A logical calculus of ideas immanent in nervous activity, *Bulletin of Mathematical Biophysics*, **5**, pp. 115–133.

Merritt, H.E. (1967) *Hydraulic Control Systems*, John Wiley and Sons, New York.

Minorski, N. (1922) Directional stability of automatically steered bodies, *J. American Society of Naval Engineers*, **34**, pp. 280–309.

Minorski, N. (1941) Note on the angular motion of ships. *Trans. American Society of Mech. Eng.*, **63**, pp. 111–120.

Morari, M. and Zafiriou, E. (1989) *Robust Process Control*, Prentice-Hall, Inc., Englewood Cliffs, NJ.

Mościński, J. and Ogonowski, Z. (eds.) (1995) *Advanced Control with MATLAB and SIMU-LINK*, Ellis Horwood, Hemel Hempstead, UK.

Nyquist, H. (1932) Regeneration Theory, *Bell System Technical Journal*, **11**, pp. 126–147.

Ogata, K. (1995) *Discrete-Time Control Systems*, 2nd ed., Prentice-Hall, Inc., Upper Saddle River, NJ.

Ogata, K. (1997) *Modern Control Engineering*, 3nd ed., Prentice-Hall, Inc., Upper Saddle River, NJ.

Payne, H.J. and Silverman, L.M. (1973) On the Discrete Time Algebraic Riccati Equation, *IEEE Trans. Automatic Control*, **AC-18**, pp. 226–234.

Pearson, A.R., Sutton, R., Burns, R.S. and Robinson, P. (2000) A Kalman Filter Approach to Fault Tolerance Control in Autonomous Underwater Vehicles. In: *Proc. 14th International Conference on Systems Engineering*, Coventry, 12–14 September, **2**, pp. 458–463.

Phillips, C.L. and Harbor, R.D. (2000) *Feedback Control Systems*, 4th ed., Prentice-Hall, Inc., Upper Saddle River, NJ.

Polkinghorne, M.N. (1994) A Self-Organising Fuzzy Logic Autopilot for Small Vessels, PhD Thesis, School of Manufacturing, Materials and Mechanical Engineering, University of Plymouth, UK.

Polkinghorne, M.N., Roberts, G.N. and Burns, R.S. (1994) The Implementation of a Fuzzy Logic Marine Autopilot. In: *Proc. IEE Control 94*, Warwick, UK, March **2**, pp. 1572–1577.

Pontryagin, L.S., Boltyanskii, V.G., Gamkrelidze, R.V. and Mishchenko, E.F. (1962) *The Mathematical Theory of Optimal Processes*, John Wiley & Sons, New York.

Postlethwaite, I., Edward J.M. and MacFarlane, A.G.J. (1981) Principal gains and principal phases in the analysis of linear mulltivariable feedback systems, *IEEE Transactions on Automatic Control*, **AC-26**, pp. 32–46.

Raven, F. (1990) *Automatic Control Engineering*, 2nd ed., McGraw-Hill, New York.

Rechenburg, I. (1965) *Cybernetic solution path of an experimental problem*, Royal Aircraft Establishment Translation No. 1122, B.F. Toms (trans.), Farnborough, Hants, Ministry of Aviation, Royal Aircraft Establishment.

Rezevski, G. (ed.) (1995) *Designing Intelligent Machines, Vol. 1, Perception, Cognition and Execution*, Butterworth-Heineman (in association with The Open University), Oxford, UK.

Riccati, J.F. (1724) Animadversiones In Aequationes Differentialis, *Acta Eruditorum Lipsiae*. Re-printed by Bittanti, S. (ed.) (1989) *Count Riccati and the Early Days of the Riccati Equation*, Pitagora Editrice, Bologna, Milano.

Richter, R. (2000) A Predictive Fuzzy-Neural Autopilot for the Guidance of Small Motorised Marine Craft, PhD Thesis, Department of Mechanical and Marine Engineering, University of Plymouth, UK.

Richter, R., Burns, R.S., Polkinghorne, M.N. and Nurse, P. (1997) A Predictive Ship Control using a Fuzzy-Neural Autopilot. In: *Eleventh Ship Control Systems Symposium, South-ampton, UK*, 14–18 April, **1**, pp. 161–172.

Rosenblatt, F. (1961) *Principles of Neurodynamics: Perceptrons and the Theory of Brain Mechanisms*, Spartan Press, Washington, DC.

Rosenbrock, H.H. (1974) *Computer Aided Control System Design*, Academic Press, New York.

Routh, E.J. (1905) *Dynamics of a System of Rigid Bodies*, Macmillan & Co., London.

Rumelhart, D.E., Hinton, G.E. and Williams, R.J. (1986) Learning internal representations by error propagation. In: *Parallel Distributed Processing*, Rumelhart, D.E. and McClelland, J.L. (eds.), M.I.T. Press, Cambridge, Mass.

Safanov, M.G. (1980) *Stability and Robustness of Multivariable Feedback Systems*, M.I.T. Press, Cambridge, Mass.

Schwarzenbach, J. and Gill, K.F. (1979) *System Modelling and Control*, Edward Arnold, London.

Shannon, C.E. and Weaver, W. (1949) *The mathematical theory of communication*, University of Illinois Press, Urbana.

Smith, O.J.M. (1957) Closer Control of Loops with Dead Time, *Chem. Eng. Progress*, **53**(5), pp. 217–219.

Sperry, E.A. (1922) Automatic Steering. *Trans. Soc. Naval Arch. & Marine Eng.*, **XXX**, pp.53–57.

Sugeno, M. (ed.) (1985) *Industrial Applications of Fuzzy Control*, Elsevier Science Publishers BV, North-Holland.

Sutton, R. and Jess, I.M. (1991) A Design Study of a Self-Organising Fuzzy Autopilot for Ship Control. In: *IMechE, Proc. Instn. Mech. Engrs.*, **205**, pp. 35–47.

Sutton, R. and Marsden, G.D. (1997) A Fuzzy Autopilot Optimized using a Genetic Algorithm, *Journal of Navigation*, **50**(1), pp. 120–131.

Sutton, R., Burns, R.S. and Craven, P.J. (2000) Intelligent Steering Control of an Autonomous Underwater Vehicle, *Journal of Navigation*, **53**(3), pp. 511–525.

The MathWorks Inc. (1993) *SIMULINK Numerical Simulation Software – Reference Guide*, The MathWorks Inc., Natick, Mass.

Tong, R.M. (1978) Synthesis of fuzzy models for industrial processes, *Int. J. General Systems*, **4**, pp. 143–162.

Van Dyke Parunak, H. (1990) Focus on Intelligent Control, *Inter. J. of Integrated Manufacturing*, pp. 1–5.

Werbos, P. (1974) Beyond Regression: New Tools for Prediction and Analysis in the Behavioural Sciences, Thesis in Applied Mathematics, Harvard University.

Widrow, B. (1987) The Original Adaptive Neural Net Broom-Balancer. In: *Proc. IEEE Int. Symp. Circuits and Systems*, pp. 351–357.

Widrow, B. and Hoff, M.E. (1960) Adaptive switching circuits. In: *IEEE WESCON Convention Record, IRE*, New York, pp. 96–104.

Widrow, B. and Smith, F.W. (1964) Pattern recognising control systems. In: *Computer and Information Sciences*, Ton, J.T. and Wilcox, R.H. (eds.), Spartan Books, Cleaver Hume Press, pp. 288–317.

Wiener, N. (1949) *The Extrapolation, Interpolation and Smoothing of Stationary Time Series*, John Wiley, New York.

Yan, J., Ryan, M. and Power, J. (1994) *Using fuzzy logic – Towards intelligent systems*, Prentice-Hall International (UK), Hemel Hempstead, UK.

Zadeh, L.A. (1965) Fuzzy Sets. In: *Information and Control*, **8**, pp. 338–353.

Zames, G. (1966) On the Input–Output Stability of Time-Varying Non-Linear Feedback Systems. Parts I and II. *IEEE Trans. on Automat. Contr.*, **AC-11**(2 & 3), pp. 228–238, 465–476.

Zames, G. (1981) Feedback and Optimal Sensitivity: Model Reference Transformations, Multiplicative Seminorms and Approximate Inverses, *IEEE Trans. on Automat. Contr.*, **AC-26**, pp. 301–320.

Ziegler, J.G. and Nichols, N.B. (1942) *Optimum Settings for Automatic Controllers*, trans. ASME, **64**, pp. 759–768.

# Index